ORCCA

Open Resources for Community College Algebra

ORCCA

Open Resources for Community College Algebra

Portland Community College Faculty

August 6, 2018

Project Leads: Ann Cary and Alex Jordan
Technology Engineer: Alex Jordan
Contributing Authors: Ann Cary, Alex Jordan, Ross Kouzes, Scot Leavitt, Cara Lee, Carl Yao, and Ralf Youtz
WeBWorK Problem Coding: Chris Hughes, Alex Jordan, Carl Yao
Other Contributors: Kara Colley
Cover Image: Ralf Youtz

Edition: 1.0

Website: pcc.edu/ORCCA

Acknowledgements

This book has been made possible through Portland Community College's Strategic Investment Funding, approved by PCC's Budget Planning Advisory Council and the Board of Directors. Without significant funding to provide the authors with the adequate time, an ambitious project such as this one would not be possible.

The technology that makes it possible to create synced print, eBook, and WeBWorK content is PreTeXt, created by Rob Beezer. David Farmer and the American Institute of Mathematics have worked to make the PreTeXt eBook layout functional, yet simple. A grant from OpenOregon funded the original bridge between WeBWorK and PreTeXt.

This book uses WeBWorK to provide most of its exercises, which may be used for online homework. WeB-WorK was created by Mike Gage and Arnie Pizer, and has benefited from over 25 years of contributions from open source developers. In 2013, Chris Hughes, Alex Jordan, and Carl Yao programmed most of the WeBWorK questions in this book with a PCC curriculum development grant.

The javascript library MathJax, created and maintained by David Cervone, Volker Sorge, Christian Lawson-Perfect, and Peter Krautzberger allows math content to render nicely on screen in the eBook. Additionally, MathJax makes web accessible mathematics possible.

The print edition (PDF) is built using the typesetting software LaTeX, created by Donald Knuth and enhanced by Leslie Lamport.

Each of these open technologies, along with many that we use but have not listed here, has been enhanced by many additional contributors spanning the past 40 years. We are grateful for all of these contributions.

To All

HTML, PDF, and print This book is available as an eBook, a free PDF, or printed and bound. All versions offer the same content and are synchronized such that cross-references match across versions. They can each be found at pcc.edu/orcca.

There are some differences between the eBook, PDF, and printed versions.

- The eBook is recommended, as it offers interactive elements and easier navigation than print. It requires no more than internet access and a modern web browser.

- A PDF version can be downloaded and then accessed without the internet. Some content is in color, but most of the colorized content from the eBook has been converted to black and white to ensure adequate contrast when printing in black and white. The exceptions are the graphs generated by WeBWorK.

- Printed and bound copies are available online. Up-to-date information about purchasing a copy should be available at pcc.edu/orcca. Contact the authors if you have trouble finding the latest version online. For each online sale, all royalties go to a PCC Foundation account, where roughly half will fund student scholarships, and half will fund continued maintenance of this book and other OER.

Copying Content The graphs and other images that appear in this manual may be copied in various file formats using the eBook version. Below each image are links to .png, .eps, .svg, .pdf, and .tex files that contain the image.

Mathematical content can be copied from the eBook. To copy math content into *MS Word*, right-click or control-click over the math content, and click to Show Math As MathML Code. Copy the resulting code, and Paste Special into *Word*. In the Paste Special menu, paste it as Unformatted Text. To copy math content into LaTeX source, right-click or control-click over the math content, and click to Show Math As TeX Commands.

Tables can be copied from the eBook version and pasted into applications like *MS Word*. However, mathematical content within tables will not always paste correctly without a little extra effort as described above.

Accessibility The HTML version is intended to meet or exceed web accessibility standards. If you encounter an accessibility issue, please report it.

- All graphs and images should have meaningful alt text that communicates what a sighted person would see, without necessarily giving away anything that is intended to be deduced from the image.

- All math content is rendered using MathJax. MathJax has a contextual menu that can be accessed in several ways, depending on what operating system and browser you are using. The most common way is to right-click or control-click on some piece of math content.

- In the MathJax contextual menu, you may set options for triggering a zoom effect on math content, and also by what factor the zoom will be. Also in the MathJax contextual menu, you can enable the Explorer, which allows for sophisticated navigation of the math content.

- A screen reader will generally have success verbalizing the math content from MathJax. With certain screen reader and browser combinations, you may need to set some configuration settings in the MathJax contextual menu.

Tablets and Smartphones PreTeXt documents like this book are "mobile-friendly." When you view the HTML version, the display adapts to whatever screen size or window size you are using. A math teacher will always recommend that you do not study from the small screen on a phone, but if it's necessary, the eBook gives you that option.

WeBWorK for Online Homework Most exercises are available in a ready-to-use collection of WeBWorK problem sets. Visit webwork.pcc.edu/webwork2/orcca-demonstration to see a demonstration WeBWorK course where guest login is enabled. Anyone interested in using these problem sets should contact the project leads.

Odd Answers The answers to the odd homework exercises at the end of each section are not contained in the PDF or print versions. As the eBook evolves, they may or may not be contained in an appendix there. In any case, odd answers are available *somewhere*. Check pcc.edu/orcca to see where.

Interactive and Static Examples Traditionally, a math textbook has examples throughout each section. This textbook uses two types of "example":

Static These are labeled "Example." Static examples may or may not be subdivided into a "statement" followed by a walk-through solution. This is basically what traditional examples from math textbooks do.

Active These are labeled "⬚ Checkpoint," not to be confused with the exercises that come at the end of a section that might be assigned for homework, etc. In the HTML output, active examples have WeBWorK answer blanks where a reader could try submitting an answer. In the PDF output, active examples are almost indistinguishable from static examples, but there is a WeBWorK icon indicating that a reader could interact more actively using the eBook. Generally, a walk-through solution is provided immediately following the answer blank.

Some HTML readers will skip the opportunity to try an active example and go straight to its solution. Some readers will try an active example once and then move on to the solution. Some readers will tough it out for a period of time and resist reading the solution.

For readers of the PDF, it is expected that they would read the example and its solution just as they would read a static example.

A reader is *not* required to try submitting an answer to an active example before moving on. A reader *is* expected to read the solution to an active example, even if they succeed on their own at finding an answer.

Interspersed through a section there are usually several exercises that are intended as active reading exercises. A reader can work these examples and submit answers to WeBWorK to see if they are correct. The important thing is to keep the reader actively engaged instead of providing another static written example. In most cases, it is expected that a reader will read the solutions to these exercises just as they would be expected to read a more traditional static example.

Pedagogical Decisions

The authors and the greater PCC faculty have taken various stances on certain pedagogical and notational questions that arise in basic algebra instruction. We attempt to catalog these decisions here, although this list is certainly incomplete. If you find something in the book that runs contrary to these decisions, please let us know.

- Interleaving is our preferred approach, compared to a proficiency-based approach. To us, this means that once the book covers a topic, that topic will be appear in subsequent sections and chapters in indirect ways.

- Chapter 1 is mostly written as a *review*, and is not intended to teach all of these topics from first principles.

- We round decimal results to four significant digits, or possibly fewer leaving out trailing zeros. We do this to maintain consistency with the most common level of precision that WeBWorK uses to assess decimal answers. We *round*, not *truncate*. And we use the $\approx$ symbol. For example $\pi \approx 3.142$ and Portland's population is ≈ 609500.

- We offer *alternative* video lessons associated with each section, found in most sections in the eBook. We hope these videos provide readers with an alternative to whatever is in the reading, but there may be discrpancies here and there between the video content and reading content.

- We believe in always trying to open a topic with some level of application rather than abstract examples. From applications and practical questions, we move to motivate more abstract definitions and notation. This approach is perhaps absent in the first chapter, which is intended to be a review only. At first this may feel backwards to some instructors, with some "easier" examples (with no context) appearing after "more difficult" contextual examples.

- Linear inequalities are not strictly separated from linear equations. The section that teaches how to solve $2x + 3 = 8$ is immediately followed by the section teaching how to solve $2x + 3 < 8$.

  Our aim is to not treat inequalities as an add-on optional topic, but rather to show how intimately related they are to corresponding equations.

- When issues of "proper formatting" of student work arise, we value that the reader understand *why* such things help the reader to communicate outwardly. We believe that mathematics is about more than understanding a topic, but also about understanding it well enough to communicate results to others.

  For example we promote progression of equations like

  $$1 + 1 + 1 = 2 + 1$$
  $$= 3$$

  instead of

  $$1 + 1 + 1 = 2 + 1 = 3.$$

  And we want students to *understand* that the former method makes their work easier for a reader to read. It is not simply a matter of "this is the standard and this is how it's done."

- When solving equations (or systems of linear equations), most examples should come with a check, intended to communicate to students that checking is part of the process. In Chapters 1–4, these checks will be complete simplifications using order of operations one step at a time. The later sections will often have more summary checks where either order of operations steps are skipped in groups, or we promote entering expressions into a calculator. Occasionally in later sections the checks will still have finer details, especially when there are issues like with negative numbers squared.

- Within a section, any first example of solving some equation (or system) should summarize with some variant of both "the solution is…" and "the solution set is…." Later examples can mix it up, but always offer at least one of these.

- There is a section on very basic arithmetic (five operations on natural numbers) in an appendix, not in the first chapter. This appendix is only available in the eBook.

- With applications of linear equations (as opposed to linear systems), we limit applications to situations where the setup will be in the form $x + f(x) = C$ and also certain rate problems where the setup will be in the form $5t + 4t = C$. There are other classes of application problem (mixing problems, interest problems, …) which can be handled with a system of two equations, and we reserve these until linear systems are covered.

- With simplifications of rational expressions in one variable, we always include domain restrictions that are lost in the simplification. For example, we would write $\frac{x(x+1)}{x+1} = x$, for $x \neq -1$. With multivariable rational expressions, we are content to ignore domain restrictions lost during simplification.

Entering WeBWorK Answers

This preface offers some guidance with syntax for WeBWorK answers. WeBWorK answer blanks appear in the active reading examples (called "checkpoints") in the HTML version of the book. If you are using WeBWorK for online homework, then you will also enter answers into WeBWorK answer blanks there.

Basic Arithemtic The five basic arithmetic operations are: addition, subtraction, multiplication, and raising to a power. The symbols for addition and subtraction are + and −, and both of these are directly avialable on most keyboards as + and -.

On paper, multiplication is sometimes written using × and sometimes written using · (a centered dot). Since these symbols are not available on most keyboards, WeBWorK uses * instead, which is often shift-8 on a full keyboard.

On paper, division is sometimes written using ÷, sometimes written using a fraction layout like $\frac{4}{2}$, and sometimes written just using a slash, /. The slash is available on most full keyboards, near the question mark. WeBWorK uses / to indicate division.

On paper, raising to a power is written using a two-dimensional layout like 4^2. Since we don't have a way to directly type that with a simple keyboard, calculators and computers use the caret character, ^, as in 4^2. The character is usually shift-6.

Roots and Radicals On paper, a square root is represented with a radical symbol like $\sqrt{}$. Since a keyboard does not usually have this symbol, WeBWorK and many computer applications use sqrt( ) instead. For example, to enter $\sqrt{17}$, type sqrt(17).

Higher-index radicals are written on paper like $\sqrt[4]{12}$. Again we have no direct way to write this using most keyboards. In *some* WeBWorK problems it is possible to type something like root(4, 12) for the fourth root of twelve. However this is not enabled for all WeBWorK problems.

As an alternative that you may learn about in a later chapter, $\sqrt[4]{12}$ is mathematically equal to $12^{1/4}$, so it can be typed as 12^(1/4). Take note of the parentheses, which very much matter.

Common Hiccups with Grouping Symbols Suppose you wanted to enter $\frac{x+1}{2}$. You might type x+1/2, but this is not right. The computer will use the order of operations (see Section 1.4) and do your division first, dividing 1 by 2. So the computer will see $x + \frac{1}{2}$. To address this, you would need to use grouping symbols like parentheses, and type something like (x+1)/2.

Suppose you wanted to enter $6^{1/4}$, and you typed 6^1/4. This is not right. The order of operations places a higher priority on exponentiation than division, so it calculates 6^1 first and then divides the result by 4. That is simply not the same as raising 6 to the $\frac{1}{4}$ power. Again the way to address this is to use grouping symbols, like 6^(1/4).

Entering Decimal Answers Often you will find a decimal answer with decimal places that go on and on. You are allowed to round, but not by too much. WeBWorK generally looks at how many *significant digits* you use, and generally expects you to use *four or more* correct significant digits.

"Significant digits" and "places past the decimal" are not the same thing. To count significant digits, read the number left to right and look for the first nonzero digit. Then count all the digits to the right including that first one.

The number 102.3 has four significant digits, but only one place past the decimal. This number could be a correct answer to a WeBWorK question. The number 0.0003 has one significant digit and four places past the decimal. This number might cause you trouble if you enter it, because maybe the "real" answer was 0.0003091, and rounding to 0.0003 was too much rounding.

Special Symbols   There are a handful of special symbols that are easy to write on paper, but it's not clear how to type them. Here are WeBWorK's expectations.

Symbol	Name	How to Type
∞	infinity	infinity or inf
π	pi	pi
∪	union	U
ℝ	the real numbers	R
		such that	| (shift-\, where \ is above the enter key)
≤	less than or equal to	<=
≥	greater than or equal to	>=
≠	not equal to	!=

Contents

Acknowledgements   v

To All   vii

Pedagogical Decisions   ix

Entering WeBWorK Answers   xi

1 Basic Math Review   1
    1.1 Arithmetic with Negative Numbers 1
    1.2 Fractions and Fraction Arithmetic 12
    1.3 Absolute Value and Square Root 27
    1.4 Order of Operations 32
    1.5 Set Notation and Types of Numbers 44
    1.6 Comparison Symbols and Notation for Intervals 53
    1.7 Basic Math Chapter Review 61

2 Variables, Expressions, and Equations   69
    2.1 Variables and Evaluating Expressions 69
    2.2 Geometry Formulas 80
    2.3 Combining Like Terms 94
    2.4 Equations and Inequalities as True/False Statements 99
    2.5 Solving One-Step Equations 109
    2.6 Solving One-Step Inequalities 123
    2.7 Percentages 131
    2.8 Modeling with Equations and Inequalities 143
    2.9 Introduction to Exponent Rules 153
    2.10 Simplifying Expressions 158
    2.11 Variables, Expressions, and Equations Chapter Review 173

3 Linear Equations and Inequalities   185
    3.1 Solving Multistep Linear Equations 185
    3.2 Solving Multistep Linear Inequalities 201
    3.3 Linear Equations and Inequalities with Fractions 209
    3.4 Isolating a Linear Variable 220
    3.5 Ratios and Proportions 226
    3.6 Special Solution Sets 239
    3.7 Linear Equations and Inequalities Chapter Review 244

4 Graphing Lines   253
    4.1 Cartesian Coordinates 253
    4.2 Graphing Equations 260
    4.3 Exploring Two-Variable Data and Rate of Change 269
    4.4 Slope 279
    4.5 Slope-Intercept Form 295
    4.6 Point-Slope Form 314

Contents

4.7   Standard Form 331
4.8   Horizontal, Vertical, Parallel, and Perpendicular Lines 345
4.9   Summary of Graphing Lines 361
4.10 Linear Inequalities in Two Variables 370
4.11 Graphing Lines Chapter Review 375

5 Systems of Linear Equations   393
5.1   Solving Systems of Linear Equations by Graphing 393
5.2   Substitution 405
5.3   Elimination 423
5.4   Systems of Linear Equations Chapter Review 437

6 Exponents and Polynomials   447
6.1   Exponent Rules 447
6.2   Scientific Notation 462
6.3   Adding and Subtracting Polynomials 471
6.4   Multiplying Polynomials 483
6.5   Special Cases of Multiplying Polynomials 494
6.6   Dividing by a Monomial 503
6.7   Exponents and Polynomials Chapter Review 507

7 Factoring   517
7.1   Factoring out the Common Factor 517
7.2   Factoring by Grouping 524
7.3   Factoring Trinomials with Leading Coefficient One 529
7.4   Factoring Trinomials with a Nontrivial Leading Coefficient 538
7.5   Factoring Special Polynomials 546
7.6   Factoring Strategies 556
7.7   Factoring Chapter Review 564

8 Solving Quadratic Equations   577
8.1   Solving Quadratic Equations by Factoring 577
8.2   Square Root Properties 589
8.3   Solving Quadratic Equations by Using a Square Root 602
8.4   The Quadratic Formula 613
8.5   Complex Solutions to Quadratic Equations 624
8.6   Strategies for Solving Quadratic Equations 628
8.7   Solving Quadratic Equations Chapter Review 633

9 Graphs of Quadratic Functions   649
9.1   Introduction to Functions 649
9.2   Properties of Quadratic Functions 675
9.3   Graphing Quadratic Functions 694
9.4   Graphs of Quadratic Functions Chapter Review 710

Index   731

CHAPTER 5

Systems of Linear Equations

5.1 Solving Systems of Linear Equations by Graphing

We have learned how to graph a line given its equation. In this section, we will learn what a *system* of *two* linear equations is, and how to use graphing to solve such a system.

5.1.1 Solving Systems of Equations by Graphing

Example 5.1.2

Fabiana and David are running at constant speeds in parallel lanes on a track. David starts out ahead of Fabiana, but Fabiana is running faster. We want to determine when Fabiana will catch up with David. Let's start by looking at the graph of each runner's distance over time, in Figure 5.1.3.

Each of the two lines has an equation, as discussed in Chapter 4. The line representing David appears to have y-intercept $(0, 4)$ and slope $\frac{4}{3}$, so its equation is $y = \frac{4}{3}t + 4$. The line representing Fabiana appears to have y-intercept $(0, 0)$ and slope 2, so its equation is $y = 2t$.

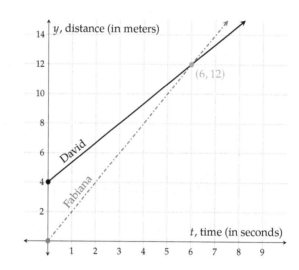

Figure 5.1.3: David and Fabiana's distances.

When these two equations are together as a package, we have what is called a **system of linear equations**:

$$\begin{cases} y = \frac{4}{3}t + 4 \\ y = 2t \end{cases}$$

The large left brace indicates that this is a collection of two distinct equations, not one equation that was somehow algebraically manipulated into an equivalent equation.

As we can see in Figure 5.1.3, the graphs of the two equations cross at the point $(6, 12)$. We refer to the point $(6, 12)$ as the **solution** to this system of linear equations. To denote the **solution set**, we write

{(6, 12)}. But it's much more valuable to interpret these numbers in context whenever possible: it took 6 seconds for the two runners to meet up, and when they met they were 12 meters up the track.

Remark 5.1.4. In Example 5.1.2, we stated that the solution was the point (6, 12). It makes sense to write this as an ordered pair when we're given a graph. In some cases when we have no graph, particularly when our variables are not x and y, it might not be clear which variable "comes first" and we won't be able to write an ordered pair. Nevertheless, given the context we can write meaningful summary statements.

Example 5.1.5 Determine the solution to the system of equations graphed in Figure 5.1.6.

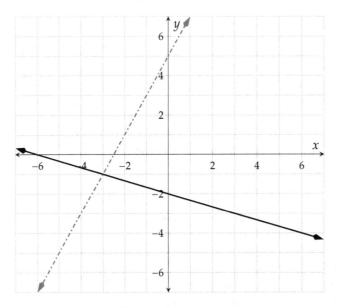

Figure 5.1.6: Graph of a System of Equations

Explanation. The two lines intersect where $x = -3$ and $y = -1$, so the solution is the point $(-3, -1)$. We write the solution set as $\{(-3, -1)\}$.

Checkpoint 5.1.7. Determine the solution to the system of equations graphed below.

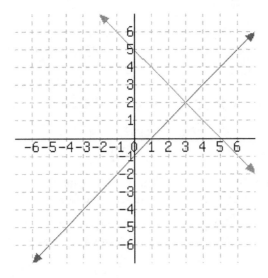

The solution is the point [   ].

Explanation. The two lines intersect where $x = 3$ and $y = 2$, so the solution is the point $(3, 2)$. We write the solution set as $\{(3, 2)\}$.

Now let's look at an example where *we* need to make a graph to find the solution.

> **Example 5.1.8** Solve the following system of equations by graphing:
>
> $$\begin{cases} y = \frac{1}{2}x + 4 \\ y = -x - 5 \end{cases}$$
>
> Notice that each of these equations is written in slope-intercept form. The first equation, $y = \frac{1}{2}x + 4$, is a linear equation with a slope of $\frac{1}{2}$ and a y-intercept of $(0, 4)$. The second equation, $y = -x - 5$, is a linear equation with a slope of -1 and a y-intercept of $(0, -5)$. We'll use this information to graph both lines.
>
> The two lines intersect where $x = -6$ and $y = 1$, so the solution of the system of equations is the point $(-6, 1)$. We write the solution set as $\{(-6, 1)\}$.
>
>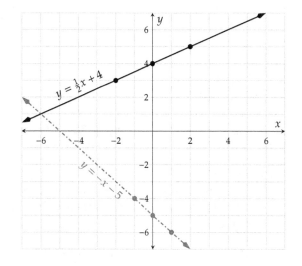
>
> **Figure 5.1.9:** $y = \frac{1}{2}x + 4$ and $y = -x - 5$.

> **Example 5.1.10** Solve the following system of equations by graphing:
>
> $$\begin{cases} x - 3y = -12 \\ 2x + 3y = 3 \end{cases}$$
>
> **Explanation.** Since both line equations are given in standard form, we'll graph each one by finding the intercepts. Recall that to find the x-intercept of each equation, replace y with 0 and solve for x. Similarly, to find the y-intercept of each equation, replace x with 0 and solve for y.
>
> For our first linear equation, we have:
>
> $$x - 3(0) = -12 \qquad 0 - 3y = -12$$
> $$x = -12 \qquad -3y = -12$$
> $$y = 4.$$
>
> So the intercepts are $(-12, 0)$ and $(0, 4)$.
>
> For our second linear equation, we have:
>
> $$2x + 3(0) = 3 \qquad 2(0) + 3y = 3$$
> $$2x = 3 \qquad 3y = 3$$
> $$x = \frac{3}{2} \qquad y = 1.$$
>
> So the intercepts are $\left(\frac{3}{2}, 0\right)$ and $(0, 1)$.

Now we can graph each line by plotting the intercepts and connecting these points:

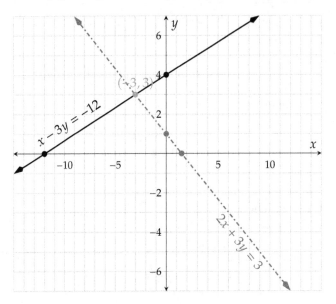

Figure 5.1.11: Graphs of $x - 3y = -12$ and $2x + 3y = 3$

It appears that the solution of the system of equations is the point of intersection of those two lines, which is $(-3, 3)$. It's important to check this is correct, because when making a hand-drawn graph, it would be easy to be off by a little bit. To check, we can substitute the values of x and y from the point $(-3, 3)$ into each equation:

$$x - 3y = -12 \qquad\qquad\qquad 2x + 3y = 3$$
$$-3 - 3(3) \overset{?}{=} -12 \qquad\qquad 2(-3) + 3(3) \overset{?}{=} 3$$
$$-3 - 9 \overset{\checkmark}{=} -12 \qquad\qquad -6 + 9 \overset{\checkmark}{=} 3$$

So we have checked that $(-3, 3)$ is indeed the solution for the system. We write the solution set as $\{(-3, 3)\}$.

Example 5.1.12 A college has a north campus and a south campus. The north campus has 18,000 students, and the south campus has 4,000 students. In the past five years, the north campus lost 4,000 students, and the south campus gained 3,000 students. If these trends continue, in how many years would the two campuses have the same number of students? Write and solve a system of equations modeling this problem.

Explanation. Since all the given student counts are in the thousands, we make the decision to measure student population in thousands. So for instance, the north campus starts with a student population of 18 (thousand students).

The north campus lost 4 thousand students in 5 years. So it is losing students at a rate of $\frac{4\,\text{thousand}}{5\,\text{year}}$, or $\frac{4}{5}\frac{\text{thousand}}{\text{year}}$. This rate of change should be interpreted as a negative number, because the north campus is losing students over time. So we have a linear model with starting value 18 thousand students, and a slope of $-\frac{4}{5}$ thousand students per year. In other words,

$$y = -\frac{4}{5}t + 18,$$

where y stands for the number of students in thousands, and t stands for the number of years into the future.

Similarly, the number of students at the south campus can be modeled by $y = \frac{3}{5}t + 4$. Now we have a system of equations:

$$\begin{cases} y = -\dfrac{4}{5}t + 18 \\ y = \dfrac{3}{5}t + 4 \end{cases}$$

We will graph both lines using their slopes and y-intercepts.

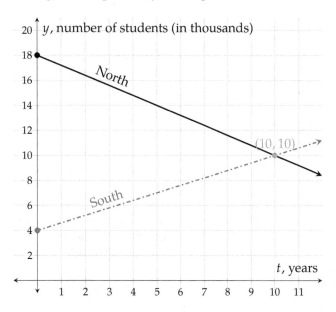

Figure 5.1.13: Number of Students at the South Campus and North Campus

According to the graph, the lines intersect at $(10, 10)$. So if the trends continue, both campuses will have 10,000 students 10 years from now.

Example 5.1.14 Solve the following system of equations by graphing:

$$\begin{cases} y = 3(x - 2) + 1 \\ y = -\frac{1}{2}(x + 1) - 1 \end{cases}$$

Explanation. Since both line equations are given in point-slope form, we can start by graphing the point indicated in each equation and use the slope to determine the rest of the line.

For our first equation, $y = 3(x - 2) + 1$, the point indicated in the equation is $(2, 1)$ and the slope is 3.

For our second equation, $y = -\frac{1}{2}(x + 1) - 1$, the point indicated in the equation is $(-1, -1)$ and the slope is $-\frac{1}{2}$.

Now we can graph each line by plotting the points and using their slopes.

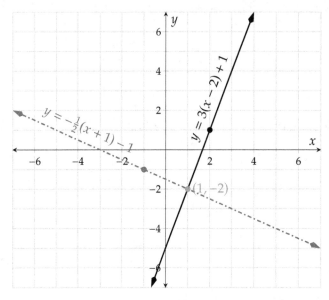

Figure 5.1.15: Graphs of $y = 3(x - 2) + 1$ and $y = -\frac{1}{2}(x + 1) - 1$

It appears that the solution of the system of equations is the point of intersection of those two lines, which is $(1, -2)$. It's important to check this is correct, because when making a hand-drawn graph, it would be easy to be off by a little bit. To check, we can substitute the values of x and y from the point $(1, -2)$ into each equation:

$$y = 3(x - 2) + 1$$
$$-2 \stackrel{?}{=} 3(1 - 2) + 1$$
$$-2 \stackrel{?}{=} 3(-1) + 1$$
$$-2 \stackrel{\checkmark}{=} -3 + 1$$

$$y = -\frac{1}{2}(x + 1) - 1$$
$$-2 \stackrel{?}{=} -\frac{1}{2}(1 + 1) - 1$$
$$-2 \stackrel{?}{=} -\frac{1}{2}(2) - 1$$
$$-2 \stackrel{\checkmark}{=} -1 - 1$$

So we have checked that $(2, -1)$ is indeed the solution for the system. We write the solution set as $\{(2, -1)\}$.

5.1.2 Special Systems of Equations

Recall that when we solved linear equations in one variable, we had two special cases. In one special case there was no solution and in the other case, there were infinitely many solutions. When solving systems of equations in two variables, we have two similar special cases.

> **Example 5.1.16 Parallel Lines.** Let's look at the graphs of two lines with the same slope, $y = 2x - 4$ and $y = 2x + 1$:
>
>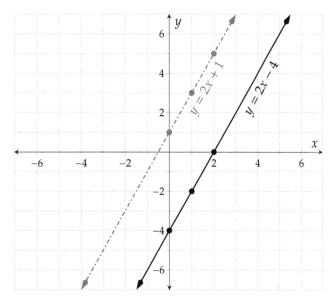
>
> **Figure 5.1.17:** Graphs of $y = 2x - 4$ and $y = 2x + 1$

For this system of equations, what is the solution? Since the two lines have the same slope they are **parallel lines** and will never intersect. This means that there is *no solution* to this system of equations. We write the solution set as $\emptyset$.

The symbol $\emptyset$ is a special symbol that represents the **empty set**, a *set* that has no numbers in it. This symbol is *not* the same thing as the number zero. The *number* of eggs in an empty egg carton is zero whereas the empty carton itself could represent the empty set. The symbols for the empty set and the number zero may look similar depending on how you write the number zero. Try to keep the concepts separate.

> **Example 5.1.18 Coinciding Lines.** Next we'll look at the other special case. Let's start with this system of equations:
> $$\begin{cases} y = 2x - 4 \\ 6x - 3y = 12 \end{cases}$$

To solve this system of equations, we want to graph each line. The first equation is in slope-intercept form and can be graphed easily using its slope of 2 and its y-intercept of $(0, -4)$.

The second equation, $6x - 3y = 12$, can either be graphed by solving for y and using the slope-intercept form or by finding the intercepts. If we use the intercept method, we'll find that this line has an x-intercept of $(2, 0)$ and a y-intercept of $(0, -4)$. When we graph both lines we get Figure 5.1.19.

Now we can see these are actually the *same* line, or **coinciding lines**. To determine the solution to this system, we'll note that they overlap everywhere. This means that we have an infinite number of solutions: *all* points that fall on the line. It may be enough to report that there are infinitely many solutions. In order to be more specific, all we can do is say that any ordered pair (x, y) satisfying the line equation is a solution. In set-builder notation, we would write $\{(x, y) \mid y = 2x - 4\}$.

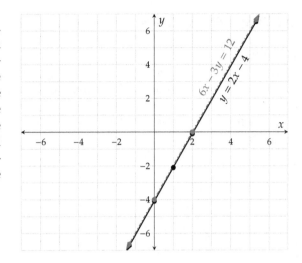

Figure 5.1.19: Graphs of $y = 2x - 4$ and $6x - 3y = 12$

Remark 5.1.20. In Example 5.1.18, what would have happened if we had decided to convert the second line equation into slope-intercept form?

$$6x - 3y = 12$$
$$6x - 3y - 6x = 12 - 6x$$
$$-3y = -6x + 12$$
$$-\frac{1}{3} \cdot (-3y) = -\frac{1}{3} \cdot (-6x + 12)$$
$$y = 2x - 4$$

This is the literally the same as the first equation in our system. This is a different way to show that these two equations are equivalent and represent the same line. Any time we try to solve a system where the equations are equivalent, we'll have an infinite number of solutions.

Warning 5.1.21. Notice that for a system of equations with infinite solutions like Example 5.1.18, we didn't say that *every* point was a solution. Rather, every point that falls on that line is a solution. It would be incorrect to state this solution set as "all real numbers" or as "all ordered pairs."

Intersecting Lines: If two linear equations have different slopes, the system has one solution.

Parallel Lines: If the linear equations have the same slope with different y-intercepts, the system has no solution.

Coinciding Lines: If two linear equations have the same slope and the same y-intercept (in other words, they are equivalent equations), the system has infinitely many solutions. This solution set consists of all ordered pairs on that line.

List 5.1.22: A summary of the three types of systems of equations and their solution sets:

Exercises

Warmup and Review   Find the line's slope and y-intercept.

1. A line has equation $y = 9x + 3$.

This line's slope is [   ].

This line's y-intercept is [   ].

2. A line has equation $y = 10x + 9$.

This line's slope is [   ].

This line's y-intercept is [   ].

3. A line has equation $y = -x - 9$.

This line's slope is [   ].

This line's y-intercept is [   ].

4. A line has equation $y = -x - 6$.

This line's slope is [   ].

This line's y-intercept is [   ].

5. A line has equation $y = -\dfrac{4x}{9} + 3$.

This line's slope is [   ].

This line's y-intercept is [   ].

6. A line has equation $y = -\dfrac{4x}{5} - 9$.

This line's slope is [   ].

This line's y-intercept is [   ].

7. A line has equation $y = \dfrac{x}{6} - 7$.

This line's slope is [   ].

This line's y-intercept is [   ].

8. A line has equation $y = \dfrac{x}{8} + 6$.

This line's slope is [   ].

This line's y-intercept is [   ].

9. Graph the equation $y = -3x$.

10. Graph the equation $y = \frac{1}{4}x$.

11. Graph the equation $y = \frac{2}{3}x + 4$.

12. Graph the equation $y = -2x + 5$.

Solve the linear equation for y.

13.
$$12x - 4y = 44$$

14.
$$-25x - 5y = -85$$

15.
$$3x + 9y = 27$$

16.
$$-6x - 2y = -6$$

Checking Solutions for System of Equations

17. Decide whether $(-3, -5)$ is a solution to the system of equations:

$$\begin{cases} 5x - 4y = 5 \\ x + 4y = -25 \end{cases}$$

The point $(-3, -5)$ ($\square$ is $\square$ is not) a solution.

18. Decide whether $(-2, 2)$ is a solution to the system of equations:

$$\begin{cases} -x + 4y = 10 \\ x + 5y = 8 \end{cases}$$

The point $(-2, 2)$ ($\square$ is $\square$ is not) a solution.

19. Decide whether $(-1, -1)$ is a solution to the system of equations:

$$\begin{cases} 4x + y = -8 \\ y = x \end{cases}$$

The point $(-1, -1)$ ($\square$ is $\square$ is not) a solution.

20. Decide whether $(0, -5)$ is a solution to the system of equations:

$$\begin{cases} -3x - 3y = 15 \\ y = -5x - 6 \end{cases}$$

The point $(0, -5)$ ($\square$ is $\square$ is not) a solution.

21. Decide whether $\left(\frac{7}{4}, \frac{9}{4}\right)$ is a solution to the system of equations:

$$\begin{cases} -8x - 12y = -41 \\ 8x - 12y = -13 \end{cases}$$

The point $\left(\frac{7}{4}, \frac{9}{4}\right)$ ($\square$ is $\square$ is not) a solution.

22. Decide whether $\left(\frac{7}{4}, \frac{5}{4}\right)$ is a solution to the system of equations:

$$\begin{cases} 12x - 4y = 16 \\ -4x - 12y = -19 \end{cases}$$

The point $\left(\frac{7}{4}, \frac{5}{4}\right)$ ($\square$ is $\square$ is not) a solution.

Using a Graph to Solve a System Use a graph to solve the system of equations.

23.
$$\begin{cases} y = -\frac{7}{2}x - 8 \\ y = 5x + 9 \end{cases}$$

24.
$$\begin{cases} y = \frac{2}{3}x + 5 \\ y = -2x - 11 \end{cases}$$

25.
$$\begin{cases} y = 12x + 7 \\ 3x + y = -8 \end{cases}$$

26.
$$\begin{cases} y = -3x + 5 \\ 4x + y = 8 \end{cases}$$

27.
$$\begin{cases} x + y = 0 \\ 3x - y = 8 \end{cases}$$

28.
$$\begin{cases} 4x - 2y = 4 \\ x + 2y = 6 \end{cases}$$

29.

$$\begin{cases} y = 4x - 5 \\ y = -1 \end{cases}$$

30.

$$\begin{cases} 3x - 4y = 12 \\ y = 3 \end{cases}$$

31.

$$\begin{cases} x + y = -1 \\ x = 2 \end{cases}$$

32.

$$\begin{cases} x - 2y = -4 \\ x = -4 \end{cases}$$

33.

$$\begin{cases} y = 2(x + 3) - 5 \\ y = -\dfrac{4}{3}(x - 4) - 1 \end{cases}$$

34.

$$\begin{cases} y = -\dfrac{2}{3}(x - 6) - 2 \\ y = -\dfrac{1}{2}(x - 1) + 2 \end{cases}$$

35.

$$\begin{cases} y = -\dfrac{1}{2}(x - 6) + 4 \\ y = 4(x + 1) - 6 \end{cases}$$

36.

$$\begin{cases} y = \dfrac{5}{6}(x - 6) + 4 \\ y = 2(x + 1) + 4 \end{cases}$$

37.

$$\begin{cases} y = -\dfrac{4}{5}x + 8 \\ 4x + 5y = -35 \end{cases}$$

38.

$$\begin{cases} 2x - 7y = 28 \\ y = \dfrac{2}{7}x - 3 \end{cases}$$

39.

$$\begin{cases} -10x + 15y = 60 \\ 6x - 9y = 36 \end{cases}$$

40.

$$\begin{cases} 6x - 8y = 32 \\ 9x - 12y = 12 \end{cases}$$

41.

$$\begin{cases} y = -\dfrac{3}{5}x + 7 \\ 9x + 15y = 105 \end{cases}$$

42.

$$\begin{cases} 9y - 12x = 18 \\ y = \dfrac{4}{3}x + 2 \end{cases}$$

Determining the Number of Solutions in a System of Equations

43. Simply by looking at this system of equations, decide the number of solutions it has.

$$\begin{cases} y = -6x \\ y = -6x \end{cases}$$

The system has   (□ no solution   □ one solution   □ infinitely many solutions)   .

44. Simply by looking at this system of equations, decide the number of solutions it has.

$$\begin{cases} y = \dfrac{2}{5}x - 4 \\ y = \dfrac{2}{5}x - 4 \end{cases}$$

The system has   (□ no solution   □ one solution   □ infinitely many solutions)   .

45. Without graphing this system of equations, decide the number of solutions it has.

$$\begin{cases} y = \dfrac{3}{2}x + 4 \\ 9x - 6y = 0 \end{cases}$$

The system has   (□ no solution   □ one solution   □ infinitely many solutions)   .

46. Without graphing this system of equations, decide the number of solutions it has.

$$\begin{cases} y = 2x + 1 \\ 8x - 4y = 4 \end{cases}$$

The system has   (□ no solution   □ one solution   □ infinitely many solutions)   .

47. Without graphing this system of equations, decide the number of solutions it has.

$$\begin{cases} 3x + 12y = -24 \\ 4x + 16y = -16 \end{cases}$$

The system has   (□ no solution   □ one solution   □ infinitely many solutions)   .

48. Without graphing this system of equations, decide the number of solutions it has.

$$\begin{cases} 12x + 4y = -20 \\ 8x + 2y = 10 \end{cases}$$

The system has   (□ no solution   □ one solution   □ infinitely many solutions)   .

49. Simply by looking at this system of equations, decide the number of solutions it has.

$$\begin{cases} x = 4 \\ y = -5 \end{cases}$$

The system has   (□ no solution   □ one solution   □ infinitely many solutions)   .

50. Simply by looking at this system of equations, decide the number of solutions it has.

$$\begin{cases} x = 1 \\ y = -1 \end{cases}$$

The system has   (□ no solution   □ one solution   □ infinitely many solutions)   .

5.2 Substitution

In Section 5.1, we focused on solving systems of equations by graphing. In addition to being time consuming, graphing can be an awkward method to determine the exact solution when the solution has large numbers or fractions. There are two symbolic methods for solving systems of linear equations, and in this section we will use one of them: **substitution**.

5.2.1 Solving Systems of Equations Using Substitution

Example 5.2.2 The Interview. In 2014, the New York Times[a] posted the following about the movie, "The Interview":

> "The Interview" generated roughly $15 million in online sales and rentals during its first four days of availability, Sony Pictures said on Sunday.

> Sony did not say how much of that total represented $6 digital rentals versus $15 sales. The studio said there were about two million transactions overall.

A few days later, Joey Devilla cleverly pointed out in his blog[b], that there is enough information given to find the amount of sales versus rentals. Using algebra, we can write a system of equations and solve it to find the two quantities.[c]

First, we will define variables. We need two variables, because there are two unknown quantities: how many sales there were and how many rentals there were. Let r be the number of rental transactions and let s be the number of sales transactions.

If you are unsure how to write an equation from the background information, use the units to help you. The units of each term in an equation must match because we can only add like quantities. Both r and s are in transactions. The article says that the total number of transactions is 2 million. So our first equation will add the total number of rental and sales transactions and set that equal to 2 million. Our equation is:

$$(r \text{ transactions}) + (s \text{ transactions}) = 2{,}000{,}000 \text{ transactions}$$

Without the units:

$$r + s = 2{,}000{,}000$$

The price of each rental was $6. That means the problem has given us a *rate* of $6 \frac{\text{dollars}}{\text{transaction}}$ to work with. The rate unit suggests this should be multiplied by something measured in transactions. It makes sense to multiply by r, and then the number of dollars generated from rentals was $6r$. Similarly, the price of each sale was $15, so the revenue from sales was $15s$. The total revenue was $15 million, which we can represent with this equation:

$$\left(6 \tfrac{\text{dollars}}{\text{transaction}}\right)(r \text{ transactions}) + \left(15 \tfrac{\text{dollars}}{\text{transaction}}\right)(s \text{ transactions}) = \$15{,}000{,}000$$

Without the units:

$$6r + 15s = 15{,}000{,}000$$

Here is our system of equations:

$$\begin{cases} r + s = 2{,}000{,}000 \\ 6r + 15s = 15{,}000{,}000 \end{cases}$$

To solve the system, we will use the **substitution method**. The idea is to use *one* equation to find an expression that is equal to r but, cleverly, does not use the variable "r." Then, substitute this for r into the *other* equation. This leaves you with *one* equation that only has *one* variable.

The first equation from the system is an easy one to solve for r:

$$r + s = 2{,}000{,}000$$
$$r = 2{,}000{,}000 - s$$

This tells us that the expression $2{,}000{,}000 - s$ is equal to r, so we can *substitute* it for r in the second equation:

$$6r + 15s = 15{,}000{,}000$$
$$6(2{,}000{,}000 - s) + 15s = 15{,}000{,}000$$

Now we have an equation with only one variable, s, which we will solve for:

$$6(2{,}000{,}000 - s) + 15s = 15{,}000{,}000$$
$$12{,}000{,}000 - 6s + 15s = 15{,}000{,}000$$
$$12{,}000{,}000 + 9s = 15{,}000{,}000$$
$$9s = 3{,}000{,}000$$
$$\frac{9s}{9} = \frac{3{,}000{,}000}{9}$$
$$s = 333{,}333.\overline{3}$$

At this point, we know that $s = 333{,}333.\overline{3}$. This tells us that out of the 2 million transactions, roughly 333,333 were from online sales. Recall that we solved the first equation for r, and found $r = 2{,}000{,}000 - s$.

$$r = 2{,}000{,}000 - s$$
$$r = 2{,}000{,}000 - 333{,}333.\overline{3}$$
$$r = 1{,}666{,}666.\overline{6}$$

To check our answer, we will see if $s = 333{,}333.\overline{3}$ and $r = 1{,}666{,}666.\overline{6}$ make the original equations true:

$$r + s = 2{,}000{,}000$$
$$1{,}666{,}666.\overline{6} + 333{,}333.\overline{3} \overset{?}{=} 2{,}000{,}000$$
$$2{,}000{,}000 \overset{\checkmark}{=} 2{,}000{,}000$$

$$6r + 15s = 15{,}000{,}000$$
$$6\left(1{,}666{,}666.\overline{6}\right) + 15\left(333{,}333.\overline{3}\right) \overset{?}{=} 15{,}000{,}000$$
$$10{,}000{,}000 + 5{,}000{,}000 \overset{\checkmark}{=} 15{,}000{,}000$$

In summary, there were roughly 333,333 copies sold and roughly 1,666,667 copies rented.

---

[a](nyti.ms/2pupebT)

[b]http://www.joeydevilla.com/2014/12/31/

[c]Although since the given information uses approximate values, the solutions we will find will only be approximations too.

Remark 5.2.3. In Example 5.2.2, we *chose* to solve the equation $r + s = 2{,}000{,}000$ for r. We could just as easily have instead solved for s and substituted that result into the second equation instead. The summary conclusion would have been the same.

Remark 5.2.4. In Example 5.2.2, we rounded the solution values because only whole numbers make sense in the context of the problem. It was OK to round, because the original information we had to work with were rounded. In fact, it would be OK to round even more to $s = 330{,}000$ and $r = 1{,}700{,}000$, as long as we communicate clearly that we rounded and our values are rough.

In other exercises where there is no context and nothing suggests the given numbers are approximations, it is not OK to round and all answers should be communicated with their exact values.

Example 5.2.5 Solve the system of equations using substitution:

$$\begin{cases} x + 2y = 8 \\ 3x - 2y = 8 \end{cases}$$

Explanation. To use substitution, we need to solve for *one* of the variables in *one* of our equations. Looking at both equations, it will be easiest to solve for x in the first equation:

$$x + 2y = 8$$
$$x = 8 - 2y$$

Next, we replace x in the second equation with $8 - 2y$, giving us a linear equation in only one variable, y, that we may solve:

$$3x - 2y = 8$$
$$3(8 - 2y) - 2y = 8$$
$$24 - 6y - 2y = 8$$
$$24 - 8y = 8$$
$$-8y = -16$$
$$y = 2$$

Now that we have the value for y, we need to find the value for x. We have already solved the first equation for x, so that is the easiest equation to use.

$$x = 8 - 2y$$
$$x = 8 - 2(2)$$
$$x = 8 - 4$$
$$x = 4$$

To check this solution, we replace x with 4 and y with 2 in each equation:

$$x + 2y = 8 \qquad\qquad\qquad 3x - 2y = 8$$
$$4 + 2(2) \stackrel{?}{=} 8 \qquad\qquad\qquad 3(4) - 2(2) \stackrel{?}{=} 8$$

$$4 + 4 \overset{\checkmark}{=} 8 \qquad\qquad\qquad 12 - 4 \overset{\checkmark}{=} 8$$

We conclude then that this system of equations is true when $x = 4$ and $y = 2$. Our solution is the point $(4, 2)$ and we write the solution set as $\{(4, 2)\}$.

Checkpoint 5.2.6. Try a similar exercise.

Solve the following system of equations.

$$\begin{cases} 5x + y = -3 \\ \quad\ \ 0 = -1 + 4x + y \end{cases}$$

Explanation. These equations have no fractions; let's try to keep it that way.

$$\begin{cases} 5x + y = -3 \\ \quad\ \ 0 = -1 + 4x + y \end{cases}$$

Since one of the coefficients of y is 1, it is wise to solve for y in terms of the other variable and then use substitution to complete the problem.

$$y = -5x - 3 \quad \text{(from the first equation)}$$

which we can substitute in for y into the second equation:

$$0 = 4x - 1 + (-5x) - 3 \quad \text{(from the second equation)}$$
$$0 = -x - 4$$
$$x = -4$$
$$x = -4$$

We can substitute this back for x into the first equation to find y.

$$y = -5(-4) - 3 \quad \text{(from the first equation, after we had solved for } y \text{ in terms of } x)$$
$$y = 20 + (-3)$$
$$y = 17$$

So the solution is $x = -4$, $y = 17$.

Example 5.2.7 Solve this system of equations using substitution:

$$\begin{cases} 3x - 7y = \ \ 5 \\ -5x + 2y = 11 \end{cases}$$

Explanation. We need to solve for *one* of the variables in *one* of our equations. Looking at both equations, it will be easiest to solve for y in the second equation. The coefficient of y in that equation is smallest.

$$-5x + 2y = 11$$

$$2y = 11 + 5x$$

$$\frac{2y}{2} = \frac{11 + 5x}{2}$$

$$y = \frac{11}{2} + \frac{5}{2}x$$

Note that in this example, there are fractions once we solve for y. We should take care with the steps that follow that the fraction arithmetic is correct.

Replace y in the first equation with $\frac{11}{2} + \frac{5}{2}x$, giving us a linear equation in only one variable, x, that we may solve:

$$3x - 7y = 5$$

$$3x - 7\left(\frac{11}{2} + \frac{5}{2}x\right) = 5$$

$$3x - 7 \cdot \frac{11}{2} - 7 \cdot \frac{5}{2}x = 5$$

$$3x - \frac{77}{2} - \frac{35}{2}x = 5$$

$$\frac{6}{2}x - \frac{77}{2} - \frac{35}{2}x = 5$$

$$-\frac{29}{2}x - \frac{77}{2} = 5$$

$$-\frac{29}{2}x = \frac{10}{2} + \frac{77}{2}$$

$$-\frac{29}{2}x = \frac{87}{2}$$

$$-\frac{2}{29} \cdot \left(-\frac{29}{2}x\right) = -\frac{2}{29} \cdot \left(\frac{87}{2}\right)$$

$$x = -3$$

Now that we have the value for x, we need to find the value for y. We have already solved the second equation for y, so that is the easiest equation to use.

$$y = \frac{11}{2} + \frac{5}{2}x$$

$$y = \frac{11}{2} + \frac{5}{2}(-3)$$

$$y = \frac{11}{2} - \frac{15}{2}$$

$$y = -\frac{4}{2}$$

$$y = -2$$

To check this solution, we replace x with -3 and y with -2 in each equation:

$$3x - 7y = 5 \qquad\qquad\qquad -5x + 2y = 11$$

$$3(-3) - 7(-2) \overset{?}{=} 5 \qquad\qquad -5(-3) + 2(-2) \overset{?}{=} 11$$

$$-9 + 14 \stackrel{\checkmark}{=} 5 \qquad\qquad 15 - 4 \stackrel{\checkmark}{=} 11$$

We conclude then that this system of equations is true when $x = -3$ and $y = -2$. Our solution is the point $(-3, -2)$ and we write the solution set as $\{(-3, -2)\}$.

Example 5.2.8 Clearing Fraction Denominators Before Solving. Solve the system of equations using the substitution method:

$$\begin{cases} \dfrac{x}{3} - \dfrac{1}{2}y = \dfrac{5}{6} \\ \dfrac{1}{4}x = \dfrac{y}{2} + 1 \end{cases}$$

Explanation. When a system of equations has fraction coefficients, it can be helpful to take steps that replace the fractions with whole numbers. With each equation, we may multiply each side by the least common multiple of all the denominators.

In the first equation, the least common multiple of the denominators is 6, so:

$$\frac{x}{3} - \frac{1}{2}y = \frac{5}{6}$$
$$6 \cdot \left(\frac{x}{3} - \frac{1}{2}y\right) = 6 \cdot \frac{5}{6}$$
$$6 \cdot \frac{x}{3} - 6 \cdot \frac{1}{2}y = 6 \cdot \frac{5}{6}$$
$$2x - 3y = 5$$

In the second equation, the least common multiple of the denominators is 4, so:

$$\frac{1}{4}x = \frac{y}{2} + 1$$
$$4 \cdot \frac{1}{4}x = 4 \cdot \frac{y}{2} + 4 \cdot 1$$
$$4 \cdot \frac{1}{4}x = 4 \cdot \frac{y}{2} + 4 \cdot 1$$
$$x = 2y + 4$$

Now we have this system that is equivalent to the original system of equations, but there are no fraction coefficients:

$$\begin{cases} 2x - 3y = 5 \\ x = 2y + 4 \end{cases}$$

The second equation is already solved for x, so we will substitute x in the first equation with $2y + 4$, and we have:

$$2x - 3y = 5$$
$$2(2y + 4) - 3y = 5$$
$$4y + 8 - 3y = 5$$
$$y + 8 = 5$$
$$y = -3$$

And we have solved for y. To find x, we know $x = 2y + 4$, so we have:

$$x = 2y + 4$$
$$x = 2(-3) + 4$$
$$x = -6 + 4$$
$$x = -2$$

The solution is $(-2, -3)$. Checking this solution is left as an exercise.

Checkpoint 5.2.9. Try a similar exercise.

Solve the following system of equations.

$$\begin{cases} -3 = -m + \dfrac{1}{2}r \\ -m + \dfrac{4}{3} = -r \end{cases}$$

Explanation. If an equation involves fractions, it is helpful to clear denominators by multiplying both sides of the equation by a common multiple of the denominators.

$$\begin{cases} 2(-3) = 2\left(-m + \dfrac{1}{2}r\right) \\ 3\left(-m + \dfrac{4}{3}\right) = 3(-r) \end{cases}$$

$$\begin{cases} -6 = -2m + r \\ -3m + 4 = -3r \end{cases}$$

Since one of the coefficients of r is 1, it is wise to solve for r in terms of the other variable and then use substitution to complete the problem.

$$2m - 6 = r \quad \text{(from the first equation)}$$

which we can substitute in for r into the second equation:

$$4 - 3m = -3(2m - 6) \quad \text{(from the second equation)}$$
$$4 - 3m = 18 - 6m$$
$$3m = 14$$
$$m = \frac{14}{3}$$

We can substitute this back for m into the first equation to find r.

$$2\left(\frac{14}{3}\right) - 6 = r \quad \text{(from the first equation, after we had solved for } r \text{ in terms of } m)$$
$$\frac{28}{3} + (-6) = r$$
$$\frac{10}{3} = r$$

So the solution is $m = \frac{14}{3}, r = \frac{10}{3}$.

5.2.2 Applications of Systems of Equations

In Example 5.2.2, we set up and solved a system of linear equations for a real-world application. The quantities in that problem included rate units (dollars per transaction). Here are some more scenarios that we can model with systems of linear equations.

Example 5.2.10 Two Different Interest Rates. Notah made some large purchases with his two credit cards one month and took on a total of $8,400 in debt from the two cards. He didn't make any payments the first month, so the two credit card debts each started to accrue interest. That month, his Visa card charged 2% interest and his Mastercard charged 2.5% interest. Because of this, Notah's total debt grew by $178. How much money did Notah charge to each card?

Explanation. To start, we will define two variables based on our two unknowns. Let v be the amount charged to the Visa card (in dollars) and let m be the amount charged to the Mastercard (in dollars).

To determine our equations, notice that we are given two different totals. We will use these to form our two equations. The total amount charged is $8,400 so we have:

$$(v \text{ dollars}) + (m \text{ dollars}) = \$8400$$

Or without units:

$$v + m = 8400$$

The other total we were given is the total amount of interest, $178, which is also in dollars. The Visa had v dollars charged to it and accrues 2% interest. So $0.02v$ is the dollar amount of interest that comes from using this card. Similarly, $0.025m$ is the dollar amount of interest from using the Mastercard. Together:

$$0.02(v \text{ dollars}) + 0.025(m \text{ dollars}) = \$178$$

Or without units:

$$0.02v + 0.025m = 178$$

As a system, we write:

$$\begin{cases} v + m = 8400 \\ 0.02v + 0.025m = 178 \end{cases}$$

To solve this system by substitution, notice that it will be easier to solve for one of the variables in the first equation. We'll solve that equation for v:

$$v + m = 8400$$
$$v = 8400 - m$$

Now we will substitute $8400 - m$ for v in the second equation:

$$0.02v + 0.025m = 178$$
$$0.02(8400 - m) + 0.025m = 178$$
$$168 - 0.02m + 0.025m = 178$$
$$168 + 0.005m = 178$$
$$0.005m = 10$$
$$\frac{0.005m}{0.005} = \frac{10}{0.005}$$
$$m = 2000$$

Lastly, we can determine the value of v by using the earlier equation where we isolated v:

$$v = 8400 - m$$
$$v = 8400 - 2000$$

$$v = 6400$$

In summary, Notah charged \$6400 to the Visa and \$2000 to the Mastercard. We should check that these numbers work as solutions to our original system *and* that they make sense in context. (For instance, if one of these numbers were negative, or was something small like \$0.50, they wouldn't make sense as credit card debt.)

The next two examples are called **mixture problems**, because they involve mixing two quantities together to form a combination and we want to find out how much of each quantity to mix.

Example 5.2.11 Mixing Solutions with Two Different Concentrations. LaVonda is a meticulous bartender and she needs to serve 600 milliliters of Rob Roy, an alcoholic cocktail that is 34% alcohol by volume. The main ingredients are scotch that is 42% alcohol and vermouth that is 18% alcohol. How many milliliters of each ingredient should she mix together to make the concentration she needs?

Explanation. The two unknowns are the quantities of each ingredient. Let s be the amount of scotch (in mL) and let v be the amount of vermouth (in mL).

One quantity given to us in the problem is 600 mL. Since this is the total volume of the mixed drink, we must have:

$$(s\,\text{mL}) + (v\,\text{mL}) = 600\,\text{mL}$$

Or without units:

$$s + v = 600$$

To build the second equation, we have to think about the alcohol concentrations for the scotch, vermouth, and Rob Roy. It can be tricky to think about percentages like these correctly. One strategy is to focus on the *amount* (in mL) of *alcohol* being mixed. If we have s milliliters of scotch that is 42% alcohol, then $0.42s$ is the actual *amount* (in mL) of alcohol in that scotch. Similarly, $0.18v$ is the amount of alcohol in the vermouth. And the final cocktail is 600 mL of liquid that is 34% alcohol, so it has $0.34(600) = 204$ milliliters of alcohol. All this means:

$$0.42(s\,\text{mL}) + 0.18(v\,\text{mL}) = 204\,\text{mL}$$

Or without units:

$$0.42s + 0.18v = 204$$

So our system is:

$$\begin{cases} s + v = 600 \\ 0.42s + 0.18v = 204 \end{cases}$$

To solve this system, we'll solve for s in the first equation:

$$s + v = 600$$
$$s = 600 - v$$

And then substitute s in the second equation with $600 - v$:

$$0.42s + 0.18v = 204$$
$$0.42(600 - v) + 0.18v = 204$$
$$252 - 0.42v + 0.18v = 204$$

$$252 - 0.24v = 204$$
$$-0.24v = -48$$
$$\frac{-0.24v}{-0.24} = \frac{-48}{-0.24}$$
$$v = 200$$

As a last step, we will determine s using the equation where we had isolated s:

$$s = 600 - v$$
$$s = 600 - 200$$
$$s = 400$$

In summary, LaVonda needs to combine 400 mL of scotch with 200 mL of vermouth to create 600 mL of Rob Roy that is 34% alcohol by volume.

As a check for Example 5.2.11, we will use **estimation** to see that our solution is reasonable. Since LaVonda is making a 34% solution, she would need to use more of the 42% concentration than the 18% concentration, because 34% is closer to 42% than to 18%. This agrees with our answer because we found that she needed 400 mL of the 42% solution and 200 mL of the 18% solution. This is an added check that we have found reasonable answers.

Example 5.2.12 Mixing a Coffee Blend. Desi owns a coffee shop and they want to mix two different types of coffee beans to make a blend that sells for $12.50 per pound. They have some coffee beans from Columbia that sell for $9.00 per pound and some coffee beans from Honduras that sell for $14.00 per pound. How many pounds of each should they mix to make 30 pounds of the blend?

Explanation. Before we begin, it may be helpful to try to estimate the solution. Let's compare the three prices. Since $12.50 is between the prices of $9.00 and $14.00, this mixture is possible. Now we need to estimate the amount of each type needed. The price of the blend ($12.50 per pound) is closer to the higher priced beans ($14.00 per pound) than the lower priced beans ($9.00 per pound). So we will need to use more of that type. Keeping in mind that we need a total of 30 pounds, we roughly estimate 20 pounds of the $14.00 Honduran beans and 10 pounds of the $9.00 Columbian beans. How good is our estimate? Next we will solve this exercise exactly.

To set up our system of equations we define variables, letting C be the amount of Columbian coffee beans (in pounds) and H be the amount of Honduran coffee beans (in pounds).

The equations in our system will come from the total amount of beans and the total cost. The equation for the total amount of beans can be written as:

$$(C \text{ lb}) + (H \text{ lb}) = 30 \text{ lb}$$

Or without units:

$$C + H = 30$$

To build the second equation, we have to think about the cost of all these beans. If we have C pounds of Columbian beans that cost $9.00 per pound, then $9C$ is the cost of those beans in dollars. Similarly, $14H$ is the cost of the Honduran beans. And the total cost is for 30 pounds of beans priced at $12.50 per pound, totaling $12.5(30) = 37.5$ dollars. All this means:

$$\left(9 \tfrac{\text{dollars}}{\text{lb}}\right)(C \text{ lb}) + \left(14 \tfrac{\text{dollars}}{\text{lb}}\right)(H \text{ lb}) = \left(12.50 \tfrac{\text{dollars}}{\text{lb}}\right)(30 \text{ lb})$$

Or without units and carrying out the multiplication on the right:

$$9C + 14H = 37.5$$

Now our system is:

$$\begin{cases} C + H = 30 \\ 9C + 14H = 12.50(30) \end{cases}$$

To solve the system, we'll solve the first equation for C:

$$C + H = 30$$
$$C = 30 - H$$

Next, we'll substitute C in the second equation with $30 - H$:

$$9C + 14H = 375$$
$$9(30 - H) + 14H = 375$$
$$270 - 9H + 14H = 375$$
$$270 + 5H = 375$$
$$5H = 105$$
$$H = 21$$

Since $H = 21$, we can conclude that $C = 9$.

In summary, Desi needs to mix 21 pounds of the Honduran coffee beans with 9 pounds of the Columbian coffee beans to create this blend. Our estimate at the beginning was pretty close, so we feel this answer is reasonable.

5.2.3 Solving Special Systems of Equations with Substitution

Remember the two special cases we encountered when solving by graphing in Section 5.1? If the two lines represented by a system of equations have the same slope, then they might be separate lines that never meet, meaning the system has no solutions. Or they might coincide as the same line, in which case there are infinitely many solutions represented by all the points on that line. Let's see what happens when we use the substitution method on each of the special cases.

Example 5.2.13 A System with No Solution. Solve the system of equations using the substitution method:

$$\begin{cases} y = 2x - 1 \\ 4x - 2y = 3 \end{cases}$$

Explanation. Since the first equation is already solved for y, we will substitute $2x - 1$ for y in the second equation, and we have:

$$4x - 2y = 3$$
$$4x - 2(2x - 1) = 3$$
$$4x - 4x + 2 = 3$$
$$2 = 3$$

Even though we were only intending to substitute away y, we ended up with an equation where there are no variables at all. This will happen whenever the lines have the same slope. This tells us the system represents either parallel or coinciding lines. Since $2 = 3$ is false no matter what values x and y might be, there can be no solution to the system. So the lines are parallel and *distinct*. We write the solution set using the empty set symbol: the solution set is $\emptyset$.

To verify this, re-write the second equation, $4x - 2y = 3$, in slope-intercept form:

$$4x - 2y = 3$$
$$-2y = -4x + 3$$
$$\frac{-2y}{-2} = \frac{-4x + 3}{-2}$$
$$y = \frac{-4x}{-2} + \frac{3}{-2}$$
$$y = 2x - \frac{3}{2}$$

So the system is equivalent to:

$$\begin{cases} y = 2x - 1 \\ y = 2x - \dfrac{3}{2} \end{cases}$$

Now it is easier to see that the two lines have the same slope but different y-intercepts. They are parallel and distinct lines, so the system has no solution.

Example 5.2.14 A System with Infinitely Many Solutions. Solve the system of equations using the substitution method:

$$\begin{cases} y = 2x - 1 \\ 4x - 2y = 2 \end{cases}$$

Explanation. Since $y = 2x - 1$, we will substitute $2x - 1$ for y in the second equation and we have:

$$4x - 2y = 2$$
$$4x - 2(2x - 1) = 2$$
$$4x - 4x + 2 = 2$$
$$2 = 2$$

Even though we were only intending to substitute away y, we ended up with an equation where there are no variables at all. This will happen whenever the lines have the same slope. This tells us the system represents either parallel or coinciding lines. Since $2 = 2$ is true no matter what values x and y might be, the system equations are true no matter what x is, as long as $y = 2x - 1$. So the lines *coincide*. We write the solution set as $\{(x, y) \mid y = 2x - 1\}$.

To verify this, re-write the second equation, $4x - 2y = 2$, in slope-intercept form:

$$4x - 2y = 2$$
$$-2y = -4x + 2$$

$$\frac{-2y}{-2} = \frac{-4x}{-2} + \frac{2}{-2}$$
$$y = 2x - 1$$

The system looks like:

$$\begin{cases} y = 2x - 1 \\ y = 2x - 1 \end{cases}$$

Now it is easier to see that the two equations represent the same line. Every point on the line is a solution to the system, so the system has infinitely many solutions. The solution set is $\{(x, y) \mid y = 2x - 1\}$.

Exercises

Review and Warmup   Solve the equation.

1. $\frac{7}{6} - 8q = 5$

2. $\frac{3}{2} - 5y = 5$

3. $\frac{7}{10} - \frac{1}{10}r = 2$

4. $\frac{5}{6} - \frac{1}{6}a = 10$

Solve the linear equation for y.

5.
$$9x - 3y = -6$$

6.
$$20x + 4y = -32$$

7.
$$24x - 4y = 76$$

8.
$$-8x - 4y = -36$$

9.
$$-7x - y = -9$$

10.
$$-9x - y = 5$$

Solving System of Equations Using Substitution   Solve the following system of equations.

11.
$$\begin{cases} x = -5 \\ 5 + 5x = -5n \end{cases}$$

12.
$$\begin{cases} 0 = -C \\ 0 = 1 + t + 2C \end{cases}$$

13.
$$\begin{cases} -2a = -8 \\ 4C - 8 = -3a \end{cases}$$

14.
$$\begin{cases} 0 = 5q + 10 \\ -4q - 2y = -20 \end{cases}$$

15.
$$\begin{cases} y = -16 - 4x \\ y = 4x \end{cases}$$

16.
$$\begin{cases} y = -2x + 6 \\ y = x + 6 \end{cases}$$

17.

$$\begin{cases} y = -5x - 8 \\ 2y + 2x = -16 \end{cases}$$

18.

$$\begin{cases} y = -x - 3 \\ 4y + 5x = -1 \end{cases}$$

19.

$$\begin{cases} y = -36 - 2x \\ 2x - 5y = -24 \end{cases}$$

20.

$$\begin{cases} B = -38 - 3p \\ -5B + 4p = 19 \end{cases}$$

21.

$$\begin{cases} a = 7 + 2y \\ 3y - 5a = 7 \end{cases}$$

22.

$$\begin{cases} r = -4b + 18 \\ -r + 5b = 9 \end{cases}$$

23.

$$\begin{cases} 2t - 6 = a \\ 6 + 2t = 2a \end{cases}$$

24.

$$\begin{cases} -10 + 4y = 3x \\ x + 5 = 3y \end{cases}$$

25.

$$\begin{cases} y = x - \dfrac{1}{4} \\ y = x + \dfrac{2}{5} \end{cases}$$

26.

$$\begin{cases} y = 1 - 3x \\ y = -3x - 1 \end{cases}$$

27.

$$\begin{cases} -3x - 9 = 2y \\ 3x - y = -9 \end{cases}$$

28.

$$\begin{cases} 5x + 54 = -y \\ -3x = 18 - 3y \end{cases}$$

29.

$$\begin{cases} 2 = 4a + 2m \\ 0 = -1 - 3m \end{cases}$$

30.

$$\begin{cases} 0 = -3 + 4m - 4b \\ -1 = -4m \end{cases}$$

31.

$$\begin{cases} -5a - 2 = 2q \\ -3q + a = -2 \end{cases}$$

32.

$$\begin{cases} -3m = -3B + 5 \\ B + 2m = 4 \end{cases}$$

33.

$$\begin{cases} -x + 5y = 2 \\ 2y - 4 = 4x \end{cases}$$

34.

$$\begin{cases} -3x + 4y = -4 \\ -3 = -x + 5y \end{cases}$$

35.

$$\begin{cases} -\dfrac{4}{5} = -\dfrac{2}{5}x + \dfrac{3}{5}y \\ x + \dfrac{5}{4} = 0 \end{cases}$$

36.

$$\begin{cases} x = -\dfrac{3}{5} \\ -\dfrac{4}{5}y = -x - 1 \end{cases}$$

37.

$$\begin{cases} 1 = -\dfrac{5}{4}x \\ \dfrac{3}{4} - \dfrac{1}{2}y = \dfrac{2}{3}x \end{cases}$$

38.
$$\begin{cases} y - \dfrac{1}{5}B = -\dfrac{2}{5} \\ \quad 0 = -\dfrac{5}{3}B + \dfrac{1}{2} \end{cases}$$

39.
$$\begin{cases} \dfrac{1}{4}a - 1 = -A \\ \quad \dfrac{1}{4}A = -\dfrac{2}{3} + a \end{cases}$$

40.
$$\begin{cases} 0 = \dfrac{5}{4} + \dfrac{5}{2}q + p \\ -\dfrac{4}{5} = -\dfrac{5}{3}q + 3p \end{cases}$$

41.
$$\begin{cases} 5 = -B - 3a \\ -a = -\dfrac{3}{2}B + 1 \end{cases}$$

42.
$$\begin{cases} \dfrac{2}{3} - x = -2y \\ \quad 0 = 1 - \dfrac{3}{5}y - x \end{cases}$$

43.
$$\begin{cases} -5y + 4x = 1 \\ -2x + 5y = -2 \end{cases}$$

44.
$$\begin{cases} x + 3y = -4 \\ -2y + 3x = -2 \end{cases}$$

45.
$$\begin{cases} -4x + 2y = -7 \\ -4x + 3y = -\dfrac{13}{2} \end{cases}$$

46.
$$\begin{cases} 4x - 2y = -\dfrac{37}{11} \\ \quad x + y = \dfrac{97}{22} \end{cases}$$

47.
$$\begin{cases} -\dfrac{1}{5}x + \dfrac{1}{3}y = \dfrac{13}{42} \\ -\dfrac{1}{3}x + \dfrac{1}{4}y = \dfrac{151}{840} \end{cases}$$

48.
$$\begin{cases} \dfrac{1}{5}x + \dfrac{1}{3}y = \dfrac{41}{120} \\ \dfrac{1}{2}x + \dfrac{1}{4}y = \dfrac{47}{96} \end{cases}$$

49.
$$\begin{cases} 3x + 2y = -1 \\ 5x + y = -18 \end{cases}$$

50.
$$\begin{cases} 4x + 4y = -12 \\ 2x + 4y = -8 \end{cases}$$

51.
$$\begin{cases} 4x - 4y = 32 \\ 5x + 2y = -16 \end{cases}$$

52.
$$\begin{cases} 6x + 5y = 42 \\ 6x - 5y = -18 \end{cases}$$

53.
$$\begin{cases} -4x - 4y = -12 \\ -2x - 4y = -2 \end{cases}$$

54.
$$\begin{cases} -2x - y = -5 \\ -2x - 2y = 4 \end{cases}$$

55.
$$\begin{cases} -4x + 2y = -26 \\ \quad 2x = 18 \end{cases}$$

56.
$$\begin{cases} x - y = -7 \\ \quad 2x = -18 \end{cases}$$

57.
$$\begin{cases} x + y = 3 \\ -2x - 2y = 3 \end{cases}$$

58.
$$\begin{cases} 2x + 4y = 2 \\ 6x + 12y = 2 \end{cases}$$

59.

$$\begin{cases} 2x + 2y = 2 \\ 8x + 8y = 8 \end{cases}$$

60.

$$\begin{cases} 3x + 5y = 2 \\ -6x - 10y = -4 \end{cases}$$

Applications

61. A rectangle's length is 4 feet longer than four times its width. The rectangle's perimeter is 358 feet. Find the rectangle's length and width.

The rectangle's length is _____ feet, and its width is _____ feet.

62. A school fund raising event sold a total of 208 tickets and generated a total revenue of $808.80. There are two types of tickets: adult tickets and child tickets. Each adult ticket costs $4.75, and each child ticket costs $3.35. Write and solve a system of equations to answer the following questions.

_____ adult tickets and _____ child tickets were sold.

63. Phone Company A charges a monthly fee of $38.70, and $0.05 for each minute of talk time. Phone Company B charges a monthly fee of $30.00, and $0.08 for each minute of talk time. Write and solve a system equation to answer the following questions.

These two companies would charge the same amount on a monthly bill when the talk time was ____ minutes.

64. Company A's revenue this fiscal year is $840,000, but its revenue is decreasing by $14,000 each year. Company B's revenue this fiscal year is $180,000, and its revenue is increasing by $19,000 each year. Write and solve a system of equations to answer the following question.

After _____ years, Company B will catch up with Company A in revenue.

65. A test has 23 problems, which are worth a total of 96 points. There are two types of problems in the test. Each multiple-choice problem is worth 2 points, and each short-answer problem is worth 7 points. Write and solve a system equation to answer the following questions.

This test has _____ multiple-choice problems and _____ short-answer problems.

66. Penelope invested a total of $5,500 in two accounts. One account pays 7% interest annually; the other pays 6% interest annually. At the end of the year, Penelope earned a total of $365 in interest. Write and solve a system of equations to find how much money Penelope invested in each account.

Penelope invested _____ in the 7% account and _____ in the 6% account.

67. Michael invested a total of $11,000 in two accounts. After a year, one account lost 7.2%, while the other account gained 4.5%. In total, Michael lost $441. Write and solve a system of equations to find how much money Michael invested in each account.

    Michael invested [ ] in the account with 7.2% loss and [ ] in the account with 4.5% gain.

68. Town A and Town B were located close to each other, and recently merged into one city. Town A had a population with 6% native Americans. Town B had a population with 12% native Americans. After the merge, the new city has a total of 4800 residents, with 10.5% native Americans. Write and solve a system of equations to find how many residents Town A and Town B used to have.

    Town A used to have [ ] residents, and Town B used to have [ ] residents.

69. You poured some 8% alcohol solution and some 6% alcohol solution into a mixing container. Now you have 600 grams of 6.6% alcohol solution. How many grams of 8% solution and how many grams of 6% solution did you pour into the mixing container?

    Write and solve a system equation to answer the following questions.

    You mixed [ ] grams of 8% solution with [ ] grams of 6% solution.

70. Neil invested a total of $8,000 in two accounts. One account pays 5% interest annually; the other pays 4% interest annually. At the end of the year, Neil earned a total of $355 in interest. How much money did Neil invest in each account?

    Neil invested [ ] in the 5% account and [ ] in the 4% account.

71. Hannah invested a total of $72,000 in two accounts. One account pays 3.8% interest annually; the other pays 2.4% interest annually. At the end of the year, Hannah earned a total interest of $2,092. How much money did Hannah invest in each account?

    Hannah invested [ ] in the 3.8% account and [ ] in the 2.4% account.

72. Charity invested a total of $5,500 in two accounts. One account pays 5% interest annually; the other pays 6% interest annually. At the end of the year, Charity earned the same amount of interest from both accounts. How much money did Charity invest in each account?

    Charity invested [ ] in the 5% account and [ ] in the 6% account.

73. Page invested a total of $86,000 in two accounts. One account pays 5.3% interest annually; the other pays 4.7% interest annually. At the end of the year, Page earned the same amount of interest from both accounts. How much money did Page invest in each account?

    Page invested [ ] in the 5.3% account [ ] in the 4.7% account.

74. Laurie invested a total of $10,000 in two accounts. After a year, one account had *earned* 10.1%, while the other account had *lost* 5.4%. In total, Laurie had a net gain of $855. How much money did Laurie invest in each account?

Laurie invested [   ] in the account that grew by 10.1% and [   ] in the account that fell by 5.4%.

75. You've poured some 6% (by mass) alcohol solution and some 12% alcohol solution into a large glass mixing container. Now you have 680 grams of 9% alcohol solution. How many grams of 6% solution and how many grams of 12% solution did you pour into the mixing container?

You poured [   ] grams of 6% solution and [   ] grams of 12% solution into the mixing container.

76. A store has some beans selling for $2.30 per pound, and some vegetables selling for $5.00 per pound. The store plans to use them to produce 13 pounds of mixture and sell for $3.92 per pound. How many pounds of beans and how many pounds of vegetables should be used?

To produce 13 pounds of mixture, the store should use [   ] pounds of beans and [   ] pounds of vegetables.

77. Town A and Town B were located close to each other, and recently merged into one city. Town A had a population with 6% native Americans. Town B had a population with 10% native Americans. After the merge, the new city has a total of 4000 residents, with 8.8% native Americans. How many residents did Town A and Town B used to have?

Town A used to have [   ] residents, and Town B used to have [   ] residents.

5.3 Elimination

We learned how to solve a system of linear equations using substitution in Section 5.2. In this section, we will learn a second symbolic method for solving systems of linear equations.

5.3.1 Solving Systems of Equations by Elimination

Example 5.3.2 Alicia has $1000 to give to her two grandchildren for New Year's. She would like to give the older grandchild $120 more than the younger grandchild, because that is the cost of the older grandchild's college textbooks this term. How much money should she give to each grandchild?

To answer this question, we will demonstrate a new technique. You may have a very good way for finding how much money Alicia should give to each grandchild, but right now we will try to see this new method.

Let A be the dollar amount she gives to her older grandchild, and B be the dollar amount she gives to her younger grandchild. (As always, we start solving a word problem like this by defining the variables, including their units.) Since the total she has to give is $1000, we can say that $A + B = 1000$. And since she wants to give $120 more to the older grandchild, we can say that $A - B = 120$. So we have the system of equations:

$$\begin{cases} A + B = 1000 \\ A - B = 120 \end{cases}$$

We could solve this system by substitution as we learned in Section 5.2, but there is an easier method. If we add together the *left* sides from the two equations, it should equal the sum of the *right* sides:

$$\begin{array}{c} A + B \\ + A - B \end{array} = \begin{array}{c} 1000 \\ + 120 \end{array} \quad .$$

So we have:

$$2A = 1120$$

Note that the variable B is eliminated. This happened because the " $+ B$" and the " $- B$" were perfectly in shape to cancel each other out. With only one variable left, it doesn't take much to finish:

$$2A = 1120$$
$$A = 560$$

To finish solving this system of equations, we need the value of B. For now, an easy way to find B is to substitute in our value of A into one of the original equations:

$$A + B = 1000$$
$$560 + B = 1000$$
$$B = 440$$

To check our work, substitute $A = 560$ and $B = 440$ into the original equations:

$$A + B = 1000 \qquad\qquad A - B = 120$$

$$560 + 440 \overset{?}{=} 1000 \qquad\qquad 560 - 440 \overset{?}{=} 120$$

$$1000 \overset{\checkmark}{=} 1000 \qquad\qquad 120 \overset{\checkmark}{=} 120$$

This confirms that our solution is correct. In summary, Alicia should give \$560 to her older grandchild, and \$440 to her younger grandchild.

This method for solving the system of equations in Example 5.3.2 worked because B and $-B$ add to zero. Once the B-terms were eliminated we were able to solve for A. This method is called the **elimination method**. Some textbooks call it the **addition method**, because we added the corresponding sides from the two equations to eliminate a variable.

If neither variable can be immediately eliminated, we can still use this method but it will require that we first adjust one or both of the equations. Let's look at an example where we need to adjust one of the equations.

Example 5.3.3 Scaling One Equation. Solve the system of equations using the elimination method.

$$\begin{cases} 3x - 4y = 2 \\ 5x + 8y = 18 \end{cases}$$

Explanation. To start, we want to see whether it will be easier to eliminate x or y. We see that the coefficients of x in each equation are 3 and 5, and the coefficients of y are -4 and 8. Because 8 is a multiple of 4 and the coefficients already have opposite signs, the y variable will be easier to eliminate.

To eliminate the y terms, we will multiply each side of the first equation by 2 so that we will have $-8y$. We can call this process **scaling** the first equation by 2.

$$\begin{cases} 2 \cdot (3x - 4y) = 2 \cdot (2) \\ 5x + 8y = 18 \end{cases}$$

$$\begin{cases} 6x - 8y = 4 \\ 5x + 8y = 18 \end{cases}$$

We now have an equivalent system of equations where the y-terms can be eliminated:

$$\begin{array}{r} 6x - 8y \\ + 5x + 8y \end{array} = \begin{array}{l} 4 \\ + 18 \end{array}$$

So we have:

$$11x = 22$$
$$x = 2$$

To solve for y, we will substitute 2 for x into either of the original equations or the new one. We will use the original first equation, $3x - 4y = 2$:

$$3x - 4y = 2$$
$$3(2) - 4y = 2$$
$$6 - 4y = 2$$

$$-4y = -4$$
$$y = 1$$

Our solution is $x = 2$ and $y = 1$. We will check this in both of the original equations:

$$5x + 8y = 18 \qquad\qquad\qquad 3x - 4y = 2$$
$$5(2) + 8(1) \overset{?}{=} 18 \qquad\qquad 3(2) - 4(1) \overset{?}{=} 2$$
$$10 + 8 \overset{\checkmark}{=} 18 \qquad\qquad\qquad 6 - 4 \overset{\checkmark}{=} 2$$

The solution to this system is $(2, 1)$ and the solution set is $\{(2, 1)\}$.

Checkpoint 5.3.4. Try a similar exercise.

Solve the following system of equations.

$$\begin{cases} 5x + 4y = -7 \\ 5x + 2y = -1 \end{cases}$$

Explanation.

1. We subtract the two equations, which will cancel the terms in involving x and give $4y - 2y = -7 - (-1)$.

2. This gives $y = -3$.

3. Now that we have y, we find x using either equation. Let's use the first: $5x - 12 = -7$, so $x = 1$.

4. The solution to the system is $(1, -3)$. It is left as an exercise to check. Please also note that you may have solved this problem a different way.

Here's an example where we have to scale both equations.

Example 5.3.5 Scaling Both Equations. Solve the system of equations using the elimination method.

$$\begin{cases} 2x + 3y = 10 \\ -3x + 5y = -15 \end{cases}$$

Explanation. Considering the coefficients of x (2 and -3) and the coefficients of y (3 and 5) we see that we cannot eliminate the x or the y variable by scaling a single equation. We will need to scale *both*.

The x-terms already have opposite signs, so we choose to eliminate x. The least common multiple of 2 and 3 is 6. We can scale the first equation by 3 and the second equation by 2 so that the equations have terms $6x$ and $-6x$, which will cancel when added.

$$\begin{cases} 3 \cdot (2x + 3y) = 3 \cdot (10) \\ 2 \cdot (-3x + 5y) = 2 \cdot (-15) \end{cases}$$
$$\begin{cases} 6x + 9y = 30 \\ -6x + 10y = -30 \end{cases}$$

At this point we can add the corresponding sides from the two equations and solve for y:

$$\frac{6x + 9y}{-6x + 10y} = \frac{30}{-30}$$

So we have:

$$19y = 0$$
$$y = 0$$

To solve for x, we'll replace y with 0 in $2x + 3y = 10$:

$$2x + 3y = 10$$
$$2x + 3(0) = 10$$
$$2x = 10$$
$$x = 5$$

We'll check the system using $x = 5$ and $y = 0$ in each of the original equations:

$$2x + 3y = 10 \qquad\qquad -3x + 5y = -15$$
$$2(5) + 3(0) \stackrel{?}{=} 10 \qquad\qquad -3(5) + 5(0) \stackrel{?}{=} -15$$
$$10 + 0 \stackrel{\checkmark}{=} 10 \qquad\qquad -15 + 0 \stackrel{\checkmark}{=} -15$$

So the system's solution is $(5, 0)$ and the solution set is $\{(5, 0)\}$.

Checkpoint 5.3.6. Try a similar exercise.

Solve the following system of equations.

$$\begin{cases} 3x + 4y = -26 \\ 5x + 5y = -40 \end{cases}$$

Explanation.

1. Let's multiply the *first* equation by 5 and the *second* equation by 3

$$15x + 20y = -130$$
$$15x + 15y = -120$$

2. Subtracting these two equations gives $20y - 15y = -10$, so $y = -2$.

3. Now that we have y, we can use either equation to find x; let's use the first one:

$$3x + (4) \cdot (-2) = -26$$

so $x = -6$.

4. The solution to the system is $(-6, -2)$. It is left as an exercise to check. Please also note that you may have solved this problem a different way.

Example 5.3.7 Meal Planning. Javed is on a meal plan and needs to consume 600 calories and 20 grams of fat for breakfast. A small avocado contains 300 calories and 30 grams of fat. He has bagels that contain 400 calories and 8 grams of fat. Write and solve a system of equations to determine how much bagel and avocado would combine to make his target calories and fat.

Explanation. To write this system of equations, we first need to define our variables. Let A be the number of avocados consumed and let B be the number of bagels consumed. Both A and B might be fractions. For our first equation, we will count calories from the avocados and bagels:

$$\left(300 \tfrac{\text{calories}}{\text{avocado}}\right)(A \text{ avocados}) + \left(400 \tfrac{\text{calories}}{\text{bagel}}\right)(B \text{ bagel}) = 600 \text{ calories}$$

Or, without the units:

$$300A + 400B = 600$$

Similarly, for our second equation, we will count the grams of fat:

$$\left(30 \tfrac{\text{g fat}}{\text{avocado}}\right)(A \text{ avocados}) + \left(8 \tfrac{\text{g fat}}{\text{bagel}}\right)(B \text{ bagel}) = 20 \text{ g fat}$$

Or, without the units:

$$30A + 8B = 20$$

So the system of equations is:

$$\begin{cases} 300A + 400B = 600 \\ 30A + 8B = 20 \end{cases}$$

Since none of the coefficients are equal to 1, it will be easier to use the elimination method to solve this system. Looking at the terms $300A$ and $30A$, we can eliminate the A variable if we multiply the second equation by -10 to get $-300A$:

$$\begin{cases} 300A + 400B = 600 \\ -10 \cdot (30A + 8B) = -10 \cdot (20) \end{cases}$$

$$\begin{cases} 300A + 400B = 600 \\ -300A + (-80B) = -200 \end{cases}$$

When we add the corresponding sides from the two equations together we have:

$$\begin{array}{c} 300A + 400B \\ -\,300A - 80B \end{array} = \begin{array}{c} 600 \\ -\,200 \end{array}$$

So we have:

$$320B = 400$$
$$\frac{320B}{320} = \frac{400}{320}$$
$$B = \frac{5}{4}$$

We now know that Javed should eat $\frac{5}{4}$ bagels (or one and one-quarter bagels). To determine the number of avocados, we will substitute B with $\frac{5}{4}$ in either of our original equations.

$$300A + 400B = 600$$

$$300A + 400\left(\frac{5}{4}\right) = 600$$

$$300A + 500 = 600$$

$$300A = 100$$

$$\frac{300A}{300} = \frac{100}{300}$$

$$A = \frac{1}{3}$$

To check this result, try using $B = \frac{5}{4}$ and $A = \frac{1}{3}$ in each of the original equations:

$$300A + 400B = 600 \qquad\qquad 30A + 8B = 20$$

$$300\left(\frac{1}{3}\right) + 400\left(\frac{5}{4}\right) \overset{?}{=} 600 \qquad 30\left(\frac{1}{3}\right) + 8\left(\frac{5}{4}\right) \overset{?}{=} 20$$

$$100 + 500 \overset{\checkmark}{=} 600 \qquad\qquad 10 + 10 \overset{\checkmark}{=} 20$$

In summary, Javed can eat $\frac{5}{4}$ of a bagel (so one and one-quarter bagel) and $\frac{1}{3}$ of an avocado in order to consume exactly 600 calories and 20 grams of fat.

5.3.2 Solving Special Systems of Equations with Elimination

Remember the two special cases we encountered when solving by graphing and substitution? Sometimes a system of equations has no solutions at all, and sometimes the solution set is infinite with all of the points on one line satisfying the equations. Let's see what happens when we use the elimination method on each of the special cases.

Example 5.3.8 A System with Infinitely Many Solutions. Solve the system of equations using the elimination method.

$$\begin{cases} 3x + 4y = 5 \\ 6x + 8y = 10 \end{cases}$$

Explanation. To eliminate the x-terms, we multiply each term in the first equation by -2, and we have:

$$\begin{cases} -2 \cdot (3x + 4y) = -2 \cdot 5 \\ 6x + 8y = 10 \end{cases}$$

$$\begin{cases} -6x + -8y = -10 \\ 6x + 8y = 10 \end{cases}$$

We might notice that the equations look very similar. Adding the respective sides of the equation, we have:

$$0 = 0$$

Both of the variables have been eliminated. Since the statement $0 = 0$ is true no matter what x and y are, the solution set is infinite. Specifically, you just need any (x, y) satisfying *one* of the two equations, since the two equations represent the same line. We can write the solution set as $\{(x, y) \mid 3x + 4y = 5\}$.

Example 5.3.9 A System with No Solution. Solve the system of equations using the elimination method.

$$\begin{cases} 10x + 6y = 9 \\ 25x + 15y = 4 \end{cases}$$

Explanation. To eliminate the x-terms, we will scale the first equation by -5 and the second by 2:

$$\begin{cases} -5 \cdot (10x + 6y) = -5 \cdot (9) \\ 2 \cdot (25x + 15y) = 2 \cdot (4) \end{cases}$$
$$\begin{cases} -50x + (-30y) = -45 \\ 50x + 30y = 8 \end{cases}$$

Adding the respective sides of the equation, we have:

$$0 = -37$$

Both of the variables have been eliminated. In this case, the statement $0 = -37$ is just false, no matter what x and y are. So the system has no solution.

5.3.3 Deciding to Use Substitution versus Elimination

In every example so far from this section, both equations were in standard form, $Ax + By = C$. And all of the coefficients were integers. If none of the coefficients are equal to 1 then it is usually easier to use the elimination method, because otherwise you will probably have some fraction arithmetic to do in the middle of the substitution method. If there *is* a coefficient of 1, then it is a matter of preference.

Example 5.3.10 A college used to have a north campus with 6000 students and a south campus with 15,000 students. The percentage of students at the north campus who self-identify as LGBTQ was three times the percentage at the south campus. After the merge, 5.5% of students identify as LGBTQ. What percentage of students on each campus identified as LGBTQ before the merge?

Explanation. We will define N as the percentage (as a decimal) of students at the north campus and S as the percentage (as a decimal) of students at the south campus that identified as LGBTQ. Since the percentage of students at the north campus was three times the percentage at the south campus, we have:

$$N = 3S$$

For our second equation, we will count LGBTQ students at the various campuses. At the north campus, multiply the population, 6000, by the percentage N to get $6000N$. This must be the actual number of LGBTQ students. Similarly, the south campus has $15000S$ LGBTQ students, and the combined school has $21000(0.055) = 1155$. When we combine the two campuses, we have:

$$6000N + 15000S = 1155$$

We write the system as:

$$\begin{cases} N = 3S \\ 6000N + 15000S = 1155 \end{cases}$$

Because the first equation is already solved for N, this is a good time to *not* use the elimination method. Instead we can substitute N in our second equation with $3S$ and solve for S:

$$6000N + 15000S = 1155$$
$$6000(3S) + 15000S = 1155$$
$$18000S + 15000S = 1155$$
$$33000S = 1155$$
$$\frac{33000S}{33000} = \frac{1155}{33000}$$
$$S = 0.035$$

We can determine N using the first equation:

$$N = 3S$$
$$N = 3(0.035)$$
$$N = 0.105$$

Before the merge, 10.5% of the north campus students self-identified as LGBTQ, and 3.5% of the south campus students self-identified as LGBTQ.

If you need to solve a system, and one of the equations is not in standard form, substitution may be easier. But you also may find it easier to convert the equations into standard form. Additionally, if the system's coefficients are fractions or decimals, you may take an additional step to scale the equations so that they only have integer coefficients.

Example 5.3.11 Solve the system of equations using the method of your choice.

$$\begin{cases} -\frac{1}{3}y = \frac{1}{15}x + \frac{1}{5} \\ \frac{5}{2}x - y = 6 \end{cases}$$

Explanation. First, we can cancel the fractions by using the least common multiple of the denominators in each equation, similarly to the topic of Section 3.3. We have:

$$\begin{cases} 15 \cdot -\frac{1}{3}y = 15 \cdot \left(\frac{1}{15}x + \frac{1}{5} \right) \\ 2 \cdot \left(\frac{5}{2}x - y \right) = 2 \cdot (6) \end{cases}$$

$$\begin{cases} -5y = x + 3 \\ 5x - 2y = 12 \end{cases}$$

We could put convert the first equation into standard form by subtracting x from both sides, and then use elimination. However, the x-variable in the first equation has a coefficient of 1, so the substitution method may be faster. Solving for x in the first equation we have:

$$-5y = x + 3$$
$$-5y - 3 = x + 3 - 3$$
$$-5y - 3 = x$$

Substituting $-5y - 3$ for x in the second equation we have:

$$5(-5y - 3) - 2y = 12$$
$$-25y - 15 - 2y = 12$$
$$-27y - 15 = 12$$
$$-27y = 27$$
$$y = -1$$

Using the equation where we isolated x and substituting -1 for y, we have:

$$-5(-1) - 3 = x$$
$$5 - 3 = x$$
$$2 = x$$

The solution is $(2, -1)$. Checking the solution is left as an exercise.

Example 5.3.12 A penny is made by combining copper and zinc. A chemistry reference source says copper has a density of $9 \frac{g}{cm^3}$ and zinc has a density of $7.1 \frac{g}{cm^3}$. A penny's mass is 2.5 g and its volume is $0.35 \, cm^3$. How many cm^3 each of copper and zinc go into one penny?

Explanation. Let c be the volume of copper and z be the volume of zinc in one penny, both measured in cm^3. Since the total volume is $0.35 \, cm^3$, one equation is:

$$(c \, cm^3) + (z \, cm^3) = 0.35 \, cm^3$$

Or without units:

$$c + z = 0.35.$$

For the second equation, we will examine the masses of copper and zinc. Since copper has a density of $9 \frac{g}{cm^3}$ and we are using c to represent the volume of copper, the mass of copper is $9c$. Similarly, the mass of zinc is 7.1. Since the total mass is 2.5 g, we have the equation:

$$\left(9 \tfrac{g}{cm^3}\right)(c \, cm^3) + \left(7.1 \tfrac{g}{cm^3}\right)(z \, cm^3) = 2.5 \, g$$

Or without units:

$$9c + 7.1z = 2.5.$$

So we have a system of equations:

$$\begin{cases} c + z = 0.35 \\ 9c + 7.1z = 2.5 \end{cases}$$

Since the coefficient of c (or z) in the first equation is 1, we could solve for one of these variables and use substitution to complete the problem. Some decimal arithmetic would be required. Alternatively, we can scale the equations by the right power of 10 to make all the coefficients integers:

$$\begin{cases} 100 \cdot (c + z) = 100 \cdot (0.35) \\ 10 \cdot (9c + 7.1z) = 10 \cdot (2.5) \end{cases}$$
$$\begin{cases} 100c + 100z = 35 \\ 90c + 71z = 25 \end{cases}$$

Now to set up elimination, scale each equation again to eliminate c:

$$\begin{cases} 9 \cdot (100c + 100z) = 9 \cdot (35) \\ -10 \cdot (90c + 71z) = -10 \cdot (25) \end{cases}$$
$$\begin{cases} 900c + 900z = 315 \\ -900c + (-710z) = -250 \end{cases}$$

Adding the corresponding sides from the two equations gives

$$190z = 65,$$

from which we find $z = \frac{65}{190} \approx 0.342$. So there is about $0.342 \, \text{cm}^3$ of zinc in a penny.

To solve for c, we can use one of the original equations:

$$c + z = 0.35$$
$$c + 0.342 \approx 0.35$$
$$c \approx 0.008$$

Therefore there is about $0.342 \, \text{cm}^3$ of zinc and $0.008 \, \text{cm}^3$ of copper in a penny.

To summarize, if a variable is already isolated or has a coefficient of 1, consider using the substitution method. If both equations are in standard form or none of the coefficients are equal to 1, we suggest using the elimination method. Either way, if you have fraction or decimal coefficients, it may help to scale your equations so that only integer coefficients remain.

Exercises

Review and Warmup Solve the equation.

1. $\dfrac{7}{2} - 8C = 4$

2. $\dfrac{3}{8} - 6n = 4$

3. $\dfrac{5}{4} - \dfrac{1}{4}q = 3$

4. $\dfrac{9}{10} - \dfrac{1}{10}x = 10$

5. $\dfrac{4r}{7} - 8 = -\dfrac{76}{7}$

6. $\dfrac{6t}{5} - 6 = -\dfrac{72}{5}$

Solving System of Equations by Elimination Solve the following system of equations.

7.
$$\begin{cases} x + y = 6 \\ 2x + 4y = 30 \end{cases}$$

8.
$$\begin{cases} 3x + 5y = 10 \\ 2x + y = 2 \end{cases}$$

9.
$$\begin{cases} 6x + 3y = -3 \\ -x + 5y = -27 \end{cases}$$

10.
$$\begin{cases} -2x + 3y = 19 \\ 2x + 2y = 26 \end{cases}$$

11.
$$\begin{cases} -2x - 5y = -19 \\ -5x - 2y = -37 \end{cases}$$

12.
$$\begin{cases} -5x - 2y = -33 \\ -5x - 5y = -15 \end{cases}$$

13.
$$\begin{cases} x - y = -18 \\ -4x = 40 \end{cases}$$

14.
$$\begin{cases} -5x - 4y = 35 \\ -4x = 28 \end{cases}$$

15.
$$\begin{cases} 2x + y = -8 \\ 6x + 3y = -8 \end{cases}$$

16.
$$\begin{cases} 2x + 5y = -8 \\ 8x + 20y = -8 \end{cases}$$

17.
$$\begin{cases} 3x + 3y = -8 \\ -6x - 6y = 16 \end{cases}$$

18.
$$\begin{cases} 3x + y = -8 \\ -12x - 4y = 32 \end{cases}$$

19.
$$\begin{cases} -2y = 2x + 8 \\ -5x = -8 + y \end{cases}$$

20.
$$\begin{cases} 2x = -18 + y \\ -5y + 4x + 18 = 0 \end{cases}$$

21.
$$\begin{cases} -5x - 3y = 3 \\ -4y + x = 1 \end{cases}$$

22.
$$\begin{cases} -4y + 5x = -2 \\ 2x + 3y = 3 \end{cases}$$

23.
$$\begin{cases} 4m = 2b + 30 \\ 30 - 3b = -m \end{cases}$$

24.
$$\begin{cases} -4x + 18 - r = 0 \\ 2r + 2x - 6 = 0 \end{cases}$$

25.
$$\begin{cases} 3B + C = 3 \\ -2C = 5B + 4 \end{cases}$$

26.
$$\begin{cases} -5 - 3n = q \\ 4 = -2n - 3q \end{cases}$$

27.
$$\begin{cases} -y = -5 + 4x \\ 0 = 2x + 3y + 1 \end{cases}$$

28.
$$\begin{cases} -1 = -5x + 4y \\ -y = -2x \end{cases}$$

29.
$$\begin{cases} -5 - \dfrac{2}{5}x + 4y = 0 \\ 0 = \dfrac{3}{5}x + \dfrac{1}{5} + y \end{cases}$$

30.
$$\begin{cases} 0 = \dfrac{5}{3} - 3x + \dfrac{1}{2}y \\ y = -\dfrac{4}{5} + \dfrac{3}{4}x \end{cases}$$

31.
$$\begin{cases} -B = y - 1 \\ -\dfrac{4}{5}y - \dfrac{1}{5} = -\dfrac{5}{4}B \end{cases}$$

32.
$$\begin{cases} -2t + \dfrac{5}{3} = 2A \\ -t = \dfrac{3}{2}A + 1 \end{cases}$$

33.
$$\begin{cases} 0 = 4q - 2p + 12 \\ -2p + 12 = -4q \end{cases}$$

34.
$$\begin{cases} 5 = 4b + 4A \\ -A - b - 2 = 0 \end{cases}$$

35.
$$\begin{cases} 5x + 2y = \dfrac{21}{2} \\ 4x - 5y = -\dfrac{129}{20} \end{cases}$$

36.
$$\begin{cases} -x - 4y = -\dfrac{72}{11} \\ -3x - 3y = -\dfrac{135}{22} \end{cases}$$

37.
$$\begin{cases} \dfrac{1}{5}x + \dfrac{1}{4}y = \dfrac{127}{280} \\ \dfrac{1}{3}x - \dfrac{1}{2}y = -\dfrac{29}{420} \end{cases}$$

38.
$$\begin{cases} -\dfrac{1}{5}x - \dfrac{1}{4}y = -\dfrac{19}{24} \\ \dfrac{1}{4}x - \dfrac{1}{5}y = \dfrac{11}{15} \end{cases}$$

Applications

39. A test has 20 problems, which are worth a total of 130 points. There are two types of problems in the test. Each multiple-choice problem is worth 5 points, and each short-answer problem is worth 10 points. Write and solve a system of equations to answer the following questions.

This test has ⬚ multiple-choice problems and ⬚ short-answer problems.

40. Barbara invested a total of $5,000 in two accounts. One account pays 5% interest annually; the other pays 6% interest annually. At the end of the year, Barbara earned a total of $255 in interest. Write and solve a system of equations to find how much money Barbara invested in each account.

Barbara invested ⬚ in the 5% account and ⬚ in the 6% account.

41. Diane invested a total of $10,000 in two accounts. After a year, one account lost 6.3%, while the other account gained 6.8%. In total, Diane lost $499. Write and solve a system of equations to find how much money Diane invested in each.

Diane invested [_____] in the account with 6.3% loss and [_____] in the account with 6.8% gain.

42. Town A and Town B were located close to each other, and recently merged into one city. Town A had a population with 10% Hispanics. Town B had a population with 6% Hispanics. After the merge, the new city has a total of 5000 residents, with 7.12% Hispanics. Write and solve a system of equations to find how many residents Town A and Town B used to have.

Town A used to have [_____] residents, and Town B used to have [_____] residents.

43. You poured some 8% alcohol solution and some 12% alcohol solution into a mixing container. Now you have 680 grams of 10% alcohol solution. Write and solve a system of equations to find how many grams of 8% solution and how many grams of 12% solution you poured into the mixing container.

You mixed [_____] grams of 8% solution with [_____] grams of 12% solution.

44. You will purchase some CDs and DVDs. If you purchase 13 CDs and 5 DVDs, it will cost you $85.20; if you purchase 5 CDs and 13 DVDs, it will cost you $130.80. Write and solve a system of equations to answer the following questions.

Each CD costs [_____] and each DVD costs [_____].

45. A school fund raising event sold a total of 211 tickets and generated a total revenue of $861.30. There are two types of tickets: adult tickets and child tickets. Each adult ticket costs $6.60, and each child ticket costs $2.75. Write and solve a system of equations to answer the following questions.

[_____] adult tickets and [_____] child tickets were sold.

46. Phone Company A charges a monthly fee of $35.80, and $0.03 for each minute of talk time. Phone Company B charges a monthly fee of $25.00, and $0.07 for each minute of talk time. Write and solve a system equation to answer the following questions.

These two companies would charge the same amount on a monthly bill when the talk time was [____] minutes.

47. Company A's revenue this fiscal year is $805,000, but its revenue is decreasing by $5,000 each year. Company B's revenue this fiscal year is $409,000, and its revenue is increasing by $17,000 each year. Write and solve a system of equations to answer the following question.

After [_____] years, Company B will catch up with Company A in revenue.

48. If a boat travels from Town A to Town B, it has to travel 990 mi along a river. A boat traveled from Town A to Town B along the river's current with its engine running at full speed. This trip took 27.5 hr. Then the boat traveled back from Town B to Town A, again with the engine at full speed, but this time against the river's current. This trip took 45 hr. Write and solve a system of equations to answer the following questions.

The boat's speed in still water with the engine running at full speed is ⬚.

The river current's speed was ⬚.

49. A small fair charges different admission for adults and children. It charges $3.75 for adults, and $1 for children. On a certain day, the total revenue is $384.25 and the fair admits 200 people. How many adults and children were admitted?

There were ⬚ adults and ⬚ children at the fair.

Challenge

50. Find the value of b so that the system of equations has an infinite number of solutions.

$$\begin{cases} -10x + 35y = 25 \\ 2x - by = -5 \end{cases}$$

5.4 Systems of Linear Equations Chapter Review

5.4.1 Solving Systems of Linear Equations by Graphing

In Section 5.1 we covered the definition of system of linear equations and how a solution to a system of linear equation is a point where the graphs of the two equations cross. We also considered special systems of equations that overlap or never touch.

Example 5.4.1 Solving Systems of Linear Equations by Graphing. Solve the following system of equations by graphing:

$$\begin{cases} y = -\dfrac{2}{3}x - 4 \\ y = -4x - 14 \end{cases}$$

Explanation.

The first equation, $y = -\frac{2}{3}x - 4$, is a linear equation in slope-intercept form with a slope of $-\frac{2}{3}$ and a y-intercept of $(0, -4)$. The second equation, $y = -4x - 14$, is a linear equation in slope-intercept form with a slope of -4 and a y-intercept of $(0, -14)$. We'll use this information to graph both lines in Figure 5.4.2.

The two lines intersect where $x = -3$ and $y = -2$, so the solution of the system of equations is the point $(-3, -2)$. We write the solution set as $\{(-3, -2)\}$.

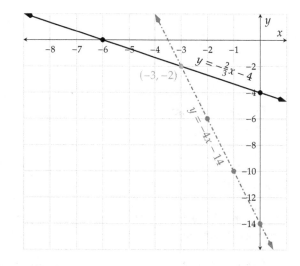

Figure 5.4.2: Graphs of $y = -\frac{2}{3}x - 4$ and $y = -4x - 14$.

Example 5.4.3 Special Systems of Equations. Solve the following system of equations by graphing:

$$\begin{cases} y = \dfrac{3}{2}(x - 1) + 4 \\ 3x - 2y = 4 \end{cases}$$

Explanation. The first equation, $y = \frac{3}{2}(x-1)+4$, is a linear equation in point-slope form with a slope of $\frac{3}{2}$ that passes through the point $(1, 4)$. The second equation, $3x - 2y = 4$, is a linear equation in standard form To graph this line, we either need to find the intercepts or put the equation into slope-intercept form. Just for practice, we will put the line in slope-intercept form.

$$3x - 2y = 4$$
$$-2y = -3x + 4$$

$$\frac{-2y}{-2} = \frac{-3x}{-2} + \frac{4}{-2}$$

$$y = \frac{3}{2}x - 2$$

We'll use this information to graph both lines:

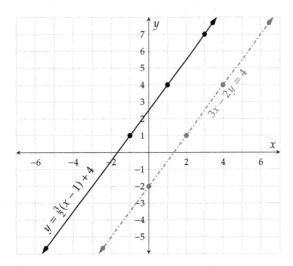

The two lines never intersect: they are parallel. So there are no solutions to the system of equations. We write the solution set as ∅.

Figure 5.4.4: Graphs of $y = \frac{3}{2}(x-1)+4$ and $3x - 2y = 4$.

5.4.2 Substitution

In Section 5.2, we covered the substitution method of solving systems of equations. We isolated one variable in one equation and then substituted into the other equation to solve for one variable.

Example 5.4.5 Solving Systems of Equations Using Substitution. Solve this system of equations using substitution:

$$\begin{cases} -5x + 6y = -10 \\ 4x - 3y = -1 \end{cases}$$

Explanation. We need to solve for *one* of the variables in *one* of our equations. Looking at both equations, it will be best to solve for y in the second equation. The coefficient of y in that equation is smallest.

$$4x - 3y = -1$$
$$-3y = -1 - 4x$$
$$\frac{-3y}{-3} = \frac{-1}{-3} - \frac{4x}{-3}$$
$$y = \frac{1}{3} + \frac{4}{3}x$$

Replace y in the first equation with $\frac{1}{3} + \frac{4}{3}x$, giving us a linear equation in only one variable, x, that we may solve:

$$-5x + 6y = -10$$
$$-5x + 6\left(\frac{1}{3} + \frac{4}{3}x\right) = -10$$
$$-5x + 2 + 8x = -10$$
$$3x + 2 = -10$$
$$3x = -12$$
$$x = -4$$

Now that we have the value for x, we need to find the value for y. We have already solved the second equation for y, so that is the easiest equation to use.

$$y = \frac{1}{3} + \frac{4}{3}x$$
$$y = \frac{1}{3} + \frac{4}{3}(-4)$$
$$y = \frac{1}{3} - \frac{16}{3}$$
$$y = -\frac{15}{3}$$
$$y = -5$$

To check this solution, we replace x with -4 and y with -5 in each equation:

$$-5x + 6y = -10$$
$$-5(-4) + 6(-5) \stackrel{?}{=} -10$$
$$20 - 30 \stackrel{\checkmark}{=} -10$$

$$4x - 3y = -1$$
$$4(-4) - 3(-4) \stackrel{?}{=} -1$$
$$-16 + 15 \stackrel{\checkmark}{=} -1$$

We conclude then that this system of equations is true when $x = -4$ and $y = -5$. Our solution is the point $(-4, -5)$ and we write the solution set as $\{(-4, -5)\}$.

Example 5.4.6 Applications of Systems of Equations. The Rusk Ranch Nature Center[a] in south-western Oregon is a volunteer run nonprofit that exists to promote the wellbeing of the local communities and conserve local nature with an emphasis on native butterflies. They sell admission tickets: $6 for adults and $4 for children. Amanda, who was working at the front desk, counted that one day she sold a total of 79 tickets and had $384 in the register from those ticket sales. She didn't bother to count how many were adult tickets and how many were child tickets because she knew she could use math to figure it out at the end of the day. So, how many of the 79 tickets were adult and how many were child?

Explanation. Let's let a represent the number of adult tickets sold and c represent the number of child tickets sold. We need to build two equations to solve a system for both variables.

The first equation we will build relates to the fact that there were 79 total tickets sold. If we combine both the number of adult tickets and child tickets, the total is 79. This fact becomes:

$$a + c = 79$$

For the second equation we need to use the per-ticket dollar amounts to generate the total cost of $384. The amount of money that was made from adult tickets is found my multiplying the number of adult tickets sold, a, by the price per ticket, $6. Similarly, the amount of money from child tickets is found my multiplying the number of child tickets sold, c, by the price per ticket, $4. These two amounts will add to be $384. This fact becomes:

$$6a + 4c = 384$$

And so, our system is

$$\begin{cases} a + c = 79 \\ 6a + 4c = 384 \end{cases}$$

To solve, we will use the substitution method and solve the first equation for the variable a.

$$a + c = 79$$
$$a = 79 - c$$

Now we will substitute $79 - c$ for a in the second equation.

$$6a + 4c = 384$$
$$6(79 - c) + 4c = 384$$
$$474 - 6c + 4c = 384$$
$$474 - 2c = 384$$
$$-2c = -90$$
$$c = 45$$

Last, we will solve for a by substituting 45 in for c in the equation $a = 79 - c$.

$$a = 79 - c$$
$$a = 79 - 45$$
$$a = 34$$

Our conclusion is that Amanda sold 34 adult tickets and 45 child tickets.

---

[a]ruskranchnaturecenter.org

Example 5.4.7 Solving Special Systems of Equations with Substitution. Solve the systems of linear equations using substitution.

a.
$$\begin{cases} 3x - 5y = 9 \\ x = \dfrac{5}{3}y + 3 \end{cases}$$

b.
$$\begin{cases} y + 7 = 4x \\ 2y - 8x = 7 \end{cases}$$

Explanation. To solve the systems using substitution, we first need to solve for one variable in one equation, then substitute into the *other* equation.

a. In this case, x is already solved for in the second equation so we can substitute $\frac{5}{3}y + 3$ everywhere we see x in the first equation. Then simplify and solve for y.

$$3x - 5y = 9$$
$$3\left(\frac{5}{3}y + 3\right) - 5y = 9$$
$$5y + 9 - 5y = 9$$
$$9 = 9$$

We will stop here since we have eliminated all of the variables in the equation and ended with a *true* statement. Since 9 always equals 9, no matter what, then any value of y must make the original equation, $3\left(\frac{5}{3}y + 3\right) - 5y = 9$ true. If you recall from the section on substitution, this means that both lines $3x - 5y = 9$ and $x = \frac{5}{3}y + 3$ are in fact the same line. Since a solution to a system of linear equations is any point where the lines touch, *all* points along both lines are solutions. We can say this in English as, "The solutions are all points on the line $3x - 5y = 9$," or in math as, "The solution set is $\{(x, y) | 3x - 5y = 9\}$."

b. We will first solve the top equation for y.

$$y + 7 = 4x$$
$$y = 4x - 7$$

Now we can substitute $4x - 7$ wherever we see y in the second equation.

$$2y - 8x = 7$$
$$2(4x - 7) - 8x = 7$$
$$8x - 14 - 8x = 7$$
$$-14 = 7$$

We will stop here since we have eliminated all of the variables in the equation and ended with a *false* statement. Since -14 never equals 7, then no values of x and y can make the original system true. If you recall from the section on substitution, this means that the lines $y + 7 = 4x$ and $2y - 8x = 7$ are parallel. Since a solution to a system of linear equations is any point where the lines touch, and parallel lines never touch, *no* points are solutions. We can say this in English as, "There are no solutions," or in math as, "The solution set is $\emptyset$."

5.4.3 Elimination

In Section 5.3, we explored a third way of solving systems of linear equations called elimination where we add two equations together to cancel a variable.

Example 5.4.8 Solving Systems of Equations by Elimination. Solve the system using elimination.

$$\begin{cases} 4x - 6y = 13 \\ 5x + 4y = -1 \end{cases}$$

Explanation. To solve the system using elimination, we first need to scale one or both of the equations so that one variable has equal but opposite coefficients in the system. In this case, we will choose to make y have opposite coefficients because the signs are already opposite for that variable in the system.

We need to multiply the first equation by 2 and the second equation by 3.

$$\begin{cases} 4x - 6y = 13 \\ 5x + 4y = -1 \end{cases}$$
$$\begin{cases} 2 \cdot (4x - 6y) = 2 \cdot (13) \\ 3 \cdot (5x + 4y) = 3 \cdot (-1) \end{cases}$$

$$\begin{cases} 8x - 12y = 26 \\ 15x + 12y = -3 \end{cases}$$

We now have an equivalent system of equations where the y-terms can be eliminated:

$$\begin{array}{r} 8x - 12y \\ + 15x + 12y \end{array} = \begin{array}{l} 26 \\ + (-3) \end{array}$$

So we have:

$$23x = 23$$
$$x = 1$$

To solve for y, we will substitute 1 for x into either of the original equations. We will use the first equation, $4x - 6y = 13$:

$$4x - 6y = 13$$
$$4(1) - 6y = 13$$
$$4 - 6y = 13$$
$$-6y = 9$$
$$\frac{-6y}{-6} = \frac{9}{-6}$$
$$y = -\frac{3}{2}$$

To verify this, we substitute the x and y values into both of the original equations.

$$4x - 6y = 13 \qquad\qquad\qquad 5x + 4y = -1$$
$$4(1) - 6\left(-\frac{3}{2}\right) \stackrel{?}{=} 13 \qquad\qquad 5(1) + 4\left(-\frac{3}{2}\right) \stackrel{?}{=} -1$$
$$4 + 9 \stackrel{\checkmark}{=} 13 \qquad\qquad\qquad 5 - 6 \stackrel{\checkmark}{=} -1$$

So the solution is the point $\left(-\frac{3}{2}, 1\right)$ and the solution set is $\left\{\left(-\frac{3}{2}, 1\right)\right\}$.

Example 5.4.9 Solving Special Systems of Equations with Elimination. Solve the system of equations using the elimination method.

$$\begin{cases} 24x + 6y = 9 \\ 8x + 2y = 2 \end{cases}$$

Explanation. To eliminate the x-terms, we will scale the second equation by -3.

$$\begin{cases} 24x + 6y = 9 \\ -3 \cdot (8x + 2y) = -3 \cdot (2) \end{cases}$$
$$\begin{cases} 24x + 6y = 9 \\ -24x - 6y = -6 \end{cases}$$

Adding the respective sides of the equation, we have:

$$0 = 3$$

Both of the variables have been eliminated. In this case, the statement $0 = 3$ is just false, no matter what x and y are. So the system has no solution. The solution set is $\emptyset$.

Example 5.4.10 Deciding to Use Substitution versus Elimination. Decide which method would be easiest to solve the systems of linear equations: substitution or elimination.

a. $\begin{cases} 2x + 3y = -11 \\ 5x - 6y = 13 \end{cases}$
b. $\begin{cases} x - 7y = 10 \\ 9x - 16y = -4 \end{cases}$
c. $\begin{cases} 6x + 30y = 15 \\ 4x + 20y = 10 \end{cases}$
d. $\begin{cases} y = 3x - 2 \\ y = 7x + 6 \end{cases}$

Explanation.

a. Elimination is probably easiest here. Multiply the first equation by 2 and eliminate the y variables. The solution to this one is $(-1, -3)$ if you want to solve it for practice.

b. Substitution is probably easiest here. Solve the first equation for x and substitute it into the second equation. We *could* use elimination if we multiplied the first equation by -9 and eliminate the x variable, but it's probably a little more work than substitution. The solution to this one is $(-4, -2)$ if you want to solve it for practice.

c. Elimination is probably easiest here. Multiply the first equation by 2 and the second equation by -3. Doing this will eliminate both variables and leave you with $0 = 0$. This should mean that all points on the line are solutions. So the solution set is $\{(x, y)|6x + 30y = 15\}$.

d. Substitution is definitely easiest here. Substituting y from one equation into y in the other equation gives you $3x - 2 = 7x + 6$. Solve this and find then find y and you should get the solution to the system to be $(-2, -8)$ if you want to solve it for practice.

Exercises

Solving Systems of Linear Equations by Graphing Use a graph to solve the system of equations.

1. $\begin{cases} x + y = 0 \\ 3x - y = 8 \end{cases}$
2. $\begin{cases} 4x - 2y = 4 \\ x + 2y = 6 \end{cases}$
3. $\begin{cases} x + y = -1 \\ x = 2 \end{cases}$

4. $\begin{cases} x - 2y = -4 \\ x = -4 \end{cases}$
5. $\begin{cases} -10x + 15y = 60 \\ 6x - 9y = 36 \end{cases}$
6. $\begin{cases} 6x - 8y = 32 \\ 9x - 12y = 12 \end{cases}$

7. $\begin{cases} y = -\dfrac{3}{5}x + 7 \\ 9x + 15y = 105 \end{cases}$
8. $\begin{cases} 9y - 12x = 18 \\ y = \dfrac{4}{3}x + 2 \end{cases}$

Substitution   Solve the following system of equations.

9.

$$\begin{cases} c = -46 - 5C \\ 2C + 5c = 0 \end{cases}$$

10.

$$\begin{cases} p = -2a \\ 4a + 2p = 0 \end{cases}$$

11.

$$\begin{cases} 2x + 2y = 18 \\ 4x + 3y = 26 \end{cases}$$

12.

$$\begin{cases} x + 3y = 8 \\ 3x + 2y = 10 \end{cases}$$

13.

$$\begin{cases} 2x + 5y = -17 \\ 2x - 4y = 28 \end{cases}$$

14.

$$\begin{cases} 4x - 4y = -12 \\ 3x + 3y = 45 \end{cases}$$

15.

$$\begin{cases} 5x + 3y = 2 \\ -20x - 12y = -8 \end{cases}$$

16.

$$\begin{cases} x + 2y = 2 \\ 3x + 6y = 6 \end{cases}$$

17. A rectangle's length is 6 feet longer than five times its width. The rectangle's perimeter is 192 feet. Find the rectangle's length and width.

The rectangle's length is [   ] feet, and its width is [   ] feet.

18. A school fund raising event sold a total of 178 tickets and generated a total revenue of $420.90. There are two types of tickets: adult tickets and child tickets. Each adult ticket costs $4.45, and each child ticket costs $1.55. Write and solve a system of equations to answer the following questions.

[   ] adult tickets and [   ] child tickets were sold.

19. A test has 21 problems, which are worth a total of 78 points. There are two types of problems in the test. Each multiple-choice problem is worth 3 points, and each short-answer problem is worth 6 points. Write and solve a system equation to answer the following questions.

This test has [   ] multiple-choice problems and [   ] short-answer problems.

20. A test has 20 problems, which are worth a total of 130 points. There are two types of problems in the test. Each multiple-choice problem is worth 3 points, and each short-answer problem is worth 10 points. Write and solve a system equation to answer the following questions.

This test has [   ] multiple-choice problems and [   ] short-answer problems.

21. Kristen invested a total of $8,000 in two accounts. One account pays 5% interest annually; the other pays 4% interest annually. At the end of the year, Kristen earned a total of $355 in interest. Write and solve a system of equations to find how much money Kristen invested in each account.

   Kristen invested [   ] in the 5% account and [   ] in the 4% account.

22. Lily invested a total of $13,000 in two accounts. After a year, one account lost 6.8%, while the other account gained 2.5%. In total, Lily lost $419. Write and solve a system of equations to find how much money Lily invested in each account.

   Lily invested [   ] in the account with 6.8% loss and [   ] in the account with 2.5% gain.

23. Town A and Town B were located close to each other, and recently merged into one city. Town A had a population with 12% African Americans. Town B had a population with 6% African Americans. After the merge, the new city has a total of 4000 residents, with 8.1% African Americans. Write and solve a system of equations to find how many residents Town A and Town B used to have.

   Town A used to have [   ] residents, and Town B used to have [   ] residents.

24. You poured some 12% alcohol solution and some 6% alcohol solution into a mixing container. Now you have 600 grams of 8% alcohol solution. How many grams of 12% solution and how many grams of 6% solution did you pour into the mixing container?

   Write and solve a system equation to answer the following questions.

   You mixed [   ] grams of 12% solution with [   ] grams of 6% solution.

Elimination   Solve the following system of equations.

25.
$$\begin{cases} x + 3y = -37 \\ 3x + 3y = -57 \end{cases}$$

26.
$$\begin{cases} 3x + 2y = -14 \\ 3x + 5y = 1 \end{cases}$$

27.
$$\begin{cases} 6x - 3y = -24 \\ 3x + 5y = -25 \end{cases}$$

28.
$$\begin{cases} 3x + 4y = -45 \\ -3x + 3y = -18 \end{cases}$$

29.
$$\begin{cases} 3x + 4y = -2 \\ -12x - 16y = -2 \end{cases}$$

30.
$$\begin{cases} 3x + 2y = -2 \\ -6x - 4y = -2 \end{cases}$$

31.
$$\begin{cases} 4x + y = -2 \\ 12x + 3y = -6 \end{cases}$$

32.
$$\begin{cases} 4x + 4y = -3 \\ 8x + 8y = -6 \end{cases}$$

33. A test has 20 problems, which are worth a total of 112 points. There are two types of problems in the test. Each multiple-choice problem is worth 5 points, and each short-answer problem is worth 7 points. Write and solve a system of equations to answer the following questions.

This test has [   ] multiple-choice problems and [   ] short-answer problems.

34. Wenwu invested a total of $5,000 in two accounts. One account pays 7% interest annually; the other pays 6% interest annually. At the end of the year, Wenwu earned a total of $345 in interest. Write and solve a system of equations to find how much money Wenwu invested in each account.

Wenwu invested [   ] in the 7% account and [   ] in the 6% account.

35. Town A and Town B were located close to each other, and recently merged into one city. Town A had a population with 6% Hispanics. Town B had a population with 8% Hispanics. After the merge, the new city has a total of 5000 residents, with 7.52% Hispanics. Write and solve a system of equations to find how many residents Town A and Town B used to have.

Town A used to have [   ] residents, and Town B used to have [   ] residents.

36. You poured some 6% alcohol solution and some 10% alcohol solution into a mixing container. Now you have 800 grams of 8.4% alcohol solution. Write and solve a system of equations to find how many grams of 6% solution and how many grams of 10% solution you poured into the mixing container.

You mixed [   ] grams of 6% solution with [   ] grams of 10% solution.

37. If a boat travels from Town A to Town B, it has to travel 797.5 mi along a river. A boat traveled from Town A to Town B along the river's current with its engine running at full speed. This trip took 27.5 hr. Then the boat traveled back from Town B to Town A, again with the engine at full speed, but this time against the river's current. This trip took 72.5 hr. Write and solve a system of equations to answer the following questions.

The boat's speed in still water with the engine running at full speed is [   ].

The river current's speed was [   ].

38. A small fair charges different admission for adults and children. It charges $3.50 for adults, and $2 for children. On a certain day, the total revenue is $5,957.50 and the fair admits 2300 people. How many adults and children were admitted?

There were [   ] adults and [   ] children at the fair.

CHAPTER 6

Exponents and Polynomials

6.1 Exponent Rules

6.1.1 Review of Exponent Rules for Products and Exponents

In Section 2.9, we introduced three basic rules involving products and exponents. We'll begin with a brief recap and explanation of these three exponent rules.

Product Rule When multiplying two expressions that have the same base, simplify the product by adding the exponents.

$$x^m \cdot x^n = x^{m+n}$$

Power to a Power Rule When a base is raised to an exponent and that expression is raised to another exponent, multiply the exponents.

$$(x^m)^n = x^{m \cdot n}$$

Product to a Power Rule When a product is raised to an exponent, apply the exponent to each factor in the product.

$$(x \cdot y)^n = x^n \cdot y^n$$

List 6.1.2: Summary of Exponent Rules

Checkpoint 6.1.3.

a. Simplify $r^{16} \cdot r^5$.

b. Simplify $\left(x^{11}\right)^{10}$.

c. Simplify $(3r)^4$.

d. Simplify $\left(3y^2\right)^2 \left(y^3\right)^5$.

Explanation.

a. We *add* the exponents because this is a product of powers with the same base:

$$r^{16} \cdot r^5 = r^{16+5}$$
$$= r^{21}$$

b. We *multiply* the exponents because this is a power being raised to a power:

$$\left(x^{11}\right)^{10} = x^{11 \cdot 10}$$
$$= x^{110}$$

c. We apply the power to each factor in the product:

$$(3r)^4 = 3^4 r^4$$
$$= 81 r^4$$

d. All three exponent rules must be used, one at a time:

$$\left(3y^2\right)^2 \left(y^3\right)^5 = 3^2 \left(y^2\right)^2 \left(y^3\right)^5$$
$$= 9 \left(y^2\right)^2 \left(y^3\right)^5$$
$$= 9 y^{2 \cdot 2} y^{3 \cdot 5}$$
$$= 9 y^4 y^{15}$$
$$= 9 y^{4+15}$$
$$= 9 y^{19}$$

6.1.2 Quotients and Exponents

Since division is a form of multiplication, it should seem natural that there are some exponent rules for division as well. Not only are there division rules, these rules for division and exponents are direct counterparts for some of the product rules for exponents.

Quotient of Powers When we multiply the same base raised to powers, we end up adding the exponents, as in $2^2 \cdot 2^3 = 2^5$ since $4 \cdot 8 = 32$. What happens when we divide the same base raised to powers?

Example 6.1.4 Simplify $\frac{x^5}{x^2}$ by first writing out what each power means.

Explanation. Without knowing a rule for simplifying this quotient of powers, we can write the expressions without exponents and simplify.

$$\frac{x^5}{x^2} = \frac{x \cdot x \cdot x \cdot x \cdot x}{x \cdot x}$$
$$= \frac{\cancel{x} \cdot \cancel{x} \cdot x \cdot x \cdot x}{\cancel{x} \cdot \cancel{x} \cdot 1}$$
$$= \frac{x \cdot x \cdot x}{1}$$
$$= x^3$$

Notice that the difference of the exponents of the numerator and the denominator (5 and 2, respectively) is 3, which is the exponent of the simplified expression.

When we divide as we've just done, we end up canceling factors from the numerator and denominator one-for-one. These common factors cancel to give us factors of 1. The general rule for this is:

Fact 6.1.5 Quotient of Powers Rule. *For any non-zero real number a and integers m and n,*

$$\frac{a^n}{a^m} = a^{n-m}$$

This rule says that when you're dividing two expressions that have the same base, you can simplify the quotient by subtracting the exponents. In Example 6.1.4, this means that we can directly compute $\frac{x^5}{x^2}$:

$$\frac{x^5}{x^2} = x^{5-2}$$
$$= x^3$$

Quotient to a Power   Another rule we have learned is the product to a power rule, which applies the outer exponent to each factor in the product inside the parentheses. We can use the rules of fractions to extend this property to a *quotient* raised to a power.

Example 6.1.6 Let y be a real number, where $y \neq 0$. Find another way to write $\left(\frac{7}{y}\right)^4$.

Explanation.   Writing the expression without an exponent and then simplifying, we have:

$$\left(\frac{7}{y}\right)^4 = \left(\frac{7}{y}\right)\left(\frac{7}{y}\right)\left(\frac{7}{y}\right)\left(\frac{7}{y}\right)$$
$$= \frac{7 \cdot 7 \cdot 7 \cdot 7}{y \cdot y \cdot y \cdot y}$$
$$= \frac{7^4}{y^4}$$
$$= \frac{2401}{y^4}$$

Similar to the product to a power rule, we essentially applied the outer exponent to the "factors" inside the parentheses—to factors of the numerator *and* factors of the denominator. The general rule is:

Fact 6.1.7 Quotient to a Power Rule. *For real numbers a and b (with b $\neq$ 0) and integer n,*

$$\left(\frac{a}{b}\right)^n = \frac{a^n}{b^n}$$

This rule says that when you raise a fraction to a power, you may separately raise the numerator and denominator to that power. In Example 6.1.6, this means that we can directly calculate $\left(\frac{7}{y}\right)^4$:

$$\left(\frac{7}{y}\right)^4 = \frac{7^4}{y^4}$$
$$= \frac{2401}{y^4}$$

Practice   Try these exercises that use the quotient rules for exponents.

Checkpoint 6.1.8.

   a. Simplify $\dfrac{3^7 x^9}{3^2 x^4}$.

   b. Simplify $\left(\dfrac{p}{2}\right)^6$.

   c. Simplify $\left(\dfrac{5^6 w^7}{5^2 w^4}\right)^9$. If you end up with a large power of a specific number, leave it written that way.

   d. Simplify $\dfrac{(2r^5)^7}{(2^2 r^8)^3}$. If you end up with a large power of a specific number, leave it written that way.

Explanation.

   a. We can use the quotient of powers rule separately on the 3s and on the xs:

$$\frac{3^7 x^9}{3^2 x^4} = 3^{7-2} x^{9-4}$$
$$= 3^5 x^5$$
$$= 243 x^5$$

   b. We can use the quotient to a power rule:

$$\left(\frac{p}{2}\right)^6 = \frac{p^6}{2^6}$$
$$= \frac{p^6}{64}$$

   c. If we stick closely to the order of operations, we should first simplify inside the parentheses and then work with the outer exponent. Going this route, we will first use the quotient rule:

$$\left(\frac{5^6 w^7}{5^2 w^4}\right)^9 = \left(5^{6-2} w^{7-4}\right)^9$$
$$= \left(5^4 w^3\right)^9$$

Now we can apply the outer exponent to each factor inside the parentheses using the product to a power rule.

$$= \left(5^4\right)^9 \cdot \left(w^3\right)^9$$

To finish, we need to use the power to a power rule.

$$= 5^{4 \cdot 9} \cdot w^{3 \cdot 9}$$
$$= 5^{36} \cdot w^{27}$$

   d. According to the order of operations, we should simplify inside parentheses first, then apply exponents, then divide. Since we cannot simplify inside the parentheses, we must apply the outer exponents to each factor inside the respective set of parentheses first:

$$\frac{\left(2r^5\right)^7}{\left(2^2 r^8\right)^3} = \frac{2^7 \left(r^5\right)^7}{\left(2^2\right)^3 \left(r^8\right)^3}$$

At this point, we need to use the power-to-a-power rule:

$$= \frac{2^7 r^{5 \cdot 7}}{2^{2 \cdot 3} r^{8 \cdot 3}}$$

$$= \frac{2^7 r^{35}}{2^6 r^{24}}$$

To finish simplifying, we'll conclude with the quotient rule:

$$= 2^{7-6} r^{35-24}$$

$$= 2^1 r^{11}$$

$$= 2 r^{11}$$

6.1.3 The Zero Exponent

So far, we have been working with exponents that are natural numbers $(1, 2, 3, \ldots)$. By the end of this chapter, we will expand our understanding to include exponents that are any integer, including 0 and negative numbers. As a first step, we will focus on understanding how 0 should behave as an exponent by considering the pattern of decreasing powers of 2 below.

power		product		value	
2^4	$=$	$2 \cdot 2 \cdot 2 \cdot 2$	$=$	16	(divide by 2)
2^3	$=$	$2 \cdot 2 \cdot 2$	$=$	8	(divide by 2)
2^2	$=$	$2 \cdot 2$	$=$	4	(divide by 2)
2^1	$=$	2	$=$	2	(divide by 2)
2^0	$=$	?	$=$	?	

Table 6.1.9: Descending Powers of 2

As we move down from one row to the row below it, we reduce the power by 1 and we remove a factor of 2. The question then becomes, "What happens when you remove the only remaining factor of 2, when you have no factors of 2?" We can see that "removing a factor of 2" really means that we're dividing the value by 2. Following that pattern, we can see that moving from 2^1 to 2^0 means that we need to divide the value 2 by 2. Since $2 \div 2 = 1$, we have:

$$2^0 = 1$$

Fact 6.1.10 The Zero Exponent Rule. *For any non-zero real number a,*

$$a^0 = 1$$

We exclude the case where $a = 0$ from this rule, because our reasoning for this rule with the table had us dividing by the base. And we cannot divide by 0.

Checkpoint 6.1.11. Simplify the following expressions. Assume all variables represent non-zero real numbers.

a. $\left(173x^4y^{251}\right)^0$ b. $(-8)^0$ c. -8^0 d. $3x^0$

Explanation. To simplify any of these expressions, it is critical that we remember an exponent only applies to what it is touching or immediately next to.

a. In the expression $\left(173x^4y^{251}\right)^0$, the exponent 0 applies to everything inside the parentheses.

$$\left(173x^4y^{251}\right)^0 = 1$$

b. In the expression $(-8)^0$ the exponent applies to everything inside the parentheses, -8.

$$(-8)^0 = 1$$

c. In contrast to the previous example, the exponent only applies to the 8. The exponent has a higher priority than negation in the order of operations. We should consider that $-8^0 = -\left(8^0\right)$, and so:

$$-8^0 = -\left(8^0\right)$$
$$= -1$$

d. In the expression $3x^0$, the exponent 0 only applies to the x:

$$3x^0 = 3 \cdot x^0$$
$$= 3 \cdot 1$$
$$= 3$$

6.1.4 Negative Exponents

In Section 2.9, we developed rules for simplifying expressions with whole number exponents, like 0, 1, 2, 3, etc. It turns out that these same rules apply even if the exponent is a negative integer, like -1, -2, -3, etc.

To consider the effects of negative integer exponents, let's extend the pattern we examined in Table 6.1.9. In that table, each time we move down a row, we reduce the power by 1 and we divide the value by 2. We can continue this pattern in the power and value columns, paying particular attention to the values for negative exponents:

Power	Value	
2^3	8	(divide by 2)
2^2	4	(divide by 2)
2^1	2	(divide by 2)
2^0	1	(divide by 2)
2^{-1}	$1/2 = 1/2^1$	(divide by 2)
2^{-2}	$1/4 = 1/2^2$	(divide by 2)
2^{-3}	$1/8 = 1/2^3$	

Note that the choice of base 2 was arbitrary, and this pattern works for all bases except 0, since we cannot divide by 0 in moving from one row to the next.

Fact 6.1.12 The Negative Exponent Rule. *For any non-zero real number a and any integer n,*

$$a^{-n} = \frac{1}{a^n}$$

Note that if we take reciprocals of both sides, we have another helpful fact:

$$\frac{1}{a^{-n}} = a^n.$$

Taken together, these facts tell us that a negative exponent power in the numerator belongs in the denominator (with a positive exponent) and a negative exponent power in the denominator belongs in the numerator (with a positive exponent). In other words, you can see a negative exponent as telling you to move things in and out of the numerator and denominator of an expression.

Remark 6.1.13. You may be expected to simplify expressions so that they do not have any negative exponents. This can always be accomplished using the negative exponent rule.

Try these exercises that involve negative exponents.

Checkpoint 6.1.14.

a. Write $4y^{-6}$ without using negative exponents.

b. Write $\dfrac{3x^{-4}}{yz^{-2}}$ without using negative exponents.

c. Simplify $\left(-5x^{-5}\right)\left(-8x^4\right)$ and write it without using negative exponents.

Explanation.

a. Always remember that an exponent only applies to what it is touching. In the expression $4y^{-6}$, only the y has an exponent of -6.

$$4y^{-6} = 4 \cdot \frac{1}{y^6}$$
$$= \frac{4}{y^6}$$

b. Negative exponents tell us to move some variables between the numerator and denominator to make the exponents positive.

$$\frac{3x^{-4}}{yz^{-2}} = \frac{3z^2}{yx^4}$$

Notice that the factors of 3 and y did not move, as both of those factors had positive exponents.

c. The product of powers rule still applies, and we can add exponents even when one or both are negative:

$$\left(-5x^{-5}\right)\left(-8x^4\right) = (-5)(-8)x^{-5}x^4$$
$$= 40x^{-5+4}$$
$$= 40x^{-1}$$
$$= \frac{40}{x^1}$$
$$= \frac{40}{x}$$

6.1.5 Summary of Exponent Rules

Now that we have some new exponent rules beyond those from Section 2.9, let's summarize.

If a and b are real numbers, and n and m are integers, then we have the following rules:

Product Rule $a^n \cdot a^m = a^{n+m}$

Power to a Power Rule $(a^n)^m = a^{n \cdot m}$

Product to a Power Rule $(ab)^n = a^n \cdot b^n$

Quotient Rule $\dfrac{a^n}{a^m} = a^{n-m}$, as long as $a \neq 0$

Quotient to a Power Rule $\left(\dfrac{a}{b}\right)^n = \dfrac{a^n}{b^n}$, as long as $b \neq 0$

Zero Exponent Rule $a^0 = 1$ for $a \neq 0$

Negative Exponent Rule $a^{-n} = \dfrac{1}{a^n}$

Negative Exponent Reciprocal Rule $\dfrac{1}{a^{-n}} = a^n$

List 6.1.15: Summary of the Rules of Exponents for Multiplication and Division

Remark 6.1.16 Why we have "$a \neq 0$" and "$b \neq 0$" for some rules. Whenever we're working with division, we have to be careful to make sure the rules we state don't ever imply that we might be dividing by zero. Dividing by zero leads us to expressions that have no meaning. For example, both $\frac{9}{0}$ and $\frac{0}{0}$ are *undefined*, meaning no one has defined what it means to divide a number by 0. Also, we established that $a^0 = 1$ using repeated division by a in table rows, so that reasoning doesn't work if $a = 0$.

Warning 6.1.17 A Common Mistake. It may be tempting to apply the rules of exponents to expressions containing addition or subtraction. However, none of the rules of exponents 6.1.15 involve addition or subtraction in the initial expression. Because whole number exponents mean repeated multiplication, not repeated addition or subtraction, trying to apply exponent rules in situations that do not use multiplication simply doesn't work.

Can we say something like $a^n + a^m = a^{n+m}$? How would that work out when $a = 2$?

$$2^3 + 2^4 \overset{?}{=} 2^{3+4}$$

$$8 + 16 \overset{?}{=} 2^7$$

$$24 \neq 128$$

As we can see, that's not even close. This attempt at a "sum rule" falls apart. In fact, without knowing values for a, n, and m, there's no way to simplify the expression $a^n + a^m$.

Checkpoint 6.1.18. Decide whether each statements is true or false.

a. $(7+8)^3 = 7^3 + 8^3$

    (□ true   □ false)

b. $(xy)^3 = x^3 y^3$

    (□ true   □ false)

c. $2x^3 \cdot 4x^2 \cdot 5x^6 = (2 \cdot 4 \cdot 5)x^{3+2+6}$

    (□ true   □ false)

d. $\left(x^3 y^5\right)^4 = x^{3+4} y^{5+4}$

    (□ true   □ false)

e. $2\left(x^2 y^5\right)^3 = 8x^6 y^{15}$

    (□ true   □ false)

f. $x^2 + x^3 = x^5$

    (□ true   □ false)

g. $x^3 + x^3 = 2x^3$

    (□ true   □ false)

h. $x^3 \cdot x^3 = 2x^6$

    (□ true   □ false)

i. $3^2 \cdot 2^3 = 6^5$

    (□ true   □ false)

j. $3^{-2} = -\frac{1}{9}$

    (□ true   □ false)

Explanation.

a. False, $(7+8)^3 \neq 7^3 + 8^3$. Following the order of operations, on the left $(7+8)^3$ would simplify as 15^3, which is 3375. However, on the right side, we have

$$7^3 + 8^3 = 343 + 512$$
$$= 855$$

Since $3375 \neq 855$, the equation is false.

b. True. As the cube applies to the product of x and y, $(xy)^3 = x^3 y^3$.

c. True. The coefficients do get multiplied together and the exponents added when the expressions are multiplied, so $2x^3 \cdot 4x^2 \cdot 5x^6 = (2 \cdot 4 \cdot 5)x^{3+2+6}$.

d. False, $\left(x^3 y^5\right)^4 \neq x^{3+4} y^{5+4}$. When we have a power to a power, we multiply the exponents rather than adding them. So

$$\left(x^3 y^5\right)^4 = x^{3 \cdot 4} y^{5 \cdot 4}$$

e. False, $2\left(x^2 y^5\right)^3 \neq 8x^6 y^{15}$. The exponent of 3 applies to x^2 and y^5, but does not apply to the 2. So

$$2\left(x^2 y^5\right)^3 = 2x^{2 \cdot 3} 6y^{5 \cdot 3}$$
$$= 2x^6 y^{15}$$

f. False, $x^2 + x^3 \neq x^5$. The two terms on the left hand side are not like terms and there is no way to combine them.

g. True. The terms x^3 and x^3 are like terms, so $x^3 + x^3 = 2x^3$.

h. False, $x^3 \cdot x^3 \neq 2x^6$. When x^3 and x^3 are multiplied, their coefficients are each 1. So the coefficient of their product is still 1, and we have $x^3 \cdot x^3 = x^6$.

i. False, $3^2 \cdot 2^3 \neq 6^5$. Note that neither the bases nor the exponents are the same. Following the order of operations, on the left $3^2 \cdot 2^3$ would simplify as $9 \cdot 8$, which is 72. However, on the right side, we have $6^5 = 7776$. Since $72 \neq 7776$, the equation is false.

j. False, $3^{-2} \neq -\frac{1}{9}$. The exponent of -2 on the number 3 does not result in a negative number. Instead, $3^{-2} = \frac{1}{3^2}$, which is $\frac{1}{9}$.

As we mentioned before, many situations we'll come across will require us to use more than one exponent rule. In these situations, we'll have to decide which rule to use first. There are often different, correct approaches we could take. But if we rely on order of operations, we will have a straightforward approach to simplify the expression correctly. To bring it all together, try these exercises.

Checkpoint 6.1.19.

a. Simplify $\dfrac{6x^3}{2x^7}$ and write it without using negative exponents.

b. Simplify $4\left(\frac{1}{5}tv^{-4}\right)^2$ and write it without using negative exponents.

c. Simplify $\left(\dfrac{3^0 y^4 \cdot y^5}{6y^2}\right)^3$ and write it without using negative exponents.

d. Simplify $\left(7^4 x^{-6} t^2\right)^{-5}\left(7x^{-2}t^{-7}\right)^4$ and write it without using negative exponents. Leave larger numbers (such as 7^{10}) in exponent form.

Explanation.

a. In the expression $\frac{6x^3}{2x^7}$, the coefficients reduce using the properties of fractions. One way to simplify the variable powers is:

$$\frac{6x^3}{2x^7} = \frac{6}{2} \cdot \frac{x^3}{x^7}$$
$$= 3 \cdot x^{3-7}$$
$$= 3 \cdot x^{-4}$$
$$= 3 \cdot \frac{1}{x^4}$$
$$= \frac{3}{x^4}$$

b. In the expression $4\left(\frac{1}{5}tv^{-4}\right)^2$, the exponent 2 applies to each factor inside the parentheses.

$$4\left(\frac{1}{5}tv^{-4}\right)^2 = 4\left(\frac{1}{5}\right)^2 (t)^2 \left(v^{-4}\right)^2$$
$$= 4\left(\frac{1}{25}\right)\left(t^2\right)\left(v^{-4 \cdot 2}\right)$$
$$= 4\left(\frac{1}{25}\right)\left(t^2\right)\left(v^{-8}\right)$$
$$= 4\left(\frac{1}{25}\right)\left(t^2\right)\left(\frac{1}{v^8}\right)$$
$$= \frac{4t^2}{25v^8}$$

c. To follow the order of operations in the expression $\left(\frac{3^0 y^4 \cdot y^5}{6y^2}\right)^3$, the numerator inside the parentheses should be dealt with first. After that, we'll simplify the quotient inside the parentheses. As a final step, we'll apply the exponent to that simplified expression:

$$\left(\frac{3^0 y^4 \cdot y^5}{6y^2}\right)^3 = \left(\frac{1 \cdot y^{4+5}}{6y^2}\right)^3$$

$$= \left(\frac{y^9}{6y^2}\right)^3$$

$$= \left(\frac{y^{9-2}}{6}\right)^3$$

$$= \left(\frac{y^7}{6}\right)^3$$

$$= \frac{(y^7)^3}{6^3}$$

$$= \frac{y^{7 \cdot 3}}{216}$$

$$= \frac{y^{21}}{216}$$

d. We'll again rely on the order of operations, and look to simplify anything inside parentheses first and then apply exponents. In this example, we will begin by applying the product to a power rule, followed by the power to a power rule.

$$\left(7^4 x^{-6} t^2\right)^{-5} \left(7x^{-2} t^{-7}\right)^4 = \left(7^4\right)^{-5} \left(x^{-6}\right)^{-5} \left(t^2\right)^{-5} \cdot (7)^4 \left(x^{-2}\right)^4 \left(t^{-7}\right)^4$$

$$= 7^{-20} x^{30} t^{-10} \cdot 7^4 x^{-8} t^{-28}$$

$$= 7^{-20+4} x^{30-8} t^{-10-28}$$

$$= 7^{-16} x^{22} t^{-38}$$

$$= \frac{x^{22}}{7^{16} t^{38}}$$

Exercises

Review and Warmup   Evaluate the following.

1. a. $2^2 =$ [   ]

    b. $3^3 =$ [   ]

    c. $(-2)^2 =$ [   ]

    d. $(-5)^3 =$ [   ]

2. a. $3^2 =$ [   ]

    b. $5^3 =$ [   ]

    c. $(-2)^2 =$ [   ]

    d. $(-3)^3 =$ [   ]

Use the properties of exponents to simplify the expression.

3. $5 \cdot 5^5$ **4.** $6 \cdot 6^2$ **5.** $7^8 \cdot 7^3$ **6.** $8^5 \cdot 8^6$

Simplifying Products and Quotients Involving Exponents Use the properties of exponents to simplify the expression.

7. $t^{19} \cdot t^3$

8. $y^2 \cdot y^{15}$

9. $\left(t^3\right)^5$

10. $\left(x^4\right)^2$

11. $\left(4x^5\right)^3$

12. $\left(2t^6\right)^2$

13. $\left(6x^{12}\right) \cdot \left(-2x^{20}\right)$

14. $\left(-9x^{14}\right) \cdot \left(3x^{13}\right)$

15. $\left(-\dfrac{x^{16}}{4}\right) \cdot \left(-\dfrac{x^7}{2}\right)$

16. $\left(\dfrac{y^{18}}{8}\right) \cdot \left(\dfrac{y^{19}}{3}\right)$

17. $-3\left(-9t^2\right)^3$

18. $-5\left(-6t^3\right)^2$

19. $(-40)^0 = \boxed{}$

20. $(-34)^0 = \boxed{}$

21. $-23^0 = \boxed{}$

22. $-29^0 = \boxed{}$

23. $34^0 + (-34)^0 = \boxed{}$

24. $40^0 + (-40)^0 = \boxed{}$

25. $45n^0 = \boxed{}$

26. $50c^0 = \boxed{}$

27. $(-792y)^0 = \boxed{}$

28. $(-571m)^0 = \boxed{}$

29. $\left(\dfrac{x^5}{4}\right)^2 = \boxed{}$

30. $\left(\dfrac{x^2}{5}\right)^3 = \boxed{}$

31. $\left(\dfrac{-7}{8x^{10}}\right)^3 = \boxed{}$

32. $\left(\dfrac{-7}{4x^5}\right)^2 = \boxed{}$

33. $\left(\dfrac{9x^8}{2}\right)^2 = \boxed{}$

34. $\left(\dfrac{3x^9}{10}\right)^3 = \boxed{}$

35. $\dfrac{45y^{12}}{15y^9} = \boxed{}$

36. $\dfrac{-18t^7}{3t^5} = \boxed{}$

37. $\dfrac{5r^{11}}{25r^{10}} = \boxed{}$

38. $\dfrac{6r^{18}}{18r^{12}} = \boxed{}$

39. $\dfrac{y^{10}}{y^9} = \boxed{}$

40. $\dfrac{r^{12}}{r^7} = \boxed{}$

41. $\dfrac{16^{12}}{16^3} = \boxed{}$

42. $\dfrac{18^{19}}{18^{10}} = \boxed{}$

43. $\left(\dfrac{x^6}{2y^{10}z^4}\right)^3 = \boxed{}$

44. $\left(\dfrac{x^3}{2y^5z^2}\right)^3 = \boxed{}$

45. $\left(\dfrac{-7x^2}{4y^9}\right)^3 = \boxed{}$

46. $\left(\dfrac{-7x^3}{10y^6}\right)^2 = \boxed{}$

47. $\dfrac{12^7 \cdot 11^{14}}{12^4 \cdot 11^5} = \boxed{}$

48. $\dfrac{14^{17} \cdot 20^{14}}{14^2 \cdot 20^{10}} = \boxed{}$

49. $\dfrac{-48x^{16}y^9z^{11}}{12x^{11}y^4z^3} = \boxed{}$

50. $\dfrac{28x^{12}y^{17}z^{18}}{14x^8y^7z^7} = \boxed{}$

51. $\dfrac{-80x^4y^8}{16x^3y^2} = \boxed{}$

52. $\dfrac{72x^{17}y^{12}}{18x^{10}y^7} = \boxed{}$

Rewrite the expression simplified and using only positive exponents.

53. $\left(\dfrac{1}{10}\right)^{-2} = \boxed{}$

54. $\left(\dfrac{1}{2}\right)^{-3} = \boxed{}$

55. $\dfrac{3^{-2}}{4^{-3}} = \boxed{}$

56. $\dfrac{4^{-3}}{2^{-2}} = \boxed{}$

57. $5^{-1} - 8^{-1} = \boxed{}$

58. $6^{-1} - 4^{-1} = \boxed{}$

59. $20x^{-9} = \boxed{}$

60. $15x^{-10} = \boxed{}$

61. $\dfrac{9}{x^{-11}} = \boxed{}$

62. $\dfrac{19}{x^{-12}} = \boxed{}$

63. $\dfrac{14x^{-3}}{x} = \boxed{}$

64. $\dfrac{8x^{-5}}{x} = \boxed{}$

65. $\dfrac{7x^{-8}}{x^{-26}} = \boxed{}$

66. $\dfrac{13x^{-10}}{x^{-17}} = \boxed{}$

67. $\dfrac{20x^{-18}}{21x^{-9}} = \boxed{}$

68. $\dfrac{11x^{-21}}{12x^{-35}} = \boxed{}$

69. $\dfrac{t^{-17}}{y^{-3}} = \boxed{}$

70. $\dfrac{t^{-6}}{x^{-17}} = \boxed{}$

71. $\dfrac{t^{-12}}{r^7} = \boxed{}$

72. $\dfrac{x^{-3}}{y^7} = \boxed{}$

73. $\dfrac{1}{7x^{-10}} = \boxed{}$

74. $\dfrac{1}{39y^{-18}} = \boxed{}$

75. $\dfrac{y^8}{y^{22}} = \boxed{}$

76. $\dfrac{r^4}{r^6} = \boxed{}$

77. $\dfrac{-27r^6}{9r^{37}} = \boxed{}$

78. $\dfrac{21t^{12}}{7t^{20}} = \boxed{}$

79. $\dfrac{-7t^3}{5t^4} = \boxed{}$

80. $\dfrac{-3t^{14}}{13t^{35}} = \boxed{}$

81. $\dfrac{x^5}{(x^6)^9} = \boxed{}$

82. $\dfrac{x^4}{(x^{12})^6} = \boxed{}$

83. $\dfrac{y^{-6}}{(y^9)^4} = \boxed{}$

84. $\dfrac{y^{-2}}{(y^5)^{10}} = \boxed{}$

85. $r^{-19} \cdot r^{10} = \boxed{}$

86. $r^{-13} \cdot r^{11} = \boxed{}$

87. $(-3t^{-7}) \cdot (3t^3) = \boxed{}$

88. $(-9t^{-19}) \cdot (8t^{14}) = \boxed{}$

89. $\left(\dfrac{7}{2}\right)^{-2} = \boxed{}$

90. $\left(\dfrac{10}{7}\right)^{-2} = \boxed{}$

91. $(-3)^{-3} = \boxed{}$

92. $(-4)^{-3} = \boxed{}$

93. $\dfrac{1}{(-5)^{-2}} = \boxed{}$

94. $\dfrac{1}{(-6)^{-3}} = \boxed{}$

95. $\dfrac{6}{(-3)^{-3}} = \boxed{}$

96. $\dfrac{-3}{(-4)^{-3}} = \boxed{}$

97. $9^{-3} = \boxed{}$

98. $10^{-2} = \boxed{}$

99. $2^{-1} + 4^{-1} = \boxed{}$

100. $3^{-1} + 8^{-1} = \boxed{}$

101. $\dfrac{1}{4^{-2}} = $ ☐

102. $\dfrac{1}{5^{-2}} = $ ☐

103. $-6^{-3} = $ ☐

104. $-7^{-2} = $ ☐

105. $\dfrac{\left(6r^{12}\right)^3}{r^{37}} = $ ☐

106. $\dfrac{\left(6t^9\right)^2}{t^{37}} = $ ☐

107. $\dfrac{\left(6t^5\right)^3}{t^{-10}} = $ ☐

108. $\dfrac{\left(6x^{12}\right)^2}{x^{-6}} = $ ☐

109. $\left(\dfrac{x^{13}}{x^7}\right)^{-2} = $ ☐

110. $\left(\dfrac{y^7}{y^6}\right)^{-4} = $ ☐

111. $\left(\dfrac{10y^{19}}{5y^7}\right)^{-3} = $ ☐

112. $\left(\dfrac{25r^{12}}{5r^9}\right)^{-2} = $ ☐

113. $\left(-2r^{-6}\right)^{-2}$ ☐

114. $\left(-3r^{-18}\right)^{-3}$ ☐

115. $\left(3t^{-12}\right)^{-3}$ ☐

116. $\left(2t^{-5}\right)^{-2}$ ☐

117. $\dfrac{9x^9 \cdot 5x^9}{2x^{14}} = $ ☐

118. $\dfrac{8x^6 \cdot 7x^4}{5x^6} = $ ☐

119. $\left(y^4\right)^4 \cdot y^{-5} = $ ☐

120. $\left(y^{12}\right)^2 \cdot y^{-23} = $ ☐

121. $\left(3r^8\right)^3 \cdot r^{-17} = $ ☐

122. $\left(3r^3\right)^4 \cdot r^{-6} = $ ☐

123. $\dfrac{\left(r^8\right)^2}{\left(r^4\right)^5} = $ ☐

124. $\dfrac{\left(t^6\right)^4}{\left(t^{14}\right)^4} = $ ☐

125. $\left(t^3\right)^{-2} = $ ☐

126. $\left(x^{15}\right)^{-4} = $ ☐

127. $\left(x^{15}y^3\right)^{-4} = $ ☐

128. $\left(y^8 t^{12}\right)^{-4} = $ ☐

129. $\left(y^{-13}r^9\right)^{-4} = $ ☐

130. $\left(r^{-6}y^5\right)^{-4} = $ ☐

131. $\left(\dfrac{r^{15}}{4}\right)^{-3} = $ ☐

132. $\left(\dfrac{r^{10}}{3}\right)^{-2} = $ ☐

133. $\left(\dfrac{t^9}{x^7}\right)^{-3} = $ ☐

134. $\left(\dfrac{t^9}{x^{10}}\right)^{-5} = $ ☐

135. $\dfrac{\left(x^4 r^{-7}\right)^{-3}}{\left(x^{-3}r^8\right)^{-3}} = $ ☐

136. $\dfrac{\left(x^6 r^{-5}\right)^{-3}}{\left(x^{-8}r^7\right)^{-3}} = $ ☐

Rewrite the expression simplified and using only positive exponents.

137. $7x^{-4}y^5 z^{-2}\left(2x^2\right)^{-4} = $ ☐

138. $5x^{-5}y^5 z^{-3}\left(3x^5\right)^{-4} = $ ☐

139. $\left(\dfrac{x^7 y^4 z^2}{x^{-3}y^{-8}z^{-4}}\right)^{-4} = $ ☐

140. $\left(\dfrac{x^5 y^4 z^5}{x^{-4}y^{-4}z^{-6}}\right)^{-2} = $ ☐

Challenge

141. Consider the exponential expression $\dfrac{x^a \cdot x^b}{x^c}$ where $a > 0, b < 0,$ and $c > 0.$

   a. Are there values for a, b, and c so that the expression equals x^6? If so, fill in the blanks below with possible values for a, b, and c. If not, fill in the blanks below with the word none.

   $a =$ _____ , $b =$ _____ , and $c =$ _____

   b. Are there values for a, b, and c so that the exponential expression equals $\frac{1}{x^{13}}$? If so, fill in the blanks below with possible values for a, b, and c. If not, fill in the blanks below with the word none.

   $a =$ _____ , $b =$ _____ , and $c =$ _____

142. Consider the exponential expression $\dfrac{x^a \cdot x^b}{x^c}$ where $a < 0, b < 0,$ and $c > 0.$

   a. Are there values for a, b, and c so that the expression equals x^7? If so, fill in the blanks below with possible values for a, b, and c. If not, fill in the blanks below with the word none.

   $a =$ _____ , $b =$ _____ , and $c =$ _____

   b. Are there values for a, b, and c so that the expression equals $\frac{1}{x^9}$? If so, fill in the blanks below with possible values for a, b, and c. If not, fill in the blanks below with the word none.

   $a =$ _____ , $b =$ _____ , and $c =$ _____

143. Consider the exponential expression $\dfrac{x^a \cdot x^b}{x^c}$ where $a > 0, b > 0,$ and $c < 0.$

   a. Are there values for a, b, and c so that the expression equals x^6? If so, fill in the blanks below with possible values for a, b, and c. If not, fill in the blanks below with the word none.

   $a =$ _____ , $b =$ _____ , and $c =$ _____

   b. Are there values for a, b, and c so that the expression equals $\frac{1}{x^{14}}$? If so, fill in the blanks below with possible values for a, b, and c. If not, fill in the blanks below with the word none.

   $a =$ _____ , $b =$ _____ , and $c =$ _____

6.2 Scientific Notation

Having just learned more about exponents, including negative exponents, we can discuss a format used for very large and very small numbers called **scientific notation**.

6.2.1 The Basics of Scientific Notation

An October 3, 2016 CBS News headline[1] read:

Federal Debt in FY 2016 Jumped $1,422,827,047,452.46—that's $12,036 Per Household.

The article also later states:

By the close of business on Sept. 30, 2016, the last day of fiscal 2016, it had climbed to $19,573,444,713,936.79.

When presented in this format, trying to comprehend the value of these numbers can be overwhelming. More commonly, such numbers would be presented in a descriptive manner:

- The federal debt climbed by 1.42 trillion dollars in 2016.

- The federal debt was 19.6 trillion dollars at the close of business on Sept. 30, 2016.

Unless we're presented with such news items, most of us deal with numbers no larger than the thousands in our daily life. In science, government, business, and many other disciplines, it's not uncommon to deal with much larger numbers. When numbers get this large, it can be hard to distinguish between a number that has nine or twelve digits. On the other hand, we have descriptive language that allows us grasp the value and not be lost in the sheer size of the number.

We have descriptive language for all numbers, based on the place value of the different digits: ones, tens, thousands, ten thousands, etc. We tend to rely upon this language more when we start dealing with larger numbers. Here's a chart for some of the most common numbers we see and use in the world around us:

Number	US English Name	Power of 10
1	one	10^0
10	ten	10^1
100	hundred	10^2
1,000	one thousand	10^3
10,000	ten thousand	10^4
100,000	one hundred thousand	10^5
1,000,000	one million	10^6
1,000,000,000	one billion	10^9

Table 6.2.2: Whole Number Powers of 10

Each number above has a corresponding power of ten and this power of ten will be important as we start to work with the content in this section.

This descriptive language also covers even larger numbers: trillion, quadrillion, quintillion, sextillion, septillion, and so on. There's also corresponding language to describe very small numbers, such as thousandth, millionth, billionth, trillionth, etc.

Through centuries of scientific progress, humanity became increasingly aware of very large numbers and very small measurements. As one example, the star that is nearest to our sun is Proxima Centauri[2]. Proxima

---

[1]http://www.cnsnews.com/news/article/terence-p-jeffrey/federal-debt-fy-2016-jumped-142282704745246
[2]imagine.gsfc.nasa.gov/features/cosmic/nearest_star_info.html

Centauri is about 25,000,000,000,000 miles from our sun. Again, many will find the descriptive language easier to digest: Proxima Centauri is about 25 trillion miles from our sun.

To make computations involving such numbers more manageable, a standardized notation called **scientific notation** was established. The foundation of scientific notation is the fact that multiplying or dividing by a power of 10 will move the decimal point of a number so many places to the right or left, respectively.

Checkpoint 6.2.3. Perform the following operations:

    a. Multiply 5.7 by 10.   b. Multiply 3.1 by 10,000.

Explanation.

    a. $5.7 \times 10 = 57$

        $10 = 10^1$ and multiplying by 10^1 moved the decimal point one place to the right.

    b. $3.1 \times 10000 = 31{,}000$

        $10{,}000 = 10^4$ and multiplying by 10^4 moved the decimal point four places to the right.

Multiplying a number by 10^n where n is a positive integer had the effect of moving the decimal point n places to the right.

Every number can be written as a product of a number between 1 and 10 and a power of 10. For example, $650 = 6.5 \times 100$. Since $100 = 10^2$, we can also write

$$650 = 6.5 \times 10^2$$

and this is our first example of writing a number in **scientific notation**.

Definition 6.2.4. A positive number is written in **scientific notation** when it has the form $a \times 10^n$ where n is an integer and $1 \le a < 10$. In other words, a has precisely one digit to the left of the decimal place. The exponent n used here is called the number's **order of magnitude**. The number a is sometimes called the **significand** or the **mantissa**.

Some conventions do not require a to be between 1 and 10, excluding both values, but that is the convention used in this book.

6.2.2 Scientific Notation for Large Numbers

To write a numbers larger than 10 in scientific notation, we write a decimal point after the first non-zero digit of the number and then count the number of places between where the decimal point originally was and where it now is. Scientific notation communicates the size of a number and the order of magnitude just as quickly, but with no need to write long strings of zeros or to try to decipher the language of quintillions, sextillions, etc.

> **Example 6.2.5** To get a sense of how scientific notation works, let's consider familiar lengths of time converted to seconds.

Length of Time	Length in Seconds	Scientific Notation
one second	1 second	1×10^0 second
one minute	60 seconds	6×10^1 seconds
one hour	3600 seconds	3.6×10^3 seconds
one month	2,628,000 seconds	2.628×10^6 seconds
ten years	315,400,000 seconds	3.154×10^8 seconds
79 years (about a lifetime)	2,491,000,000 seconds	2.491×10^9 seconds

Checkpoint 6.2.6. Write each of the following in scientific notation.

a. The federal debt at the close of business on Sept. 30, 2016: about 19,600,000,000,000 dollars.

b. The world's population in 2016: about 7,418,000,000 people.

Explanation.

a. To convert the federal debt to scientific notation, we will count the number of digits after the first non-zero digit (which happens to be a 1 here). Since there are 13 places after the first non-zero digit, we write:

$$\overbrace{1\,9{,}600{,}000{,}000{,}000}^{\text{13 places}} \text{ dollars} = 1.96 \times 10^{13} \text{ dollars}$$

b. Since there are nine places after the first non-zero digit of 7, the world's population in 2016 was about

$$\overbrace{7{,}418{,}000{,}000}^{\text{9 places}} \text{ people} = 7.418 \times 10^9 \text{ people}$$

Checkpoint 6.2.7. Convert each of the following from scientific notation to decimal notation (without any exponents).

a. The earth's diameter: about 1.27×10^7 meters.

b. As of 2013, known digits of π: 1.21×10^{13}.

Explanation.

a. To convert this number to decimal notation we will move the decimal point after the digit 1 seven places to the right, including zeros where necessary. The earth's diameter is:

$$1.27 \times 10^7 \text{ meters} = \overbrace{1\,2{,}700{,}000}^{\text{7 places}} \text{ meters.}$$

b. As of 2013 there are

$$1.21 \times 10^{13} = \overbrace{1\,2{,}100{,}000{,}000{,}000}^{\text{13 places}}$$

known digits of π.

6.2.3 Scientific Notation for Small Numbers

Scientific notation can also be useful when working with numbers smaller than 1. As we saw in Table 6.2.2, we can denote thousands, millions, billions, trillions, etc., with positive integer exponents on 10. We can similarly denote numbers smaller than 1 (which are written as tenths, hundredths, thousandths, millionths, billionths, trillionths, etc.), with *negative* integer exponents on 10. This relationship is outlined in Table 6.2.8.

Number	English Name	Power of 10
1	one	10^0
0.1	one tenth	$\frac{1}{10} = 10^{-1}$
0.01	one hundredth	$\frac{1}{100} = 10^{-2}$
0.001	one thousandth	$\frac{1}{1,000} = 10^{-3}$
0.0001	one ten thousandth	$\frac{1}{10,000} = 10^{-4}$
0.00001	one hundred thousandth	$\frac{1}{100,000} = 10^{-5}$
0.000001	one millionth	$\frac{1}{1,000,000} = 10^{-6}$
0.000000001	one billionth	$\frac{1}{1,000,000,000} = 10^{-9}$

Table 6.2.8: Negative Integer Powers of 10

To see how this works with a digit other than 1, let's look at 0.05. When we state 0.05 as a number, we say "5 hundredths." Thus $0.05 = 5 \times \frac{1}{100}$. The fraction $\frac{1}{100}$ can be written as $\frac{1}{10^2}$, which we know is equivalent to 10^{-2}. Using negative exponents, we can then rewrite 0.05 as 5×10^{-2}. This is the scientific notation for 0.05.

In practice, we won't generally do that much computation. To write a small number in scientific notation we start as we did before and place the decimal point behind the first non-zero digit. We then count the number of decimal places between where the decimal had originally been and where it now is. Keep in mind that negative powers of ten are used to help represent very small numbers (smaller than 1) and positive powers of ten are used to represent very large numbers (larger than 1). So to convert 0.05 to scientific notation, we have:

$$\overbrace{0 \ .05}^{\text{2 places}} = 5 \times 10^{-2}$$

> **Example 6.2.9** In quantum mechanics, there is an important value called the Planck Constant[a]. Written as a decimal, the value of the Planck constant (rounded to 4 significant digits) is
>
> $$0.000000000000000000000000000000006626.$$
>
> In scientific notation, this number will be $6.626 \times 10^?$. To determine the exponent, we need to count the number of places from where the decimal is when the number is written as
>
> $$0.000000000000000000000000000000006626$$
>
> to where it will be when written in scientific notation:
>
> $$0 \underbrace{.000000000000000000000000000000006}_{\text{34 places}} 626$$
>
> As a result, in scientific notation, the Planck Constant value is 6.626×10^{-34}. It will be much easier to

use 6.626×10^{-34} in a calculation, and an added benefit is that scientific notation quickly communicates both the value and the order of magnitude of the Planck constant.

[a]en.wikipedia.org/wiki/Planck_constant

Checkpoint 6.2.10. Write each of the following in scientific notation.

a. The weight of a single grain of long grain rice: about 0.029 grams.

b. The gate pitch of a microprocessor: 0.000000014 meters

Explanation.

a. To convert this weight to scientific notation, we must first move the decimal behind the first non-zero digit to obtain 2.9, which requires that we move the decimal point 2 places. Thus we have:

$$\overset{\text{2 places}}{\overparen{0\ .02}}\ 9 \text{ grams} = 2.9 \times 10^{-2} \text{ grams}$$

b. The gate pitch of a microprocessor is:

$$\overset{\text{8 places}}{0.\overparen{00000001}}\ 4 \text{ meters} = 1.4 \times 10^{-8} \text{ meters}$$

Checkpoint 6.2.11. Convert each of the following from scientific notation to decimal notation (without any exponents).

a. A download speed of 7.53×10^{-3} Gigabyte per second

b. The weight of a poppy seed: about 3×10^{-7} kilograms

Explanation.

a. To convert a download speed of 7.53×10^{-3} Gigabyte per second to decimal notation, we will move the decimal point 3 places to the left and include the appropriate number of zeros:

$$7.53 \times 10^{-3} \text{ Gigabyte per second} = \overset{\text{3 places}}{0\ .\overparen{007}}\ 53 \text{ Gigabyte per second}$$

b. The weight of a poppy seed is:

$$3 \times 10^{-7} \text{ kilograms} = \overset{\text{7 places}}{0\ .\overparen{0000003}}\ \text{kilograms}$$

Checkpoint 6.2.12. Decide if the numbers are written in scientific notation or not. Use Definition 6.2.4.

a. The number $7 \times 10^{1.9}$ is   (□ in scientific notation   □ not in scientific notation)   .

b. The number 2.6×10^{-31} is   (□ in scientific notation   □ not in scientific notation)   .

c. The number 10×7^4 is   (□ in scientific notation   □ not in scientific notation)   .

d. The number 0.93×10^3 is   (□ in scientific notation   □ not in scientific notation)   .

e. The number 4.2×10^0 is   (□ in scientific notation   □ not in scientific notation)   .

f. The number 12.5×10^{-6} is   (□ in scientific notation   □ not in scientific notation)   .

Explanation.

a. The number $7 \times 10^{1.9}$ *is not* in scientific notation. The exponent on the 10 is required to be an integer and 1.9 is not.

b. The number 2.6×10^{-31} *is* in scientific notation.

c. The number 10×7^4 *is not* in scientific notation. The base must be 10, not 7.

d. The number 0.93×10^3 *is not* in scientific notation. The coefficient of the 10 must be between 1 (inclusive) and 10.

e. The number 4.2×10^0 *is* in scientific notation.

f. The number 12.5×10^{-6} *is not* in scientific notation. The coefficient of the 10 must be between 1 (inclusive) and 10.

6.2.4 Multiplying and Dividing Using Scientific Notation

One main reason for having scientific notation is to make calculations involving immensely large or small numbers easier to perform. By having the order of magnitude separated out in scientific notation, we can separate any calculation into two components.

> **Example 6.2.13** On Sept. 30th, 2016, the US federal debt was about \$19,600,000,000,000 and the US population was about 323,000,000. What was the average debt per person that day?
>
> a. Calculate the answer using the numbers provided, which are not in scientific notation.
>
> b. First, confirm that the given values in scientific notation are 1.96×10^{13} and 3.23×10^8. Then calculate the answer using scientific notation.
>
> **Explanation.** We've been asked to answer the same question, but to perform the calculation using two different approaches. In both cases, we'll need to divide the debt by the population.
>
> a. We may need to be working a calculator to handle such large numbers and we have to be careful that we type the correct number of 0s.
>
> $$\frac{19600000000000}{323000000} \approx 60681.11$$
>
> b. To perform this calculation using scientific notation, our work would begin by setting up the quotient $\frac{1.96 \times 10^{13}}{3.23 \times 10^8}$. Dividing this quotient follows the same process we did with variable expressions of the same format, such as $\frac{1.96w^{13}}{3.23w^8}$. In both situations, we'll divide the coefficients and then use exponent rules to simplify the powers.
>
> $$\frac{1.96 \times 10^{13}}{3.23 \times 10^8} = \frac{1.96}{3.23} \times \frac{10^{13}}{10^8}$$
> $$\approx 0.6068111 \times 10^5$$
> $$\approx 60681.11$$

The federal debt per capita in the US on September 30th, 2016 was about \$60,681.11 per person. Both calculations give us the same answer, but the calculation relying upon scientific notation has less room for error and allows us to perform the calculation as two smaller steps.

Whenever we multiply or divide numbers that are written in scientific notation, we must separate the calculation for the coefficients from the calculation for the powers of ten, just as we simplified earlier expressions using variables and the exponent rules.

> **Example 6.2.14**
>
> a. Multiply $(2 \times 10^5)(3 \times 10^4)$.
>
> b. Divide $\dfrac{8 \times 10^{17}}{4 \times 10^2}$.
>
> **Explanation.** We will simplify the significand/mantissa parts as one step and then simplify the powers of 10 as a separate step.
>
> a.
> $$(2 \times 10^5)(3 \times 10^4) = (2 \times 3) \times (10^5 \times 10^4)$$
> $$= 6 \times 10^9$$
>
> b.
> $$\frac{8 \times 10^{17}}{4 \times 10^2} = \frac{8}{4} \times \frac{10^{17}}{10^2}$$
> $$= 2 \times 10^{15}$$

Often when we multiply or divide numbers in scientific notation, the resulting value will not be in scientific notation. Suppose we were multiplying $(9.3 \times 10^{17})(8.2 \times 10^{-6})$ and need to state our answer using scientific notation. We would start as we have previously:

$$(9.3 \times 10^{17})(8.2 \times 10^{-6}) = (9.3 \times 8.2) \times (10^{17} \times 10^{-6})$$
$$= 76.26 \times 10^{11}$$

While this is a correct value, it is not written using scientific notation. One way to covert this answer into scientific notation is to turn just the coefficient into scientific notation and momentarily ignore the power of ten:

$$= 76.26 \times 10^{11}$$
$$= 7.626 \times 10^1 \times 10^{11}$$

Now that the coefficient fits into the proper format, we can combine the powers of ten and have our answer written using scientific notation.

$$= 7.626 \times 10^1 \times 10^{11}$$
$$= 7.626 \times 10^{12}$$

> **Example 6.2.15** Multiply or divide as indicated. Write your answer using scientific notation.
>
> a. $(8 \times 10^{21})(2 \times 10^{-7})$
>
> b. $\dfrac{2 \times 10^{-6}}{8 \times 10^{-19}}$
>
> **Explanation.** Again, we'll separate out the work for the significand/mantissa from the work for the

powers of ten. If the resulting coefficient is not between 1 and 10, we'll need to adjust that coefficient to put it into scientific notation.

a.

$$(8 \times 10^{21})(2 \times 10^{-7}) = (8 \times 2) \times (10^{21} \times 10^{-7})$$
$$= 16 \times 10^{14}$$
$$= 1.6 \times 10^{1} \times 10^{14}$$
$$= 1.6 \times 10^{15}$$

We need to remember to apply the product rule for exponents to the powers of ten.

b.

$$\frac{2 \times 10^{-6}}{8 \times 10^{-19}} = \frac{2}{8} \times \frac{10^{-6}}{10^{-19}}$$
$$= 0.25 \times 10^{13}$$
$$= 2.5 \times 10^{-1} \times 10^{13}$$
$$= 2.5 \times 10^{12}$$

There are times where we will have to raise numbers written in scientific notation to a power. For example, suppose we have to find the area of a square whose radius is 3×10^{7} feet. To perform this calculation, we first remember the formula for the area of a square, $A = s^2$ and then substitute 3×10^7 for s: $A = (3 \times 10^7)^2$. To perform this calculation, we'll need to remember to use the product to a power rule and the power to a power rule:

$$A = (3 \times 10^7)^2$$
$$= (3)^2 \times (10^7)^2$$
$$= 9 \times 10^{14}$$

Exercises

Converting To and From Scientific Notation   Write the following number in scientific notation.

1. $7500 =$ [_____]   **2.** $850000 =$ [_____]   **3.** $95000 =$ [_____]

4. $1500 =$ [_____]   **5.** $0.026 =$ [_____]   **6.** $0.0036 =$ [_____]

7. $0.00046 =$ [_____]   **8.** $0.055 =$ [_____]

Write the following number in decimal notation without using exponents.

9. $6.5 \times 10^4 =$ [_____]   **10.** $7.5 \times 10^2 =$ [_____]   **11.** $8.53 \times 10^5 =$ [_____]

12. $9.53 \times 10^4 = $ _____

13. $1.52 \times 10^0 = $ _____

14. $2.51 \times 10^0 = $ _____

15. $3.5 \times 10^{-3} = $ _____

16. $4.5 \times 10^{-4} = $ _____

17. $5.51 \times 10^{-2} = $ _____

18. $6.5 \times 10^{-3} = $ _____

Arithmetic with Scientific Notation   Multiply the following numbers, writing your answer in scientific notation.

19. $(7 \times 10^2)(2 \times 10^4) = $ _____

20. $(8 \times 10^4)(5 \times 10^3) = $ _____

21. $(9 \times 10^3)(2 \times 10^5) = $ _____

22. $(2 \times 10^2)(4 \times 10^5) = $ _____

23. $(3 \times 10^4)(7 \times 10^4) = $ _____

24. $(4 \times 10^3)(3 \times 10^3) = $ _____

Divide the following numbers, writing your answer in scientific notation.

25. $\dfrac{4.5 \times 10^5}{5 \times 10^2} = $ _____

26. $\dfrac{3.5 \times 10^6}{5 \times 10^4} = $ _____

27. $\dfrac{2.4 \times 10^4}{6 \times 10^3} = $ _____

28. $\dfrac{6.3 \times 10^6}{7 \times 10^2} = $ _____

29. $\dfrac{4.8 \times 10^2}{8 \times 10^{-4}} = $ _____

30. $\dfrac{3.6 \times 10^4}{9 \times 10^{-3}} = $ _____

31. $\dfrac{1.8 \times 10^2}{2 \times 10^{-4}} = $ _____

32. $\dfrac{1.8 \times 10^3}{3 \times 10^{-5}} = $ _____

33. $\dfrac{1.2 \times 10^{-2}}{4 \times 10^5} = $ _____

34. $\dfrac{4 \times 10^{-4}}{5 \times 10^4} = $ _____

35. $\dfrac{3 \times 10^{-3}}{5 \times 10^2} = $ _____

36. $\dfrac{1.8 \times 10^{-5}}{6 \times 10^5} = $ _____

Simplify the following expression, writing your answer in scientific notation.

37. $(4 \times 10^9)^3 = $ _____

38. $(5 \times 10^6)^4 = $ _____

39. $(5 \times 10^3)^3 = $ _____

40. $(2 \times 10^9)^4 = $ _____

41. $(2 \times 10^5)^2 = $ _____

42. $(3 \times 10^2)^4 = $ _____

6.3 Adding and Subtracting Polynomials

A polynomial is a particular type of algebraic expression used for things all around us.

- A company's sales, s (in millions of dollars), can be modeled by $2.2t + 5.8$, where t stands for the number of years since 2010.

- The height of an object from the ground, h (in feet), launched upward from the top of a building can be modeled by $-16t^2 + 32t + 300$, where t represents the amount of time (in seconds) since the launch.

- The volume of an open-top box with a square base, V (in cubic inches), can be calculated by $30s^2 - \frac{1}{2}s^2$, where s stands for the length of the square base, and the box sides have to be cut from a certain square piece of metal.

All of the expressions above are **polynomials**. In this section, we will learn some basic vocabulary relating to polynomials and we'll then learn how to add and subtract polynomials.

6.3.1 Polynomial Vocabulary

Definition 6.3.2. A **polynomial** is an expression that consists of terms summed together. Each term must be the product of a number and one or more variables raised to whole number powers. Since 0 is a whole number, a term can just be a number. A polynomial may have just one term. The expression 0 is also considered a polynomial, with zero terms.

Some examples of polynomials in one variable are:

$$x^2 - 5x + 2 \qquad t^3 - 1 \qquad 7y.$$

The expression $3x^4y^3 + 7xy^2 - 12xy$ is an example of a polynomial in several variables.

Definition 6.3.3. A **term** of a polynomial is the product of a numerical coefficient and one or more variables raised to whole number powers. Since 0 is a whole number, a term can just be a number.

For example:

- the polynomial $x^2 - 5x + 3$ has three terms: x^2, $-5x$, and 3;

- the polynomial $3x^4 + 7xy^2 - 12xy$ also has three terms;

- the polynomial $t^3 - 1$ has two terms.

Definition 6.3.4. The **coefficient** (or numerical coefficient) of a term is the numerical factor in the term.

For example:

- the coefficient of the term $\frac{4}{3}x^6$ is $\frac{4}{3}$;

- the coefficient of the second term of the polynomial $x^2 - 5x + 3$ is -5;

- the coefficient of the term $\frac{y^7}{4}$ is $\frac{1}{4}$.

Remark 6.3.5. Because variables in polynomials must have whole number exponents, a polynomial will never have a variable in the denominator of a fraction or under a square root (or any other radical).

Checkpoint 6.3.6. Identify which of the following are polynomials and which are not.

a. The expression $-2x^9 - \frac{7}{13}x^3 - 1$ ($\square$ is $\quad \square$ is not) a polynomial.

b. The expression $5x^{-2} - 5x^2 + 3$ ($\square$ is $\square$ is not)   a polynomial.

c. The expression $\sqrt{2}x - \frac{3}{5}$ ($\square$ is $\square$ is not)   a polynomial.

d. The expression $5x^3 - 5^{-5}x - x^4$ ($\square$ is $\square$ is not)   a polynomial.

e. The expression $\frac{25}{x^2} + 23 - x$ ($\square$ is $\square$ is not)   a polynomial.

f. The expression $37x^6 - x + 8^{\frac{4}{3}}$ ($\square$ is $\square$ is not)   a polynomial.

g. The expression $\sqrt{7x} - 4x^3$ ($\square$ is $\square$ is not)   a polynomial.

h. The expression $6x^{\frac{3}{2}} + 1$ ($\square$ is $\square$ is not)   a polynomial.

i. The expression $6^x - 3x^6$ ($\square$ is $\square$ is not)   a polynomial.

Explanation.

a. The expression $-2x^9 - \frac{7}{13}x^3 - 1$ is a polynomial.

b. The expression $5x^{-2} - 5x^2 + 3$ is not a polynomial because it has negative exponents on a variable.

c. The expression $\sqrt{2}x - \frac{3}{5}$ is a polynomial. Note that *coefficients* can have radicals even though variables cannot, and the square root here is *only* applied to the 2.

d. The expression $5x^3 - 5^{-5}x - x^4$ is a polynomial. Note that *coefficients* can have negative exponents even though variables cannot.

e. The expression $\frac{25}{x^2} + 23 - x$ is not a polynomial because it has a variable in a denominator.

f. The expression $37x^6 - x + 8^{\frac{4}{3}}$ is a polynomial. Note that *coefficients* can have fractional exponents even though variables cannot.

g. The expression $\sqrt{7x} - 4x^3$ is not a polynomial because it has a variable inside a radical.

h. The expression $6x^{\frac{3}{2}} + 1$ is not a polynomial because a variable has a fractional exponent.

i. The expression $6^x - 3x^6$ is not a polynomial because it has a variable in an exponent.

Definition 6.3.7. When a term only has one variable, its **degree** is the exponent on that variable. When a term has more than on variable, its **degree** is the sum of the exponents on the variables in that term. When a term has no variables, its **degree** is 0.

For example:

- the degree of $5x^2$ is 2;
- the degree of $-\frac{4}{7}y^5$ is 5.
- the degree of $-4x^2y^3$ is 5.

Polynomial terms are often classified by their degree. In doing so, we would refer to $5x^2$ as a second-degree term.

Definition 6.3.8. The **degree of a polynomial** is the greatest degree that appears amongst its terms. If the polynomial is just 0, it has no terms, and we say its degree is -1.

Definition 6.3.9. The **leading term** of a polynomial is the term with the greatest degree (assuming there is one, and there is no tie).

For example, the degree of the polynomial $x^2 - 5x + 3$ is 2 because the terms have degrees 2, 1, and 0, and 2 is the largest. Its leading term is x^2. Polynomials are often classified by their degree, and we would say that

$x^2 - 5x + 3$ is a second-degree polynomial.

The coefficient of a polynomial's leading term is called the polynomial's **leading coefficient**. For example, the leading coefficient of $x^2 - 5x + 3$ is 1 (because $x^2 = 1 \cdot x^2$).

Definition 6.3.10. A term with no variable factor is called a **constant term**.

For example, the constant term of the polynomial $x^2 - 5x + 3$ is 3.

There are some special names for polynomials with certain degrees:

- A zero-degree polynomial is called a **constant polynomial** or simply a **constant**.

  An example is the polynomial 7, which has degree zero because it can be viewed as $7x^0$.

- A first-degree polynomial is called a **linear polynomial**.

  An example is $-2x + 7$.

- A second-degree polynomial is called a **quadratic polynomial**.

  An example is $4x^2 - 2x + 7$.

- A third-degree polynomial is called a **cubic polynomial**.

  An example is $x^3 + 4x^2 - 2x + 7$.

Fourth-degree and fifth-degree polynomials are called quartic and quintic polynomials, respectively. If the degree of the polynomial, n, is greater than five, we'll simply call it an nth-degree polynomial. For example, the polynomial $5x^8 - 4x^5 + 1$ is an 8th-degree polynomial.

Remark 6.3.11. To help us recognize a polynomial's degree, it is the standard convention to write a polynomial's terms in order from greatest-degree term to lowest-degree term. When a polynomial is written in this order, it is written in **standard form**. For example, it is standard practice to write $7 - 4x - x^2$ as $-x^2 - 4x + 7$ since $-x^2$ is the leading term. By writing the polynomial in standard form, we can look at the first term to determine both the polynomial's degree and leading term.

There are special names for polynomials with a small number of terms:

Definition 6.3.12.

- A polynomial with one term, such as $3x^5$ or 9, is called a **monomial**.

- A polynomial with two terms, such as $3x^5 + 2x$ or $-2x + 1$, is called a **binomial**.

- A polynomial with three terms, such as $x^2 - 5x + 3$, is called a **trinomial**.

6.3.2 Adding and Subtracting Polynomials

Example 6.3.13 Production Costs. Bayani started a company that is devoted to one product: ketchup. The company's production costs only involve two components: supplies and labor. The cost of supplies, S (in thousands of dollars), can be modeled by $S = 0.05x^2 + 2x + 30$, where x is number of thousands of jars of ketchup produced. The labor cost for his employees, L (in thousands of dollars), can be modeled by $0.1x^2 + 4x$, where x again represents the number of jars they produce (in thousands of jars). Find a model for the company's total production costs.

Since Bayani's company only has these two costs, we can find a model for the total production costs, C (in thousands of dollars), by adding the supply costs and the labor costs:

$$C = \left(0.05x^2 + 2x + 30\right) + \left(0.1x^2 + 4x\right)$$

To finish simplifying our total production cost model, we'll combine the like terms:

$$C = 0.05x^2 + 0.1x^2 + 2x + 4x + 30$$
$$= 0.15x^2 + 6x + 30$$

This simplified model can now calculate Bayani's total production costs C (in thousands of dollars) when the company produces x thousand jars of ketchup.

In short, the process of adding two or more polynomials involves recognizing and then combining the like terms.

Checkpoint 6.3.14. Add the polynomials.

$$\left(-6x^2 - 2x\right) + \left(2x^2 + 3x\right)$$

Explanation.   We combine like terms as follows

$$\left(-6x^2 - 2x\right) + \left(2x^2 + 3x\right) = \left(2x^2 - 6x^2\right) + \left(3x - 2x\right)$$
$$= -4x^2 + x$$

Example 6.3.15 Simplify the expression $\left(\frac{1}{2}x^2 - \frac{2}{3}x - \frac{3}{2}\right) + \left(\frac{3}{2}x^2 + \frac{7}{2}x - \frac{1}{4}\right)$.

Explanation.

$$\left(\frac{1}{2}x^2 - \frac{2}{3}x - \frac{3}{2}\right) + \left(\frac{3}{2}x^2 + \frac{7}{2}x - \frac{1}{4}\right)$$
$$= \left(\frac{1}{2}x^2 + \frac{3}{2}x^2\right) + \left(\left(-\frac{2}{3}x\right) + \frac{7}{2}x\right) + \left(\left(-\frac{3}{2}\right) + \left(-\frac{1}{4}\right)\right)$$
$$= \left(\frac{4}{2}x^2\right) + \left(\left(-\frac{4}{6}x\right) + \frac{21}{6}x\right) + \left(\left(-\frac{6}{4}\right) + \left(-\frac{1}{4}\right)\right)$$
$$= \left(2x^2\right) + \left(\frac{17}{6}x\right) + \left(-\frac{7}{4}\right)$$
$$= 2x^2 + \frac{17}{6}x - \frac{7}{4}$$

Example 6.3.16 Profit, Revenue, and Costs.   From Example 6.3.13, we know Bayani's ketchup company's production costs, C (in thousands of dollars), for producing x thousand jars of ketchup is modeled by $C = 0.15x^2 + 6x + 30$. The revenue, R (in thousands of dollars), from selling the ketchup can be modeled by $R = 13x$, where x stands for the number of thousands of jars of ketchup sold. The company's net profit can be calculated using the concept:

$$\text{net profit} = \text{revenue} - \text{costs}$$

Assuming all products produced will be sold, a polynomial to model the company's net profit, P (in thousands of dollars) is:

$$P = R - C$$
$$= (13x) - \left(0.15x^2 + 6x + 30\right)$$
$$= 13x - 0.15x^2 - 6x - 30$$

$$= -0.15x^2 + (13x + (-6x)) - 30$$
$$= -0.15x^2 + 7x - 30$$

The key distinction between the addition and subtraction of polynomials is that when we subtract a polynomial, we must subtract each term in that polynomial.

Remark 6.3.17. Notice that our first step in simplifying the expression in Example 6.3.16 was to subtract *every* term in the second expression. We can also think of this as distributing a factor of -1 across the second polynomial, $0.15x^2 + 6x + 30$, and then adding these terms as follows:

$$P = R - C$$
$$= (13x) - \left(0.15x^2 + 6x + 30\right)$$
$$= 13x + (-1)(0.15x^2) + (-1)(6x) + (-1)(30)$$
$$= 13x - 0.15x^2 - 6x - 30$$
$$= -0.15x^2 + (13x + (-6x)) - 30$$
$$= -0.15x^2 + 7x - 30$$

Example 6.3.18 Subtract $\left(5x^3 + 4x^2 - 6x\right) - \left(-3x^2 + 9x - 2\right)$.

Explanation. We must first subtract every term in $\left(-3x^2 + 9x - 2\right)$ from $\left(5x^3 + 4x^2 - 6x\right)$. Then we can combine like terms.

$$\left(5x^3 + 4x^2 - 6x\right) - \left(-3x^2 + 9x - 2\right)$$
$$= 5x^3 + 4x^2 - 6x + 3x^2 - 9x + 2$$
$$= 5x^3 + \left(4x^2 + 3x^2\right) + (-6x + (-9x)) + 2$$
$$= 5x^3 + 7x^2 - 15x + 2$$

Checkpoint 6.3.19. Subtract the polynomials.

$$\left(-3x^6 + 8x^4\right) - (-2x + 3)$$

Explanation. We combine like terms as follows

$$\left(-3x^6 + 8x^4\right) - (-2x + 3) = -3x^6 + 8x^4 + 2x - 3$$

Let's look at one more example involving multiple variables. Remember that like terms must have the same variable(s) with the same exponent.

Example 6.3.20 Subtract $\left(3x^2y + 8xy^2 - 17y^3\right) - \left(2x^2y + 11xy^2 + 4y^2\right)$.

Explanation. Again, we'll begin by subtracting each term in $\left(2x^2y + 11xy^2 + 4y^2\right)$. Once we've done this, we'll need to identify and combine like terms.

$$\left(3x^2y + 8xy^2 - 17y^3\right) - \left(2x^2y + 11xy^2 + 4y^2\right)$$
$$= 3x^2y + 8xy^2 - 17y^3 - 2x^2y - 11xy^2 - 4y^2$$
$$= \left(3x^2y + \left(-2x^2y\right)\right) + \left(8xy^2 + \left(-11xy^2\right)\right) + \left(-17y^3\right) + \left(-4y^2\right)$$

$$= x^2y - 3xy^2 - 17y^3 - 4y^2$$

6.3.3 Evaluating Polynomial Expressions

Evaluating expressions was introduced in Section 2.1, and involves replacing the variable(s) in an expression with specific numbers and calculating the result. Here, we will look at evaluating polynomial expressions.

Example 6.3.21 Evaluate the expression $-12y^3 + 4y^2 - 9y + 2$ for $y = -5$.

Explanation. We will replace y with -5 and simplify the result:

$$-12y^3 + 4y^2 - 9y + 2 = -12(-5)^3 + 4(-5)^2 - 9(-5) + 2$$
$$= -12(-125) + 4(25) + 45 + 2$$
$$= 1647$$

Remark 6.3.22. Recall that $(-5)^2$ and -5^2 are not the same expressions. The first expression, $(-5)^2$, represents the number -5 squared, and is $(-5)(-5) = 25$. The second expression, -5^2, is the *opposite* of the number 5 squared, and is $-5^2 = -(5 \cdot 5) = -25$.

Example 6.3.23 Evaluate the expression $C = 0.15x^2 + 6x + 30$ from Example 6.3.13 for $x = 10$ and explain what this means in context.

Explanation. We will replace x with 10:

$$C = 0.15x^2 + 6x + 30$$
$$= 0.15(10)^2 + 6(10) + 30$$
$$= 105$$

In context, we can interpret this as it costing \$105,000 to produce 10,000 jars of ketchup.

Checkpoint 6.3.24. Evaluate the following expressions.

a. Evaluate $(-y)^2$ when $y = -2$.

$(-y)^2 = \boxed{}$

b. Evaluate $(-y)^3$ when $y = -2$.

$(-y)^3 = \boxed{}$

Explanation.

a. $(-y)^2 = \left(-1(-2)\right)^2$
$ = (2)^2$
$ = 4$

b. $(-y)^3 = \left(-1(-2)\right)^3$
$ = (2)^3$
$ = 8$

Exercises

Review and Warmup

1. List the terms in each expression.

    a. $-3.4t + 0.1x + 8.2t + 3.6s$

    b. $-2.8s^2 - 0.8z$

    c. $7.5s + 7.6 - 5.4t^2$

    d. $2.9y + 4t^2 + 5.2y + 2.5s$

2. List the terms in each expression.

    a. $-1.7t^2 + 8.3z^2 + 6.2y + 2.4x$

    b. $3.2z^2 - 2.4 + 8.9t - 8z^2$

    c. $-6.6x$

    d. $-6.6z^2 + 5.6y^2 + 0.4x^2$

3. List the terms in each expression.

    a. $-0.1t$

    b. $-1.6s + 4.2s + 1.2x^2$

    c. $7.2x + 4.8 + 3x$

    d. $-7.3s^2 - 7.4$

4. List the terms in each expression.

    a. $1.5t^2 + 6.6y$

    b. $2.2z + 1.4x$

    c. $-7.9t^2$

    d. $-3.7t - 1.4x + 8.1s^2$

Simplify each expression, if possible, by combining like terms.

5.   a. $3t - 3t^2$

    b. $-6x - 7 - 7x + 8x$

    c. $-4z^2 + 9 + s^2$

    d. $-4s - 4s^2 + 8x + 7y$

6.   a. $5z^2 + 5z^2 - 2z^2$

    b. $-3y - 5 + 4z$

    c. $-4z^2 + 7y + 8y - 9$

    d. $-4t^2 - 2y^2$

7.   a. $7z + \frac{1}{2}x + \frac{5}{4}x$

    b. $\frac{9}{4}y^2 + 1$

    c. $-x - \frac{7}{5}x$

    d. $\frac{2}{7}s + \frac{1}{2}s - \frac{9}{2}s$

8.   a. $\frac{8}{9}z^2 + 4t - \frac{6}{5}t$

    b. $-2s^2 + \frac{1}{9}s + 3z - \frac{4}{3}x$

    c. $-3s + 8s^2 - \frac{6}{5}s^2$

    d. $\frac{1}{8}y^2 + \frac{2}{9}s^2 - 2s^2 - \frac{9}{7}y$

Vocabulary Questions   Is the following expression a monomial, binomial, or trinomial?

9. $3r^{14} - 17r^6$ is a   ($\Box$ monomial $\Box$ binomial

    $\Box$ trinomial) of degree ⬚

10. $-12t^{11} - 3t^2$ is a   ($\Box$ monomial $\Box$ binomial

    $\Box$ trinomial) of degree ⬚

11. 37 is a   ($\Box$ monomial $\Box$ binomial $\Box$ trino-

    mial) of degree ⬚

12. 2 is a   ($\Box$ monomial $\Box$ binomial $\Box$ trino-

    mial) of degree ⬚

13. $-18y^{11} - 9y^7 - 20y^6$ is a   (□ monomial   □ binomial   □ trinomial)   of degree [ ]

14. $-20r^{10} - 2r^9 - 10r^2$ is a   (□ monomial   □ binomial   □ trinomial)   of degree [ ]

15. $-y^4 + 4y^7 - 19y^2$ is a   (□ monomial   □ binomial   □ trinomial)   of degree [ ]

16. $-15y^6 - 20y^7 + 10y^4$ is a   (□ monomial   □ binomial   □ trinomial)   of degree [ ]

17. $12r^9$ is a   (□ monomial   □ binomial   □ trinomial)   of degree [ ]

18. $-3r^{17}$ is a   (□ monomial   □ binomial   □ trinomial)   of degree [ ]

Find the degree of the following polynomial.

19. $13x^7y^6 - 17x^4y^2 - 10x^2 + 17$

The degree of this polynomial is [ ].

20. $17x^7y^9 + 10xy^4 + 7x^2 + 6$

The degree of this polynomial is [ ].

Simplifying Polynomials   Add the polynomials.

21. $(-10x - 2) + (-9x - 3)$

22. $(-7x - 10) + (-9x - 2)$

23. $(-5x^2 + 4x) + (9x^2 + 7x)$

24. $(-3x^2 + x) + (-3x^2 + 7x)$

25. $(-6x^2 - 8x - 3) + (-4x^2 + 9x - 3)$

26. $(7x^2 + 3x - 3) + (9x^2 - 4x + 7)$

27. $(4r^3 - 8r^2 - 4) + (7r^3 - 6r^2 - 10)$

28. $(-10t^3 - 5t^2 + 7) + (-8t^3 + 9t^2 + 4)$

29. $(-7t^6 - 3t^4 - 4t^2) + (9t^6 - 3t^4 - 4t^2)$

30. $(4x^6 - 9x^4 + 7x^2) + (10x^6 + 6x^4 + 9x^2)$

31. $(0.8x^5 - 0.5x^4 + 0.2x^2 - 0.4) + (-0.4x^5 + 0.9x^3 - 0.3)$

32. $(0.1y^5 + 0.3y^4 - 0.3y^2 - 0.5) + (0.6y^5 - 0.8y^3 + 0.4)$

33. $\left(-5x^3 - 2x^2 - 4x + \dfrac{1}{2}\right) + \left(-5x^3 + 9x^2 - 5x + \dfrac{3}{2}\right)$

34. $\left(6x^3 + 4x^2 - 8x + \dfrac{1}{2}\right) + \left(4x^3 - 6x^2 + 6x + \dfrac{3}{2}\right)$

Subtract the polynomials.

35. $(2x + 1) - (9x - 6)$

36. $(4x - 7) - (-3x - 6)$

37. $(6x^2 + 7x) - (5x^2 + 9x)$

38. $(9x^2 - 7x) - (3x^2 - 7x)$

39. $(-2x^5 - 7x^4) - (-5x^2 + 4)$

40. $(-8x^5 - 7x^4) - (6x + 10)$

41. $(-4x^3 + 5x^2 - 8x + (-5)) - (-9x^2 + 9x + 8)$

42. $(-5x^3 + 9x^2 - 8x + 6) - (10x^2 - 2x + 3)$

43. $(6x^2 - 4x - 8) - (-4x^2 - 3x + 1)$

44. $(-7x^2 + 8x - 8) - (-8x^2 + 7x + 10)$

45. $(-2r^6 - 8r^4 - 5r^2) - (-4r^6 - 2r^4 + r^2)$

46. $(8t^6 - 5t^4 + 6t^2) - (-4t^6 + 4t^4 - 10t^2)$

Add or subtract the given polynomials as indicated.

47. $\left[4t^{13} - 2t^7 + 2t^6 - (-10t^{13} + 3t^7 - 7t^6)\right] - (-3t^{13} - 5t^7 - 7t^6)$

48. $\left[x^6 - 7x^4 + 9x^3 - (-10x^6 + 8x^4 - 6x^3)\right] - (-8x^6 - 2x^4 - 3x^3)$

49. $\left[7x^5 - 4x^4 + 7x^3 - (-10x^5 + 3x^4 - 5x^3)\right] - \left[-5x^5 - 8x^4 + 3x^3 + (-2x^5 - 6x^4 - 5x^3)\right]$

50. $\left[4x^6 + 4x^5 - (-10x^6 - 4x^5)\right] - \left[-10x^6 + 8x^5 + (-6x^6 - 6x^5)\right]$

51. $(5x^8y^4 - 7xy) + (-2x^8y^4 - 8xy)$

52. $(6x^3y^7 + 2xy) + (10x^3y^7 + 9xy)$

53. $(-5x^4y^8 + 4xy - 9) + (-8x^4y^8 - 5xy - 6)$

54. $(8x^3y^6 - 9xy + 8) + (10x^3y^6 + 2xy - 2)$

55. $(9x^9y^8 + 4x^2y^2 + 2xy) + (-10x^9y^8 - 3x^2y^2 - 8xy)$

56. $(-10x^8y^9 + 8x^5y^3 - 4xy) + (-3x^8y^9 - 5x^5y^3 + 9xy)$

57. $(2x^8 - 2xy + 7y^6) - (-10x^8 + 4xy - 10y^6)$

58. $(3x^7 - 6xy - 10y^9) - (10x^7 - 5xy - 7y^9)$

59. $(-4x^8y^9 + 10x^4y^3 + 4xy) - (-10x^8y^9 + 6x^4y^3 - 3xy)$

60. $\left(5x^7y^9 + 5x^2y^4 - 7xy\right) - \left(10x^7y^9 - 7x^2y^4 - 9xy\right)$

61. $\left(-6x^4 - 8y^8\right) - \left(-9x^4 + 7x^7y^8 + 9x^4y^8 - 6y^8\right)$ **62.** $\left(-7x^3 - 3y^8\right) - \left(9x^3 + 8x^4y^8 - 3x^3y^8 - 2y^8\right)$

63. Subtract $-8r^{15} - 6r^{14} - 5r^{12}$ from the sum of $8r^{15} - 5r^{14} + 8r^{12}$ and $-3r^{15} + 3r^{14} - 3r^{12}$.

64. Subtract $-4t^8 - 10t^7$ from the sum of $5t^8 + 5t^7$ and $-3t^8 - 2t^7$.

65. Subtract $-10x^3y^5 - 5xy$ from $-9x^3y^5 - 2xy$

66. Subtract $2x^5y^8 + 9xy$ from $9x^5y^8 + 2xy$

Evaluating Polynomials

67. Evaluate the expression x^2:

   a. When $x = 5$, $x^2 = $ _____

   b. When $x = -4$, $x^2 = $ _____

68. Evaluate the expression x^2:

   a. When $x = 2$, $x^2 = $ _____

   b. When $x = -7$, $x^2 = $ _____

69. Evaluate the expression $-y^2$:

   a. When $y = 4$, $-y^2 = $ _____

   b. When $y = -2$, $-y^2 = $ _____

70. Evaluate the expression $-y^2$:

   a. When $y = 3$, $-y^2 = $ _____

   b. When $y = -4$, $-y^2 = $ _____

71. Evaluate the expression r^3:

   a. When $r = 2$, $r^3 = $ _____

   b. When $r = -5$, $r^3 = $ _____

72. Evaluate the expression r^3:

   a. When $r = 4$, $r^3 = $ _____

   b. When $r = -3$, $r^3 = $ _____

73. Evaluate the following expressions.

   a. Evaluate $(-t)^2$ when $t = -4$.

   $(-t)^2 = $ _____

   b. Evaluate $(-t)^3$ when $t = -4$.

   $(-t)^3 = $ _____

74. Evaluate the following expressions.

   a. Evaluate $(-t)^2$ when $t = -4$.

   $(-t)^2 = $ _____

   b. Evaluate $(-t)^3$ when $t = -4$.

   $(-t)^3 = $ _____

75. Evaluate the expression $\frac{1}{5}(x+1)^2 - 6$ when $x = -6$.

76. Evaluate the expression $\frac{1}{8}(x+1)^2 - 4$ when $x = -9$.

77. Evaluate the expression $\frac{1}{6}(x+1)^2 - 9$ when $x = -7$.

78. Evaluate the expression $\frac{1}{3}(x+2)^2 - 6$ when $x = -5$.

79. Evaluate the expression $-16t^2 + 64t + 128$ when $t = -2$.

80. Evaluate the expression $-16t^2 + 64t + 128$ when $t = 5$.

81. Evaluate the expression $-16t^2 + 64t + 128$ when $t = 4$.

82. Evaluate the expression $-16t^2 + 64t + 128$ when $t = 2$.

Applications of Simplifying Polynomials The formula

$$y = \frac{1}{2}a\,t^2 + v_0\,t + y_0$$

gives the vertical position of an object, at time t, thrown with an initial velocity v_0, from an initial position y_0 in a place where the acceleration of gravity is a. The acceleration of gravity on earth is $-9.8\,\frac{m}{s^2}$. It is negative, because we consider the upward direction as positive in this situation, and gravity pulls down.

83. What is the height of a baseball thrown with an initial velocity of $v_0 = 95\,\frac{m}{s}$, from an initial position of $y_0 = 92$ m, and at time $t = 19$ s?

Nineteen seconds after the baseball was thrown, it was ⬚ high in the air.

84. What is the height of a baseball thrown with an initial velocity of $v_0 = 50\,\frac{m}{s}$, from an initial position of $y_0 = 74$ m, and at time $t = 4$ s?

Four seconds after the baseball was thrown, it was ⬚ high in the air.

85. An auto company's sales volume can be modeled by $4.1x^2 + 7.1x + 3.3$, and its cost can be modeled by $2.3x^2 + 2.7x + 3.3$, where x represents the number of cars produced, and y stands for money in thousand dollars. We can calculate the company's net profit by subtracting cost from sales. Find the polynomial which models the company's sales in thousands of dollars.

The company's profit can be modeled by ⬚ dollars.

86. An auto company's sales volume can be modeled by $7.9x^2 + 1.8x + 3.2$, and its cost can be modeled by $2.7x^2 - 3.2x + 3.2$, where x represents the number of cars produced, and y stands for money in thousand dollars. We can calculate the company's net profit by subtracting cost from sales. Find the polynomial which models the company's sales in thousands of dollars.

The company's profit can be modeled by ⬚ dollars.

87. A handyman is building two pig pens sharing the same side. Assume the length of the shared side is x meters. The cost of building one pen would be $33.5x^2 - 2x + 4$ dollars, and the cost of building the other pen would be $30.5x^2 + 2x - 42$ dollars. What's the total cost of building those two pens?

   A polynomial representing the total cost of building those two pens is $\boxed{}$ dollars.

88. A handyman is building two pig pens sharing the same side. Assume the length of the shared side is x meters. The cost of building one pen would be $23x^2 + 10x + 2.5$ dollars, and the cost of building the other pen would be $33.5x^2 - 10x + 30.5$ dollars. What's the total cost of building those two pens?

   A polynomial representing the total cost of building those two pens is $\boxed{}$ dollars.

89. A farmer is building fence around a triangular area. The cost of building the shortest side is $40x$ dollars, where x stands for the length of the side in feet. The cost of building the other two sides can be modeled by $8x^2 - 0.5x + 40$ dollars and $4x^3 - 3.5x + 20$ dollars, respectively. What's the total cost of building fence for all three sides?

   The cost of building fence for all three sides would be $\boxed{}$ dollars.

90. A farmer is building fence around a triangular area. The cost of building the shortest side is $45x$ dollars, where x stands for the length of the side in feet. The cost of building the other two sides can be modeled by $5x^2 + 3.5x + 35$ dollars and $4x^3 + 1.5x + 25$ dollars, respectively. What's the total cost of building fence for all three sides?

   The cost of building fence for all three sides would be $\boxed{}$ dollars.

91. An architect is designing a house on an empty plot. The area of the plot can be modeled by the polynomial $5x^4 + 5x^2 - 4.5x$, and the area of the house's base can be modeled by $6x^3 - 4.5x - 40$. The rest of the plot is the yard. What's the yard's area?

   The area of the yard can be modeled by the polynomial $\boxed{}$.

92. An architect is designing a house on an empty plot. The area of the plot can be modeled by the polynomial $6x^4 + 16x^2 + 4x$, and the area of the house's base can be modeled by $5x^3 + 4x - 5$. The rest of the plot is the yard. What's the yard's area?

   The area of the yard can be modeled by the polynomial $\boxed{}$.

6.4 Multiplying Polynomials

Previously, we have learned to multiply monomials in Section 6.1 (such as $(4xy)(3x^2)$) and to add and subtract polynomials in Section 6.3 (such as $(4x^2 - 3x) + (5x^2 + x - 2)$). In this section, we will learn how to multiply polynomials.

> **Example 6.4.2 Revenue.** Avery owns a local organic jam company that currently sells about 1500 jars a month at a price of \$13 per jar. Avery has found that every time they raise the price of a jar by 25 cents, they will sell 50 fewer jars of jam each month.
>
> In general, this company's revenue can be calculated by multiplying the cost per jar by the total number of jars of jam sold.
>
> If we let x represent the number of 25-cent increases in the price, then the price per jar will be the current price of thirteen dollars/jar plus x times 0.25 dollars/jar, or $13 + 0.25x$.
>
> Continuing with x representing the number of 25-cent increases in the price, we know the company will sell 50 fewer jars each time the price increases by 25 cents. The number of jars the company will sell will be the 1500 they currently sell each month, minus 50 jars times x, the number of price increases. This gives us the expression $1500 - 50x$ to represent how many jars the company will sell after x 25-cent price increases.
>
> Combining this, we can now write a formula for our revenue model:
>
> $$\text{revenue} = (\text{price per item})(\text{number of items sold})$$
> $$R = (13 + 0.25x)(1500 - 50x)$$
>
> To simplify the expression $(13 + 0.25x)(1500 - 50x)$, we'll need to multiply $13 + 0.25x$ by $1500 - 50x$. In this section, we'll learn how to multiply these two expressions that each have multiple terms.

6.4.1 Review of the Distributive Property

The first step in almost every polynomial multiplication exercise will be a step of distribution. Let's quickly review the distributive property from Section 2.10, which states that $a(b + c) = ab + ac$ where a, b, and c are real numbers or variable expressions.

When we multiply a monomial with a binomial, we apply this property by distributing the monomial to each term in the binomial. For example,

$$-4x(3x^2 + 5) = (-4x) \cdot (3x^2) + (-4x) \cdot (5)$$
$$= -12x^3 - 20x$$

A visual approach to the distributive property is to treat the product as finding a rectangle's area. Such rectangles are referred to as **generic rectangles** and they can be used to model polynomial multiplication.

Figure 6.4.3: A Generic Rectangle Modeling $2x(3x + 4)$

The big rectangle consists of two smaller rectangles. The big rectangle's area is $2x(3x + 4)$, and the sum of those two smaller rectangles is $2x \cdot 3x + 2x \cdot 4$. Since the sum of the areas of those two smaller rectangles is the same as the bigger rectangle's area, we have:

$$2x(3x + 4) = 2x \cdot 3x + 2x \cdot 4$$
$$= 6x^2 + 8x$$

Generic rectangles are frequently used to visualize the distributive property.

Multiplying a monomial with a polynomial involves two steps: distribution and monomial multiplication. We also need to rely on the rules of exponents 6.1.15 when simplifying.

Checkpoint 6.4.4. Simplify the expression so that the result does not have parentheses.

a. $2(xy) = \boxed{}$

b. $2(x + y) = \boxed{}$

Explanation.

a. $2(xy) = 2xy$

b. $2(x + y) = 2x + 2y$

Checkpoint 6.4.5. Multiply the polynomials.

$-3x(-6x + 6) = \boxed{}$

Explanation. We multiply the monomial by each term in the binomial, using the properties of exponents to help us.

$$-3x(-6x + 6) = 18x^2 - 18x$$

Checkpoint 6.4.6. Multiply the polynomials.

$(4a^4)(10a^3 - 5a^{10}b^6 - 8b^7) = \boxed{}$

Explanation. We multiply the polynomials using the rule $a^m \cdot a^n = a^{m+n}$ to guide us.

$$(4a^4)(10a^3 - 5a^{10}b^6 - 8b^7) = 40a^{4+3} - 20a^{4+10}b^6 - 32a^4b^7$$
$$= 40a^7 - 20a^{14}b^6 - 32a^4b^7$$

Note that we are using the *distributive* property of multiplication in this problem: $x(y + z) = xy + xz$.

Remark 6.4.7. We can use the distributive property when multiplying on either the left or the right. This means that we can state $a(b + c) = ab + ac$, or that $(b + c)a = ba + ca$, which is equivalent to $ab + ac$. As an example,

$$(3x^2 + 5)(-4x) = (3x^2) \cdot (-4x) + (5) \cdot (-4x)$$
$$= -12x^3 - 20x$$

6.4.2 Approaches to Multiplying Binomials

Multiplying Binomials Using Distribution Whether we're multiplying a monomial with a polynomial or two larger polynomials together, the first step to carrying out the multiplication is a step of distribution.

We'll start with multiplying binomials and then move to working with larger polynomials.

We know we can distribute the 3 in $(x + 2)3$ to obtain $(x + 2) \cdot 3 = x \cdot 3 + 2 \cdot 3$. We can actually distribute anything across $(x + 2)$. For example:

$$(x + 2)🐱 = x \cdot 🐱 + 2 \cdot 🐱$$

With this in mind, we can begin multiplying $(x + 2)(x + 3)$ by distributing the $(x + 3)$ across $(x + 2)$:

$$(x + 2)(x + 3) = x(x + 3) + 2(x + 3)$$

To finish multiplying, we'll continue by distributing again, but this time across $(x + 3)$:

$$\begin{aligned}
(x + 2)(x + 3) &= x(x + 3) + 2(x + 3) \\
&= x \cdot x + x \cdot 3 + 2 \cdot x + 2 \cdot 3 \\
&= x^2 + 3x + 2x + 6 \\
&= x^2 + 5x + 6
\end{aligned}$$

To multiply a binomial by another binomial, we simply had to repeat the step of distribution and simplify the resulting terms. In fact, multiplying any two polynomials will rely upon these same steps.

Multiplying Binomials Using FOIL While multiplying two binomials requires two applications of the distributive property, people often remember this distribution process using the mnemonic FOIL. FOIL refers to the pairs of terms from each binomial that end up distributed to each other.

If we take another look at the example we just completed, $(x + 2)(x + 3)$, we can highlight how the FOIL process works. FOIL is the acronym for "First, Outer, Inner, Last".

$$\begin{aligned}
(x + 2)(x + 3) &= \overbrace{(x \cdot x)}^{F} + \overbrace{(3 \cdot x)}^{O} + \overbrace{(2 \cdot x)}^{I} + \overbrace{(2 \cdot 3)}^{L} \\
&= x^2 + 3x + 2x + 6 \\
&= x^2 + 5x + 6
\end{aligned}$$

F: x^2 The x^2 term was the result of the product of *first* terms from each binomial.

O: $3x$ The $3x$ was the result of the product of the *outer* terms from each binomial. This was from the x in the front of the first binomial and the 3 in the back of the second binomial.

I: $2x$ The $2x$ was the result of the product of the *inner* terms from each binomial. This was from the 2 in the back of the first binomial and the x in the front of the second binomial.

L: 6 The constant term 6 was the result of the product of the *last* terms of each binomial.

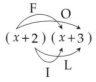

Figure 6.4.8: Using FOIL Method to multiply $(x + 2)(x + 3)$

Multiplying Binomials Using Generic Rectangles We can also approach this same example using the generic rectangle method. To use generic rectangles, we treat $x + 2$ as the base of a rectangle, and $x + 3$ as the height. Their product, $(x + 2)(x + 3)$, represents the rectangle's area. The next diagram shows how to set up generic rectangles to multiply $(x + 2)(x + 3)$.

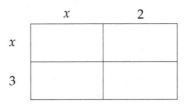

Figure 6.4.9: Setting up Generic Rectangles to Multiply $(x + 2)(x + 3)$

The big rectangle consists of four smaller rectangles. We will find each small rectangle's area in the next diagram by the formula area = base · height.

	x	2
x	x^2	$2x$
3	$3x$	6

Figure 6.4.10: Using Generic Rectangles to Multiply $(x + 2)(x + 3)$

To finish finding this product, we need to add the areas of the four smaller rectangles:

$$(x + 2)(x + 3) = x^2 + 3x + 2x + 6$$
$$= x^2 + 5x + 6$$

Notice that the areas of the four smaller rectangles are exactly the same as the four terms we obtained using distribution, which are also the same four terms that came from the FOIL method. Both the FOIL method and generic rectangles approach are different ways to represent the distribution that is occurring.

Example 6.4.11 Multiply $(2x - 3y)(4x - 5y)$ using distribution.

Explanation. To use the distributive property to multiply those two binomials, we'll first distribute the second binomial across $(2x - 3y)$. Then we'll distribute again, and simplify the terms that result.

$$(2x - 3y)(4x - 5y) = 2x(4x - 5y) - 3y(4x - 5y)$$
$$= 8x^2 - 10xy - 12xy + 15y^2$$
$$= 8x^2 - 22xy + 15y^2$$

Example 6.4.12 Multiply $(2x - 3y)(4x - 5y)$ using FOIL.

Explanation. First, Outer, Inner, Last: Either with arrows on paper or mentally in our heads, we'll pair up the four pairs of monomials and multiply those pairs together.

$$\overbrace{}^{F} \quad \overbrace{}^{O} \quad \overbrace{}^{I} \quad \overbrace{}^{L}$$

$$(2x - 3y)(4x - 5y) = (2x \cdot 4x) + (2x \cdot (-5y)) + (-3y \cdot 4x) + (-3y \cdot (-5y))$$
$$= 8x^2 - 10xy - 12xy + 15y^2$$
$$= 8x^2 - 22xy + 15y^2$$

Example 6.4.13 Multiply $(2x - 3y)(4x - 5y)$ using generic rectangles.

Explanation. We begin by drawing four rectangles and marking their bases and heights with terms in the given binomials:

Figure 6.4.14: Setting up Generic Rectangles to Multiply $(2x - 3y)(4x - 5y)$

Next, we calculate each rectangle's area by multiplying its base with its height:

Figure 6.4.15: Using Generic Rectangles to Multiply $(2x - 3y)(4x - 5y)$

Finally, we add up all rectangles' area to find the product:

$$(2x - 3y)(4x - 5y) = 8x^2 - 10xy - 12xy + 15y^2$$
$$= 8x^2 - 22xy + 15y^2$$

6.4.3 More Examples of Multiplying Binomials

When multiplying binomials, all of the approaches shown in Subsection 6.4.2 will have the same result. The FOIL method is the most direct and will be used in the examples that follow.

Checkpoint 6.4.16. Multiply the polynomials.

$(5x - 4)(x + 9) =$ ⬚

Explanation. We use the FOIL technique: *First Outside Inside Last.*

$$(5x - 4)(x + 9) = 5x^2 + 45x - 4x - 36$$
$$= 5x^2 + 41x - 36$$

Checkpoint 6.4.17. Multiply the polynomials.

$$\left(4x^2 - 5\right)\left(6x^2 - 2\right) = \boxed{}$$

Explanation. We use the FOIL technique: *First Outside Inside Last.*

$$\left(4x^2 - 5\right)\left(6x^2 - 2\right) = 24x^4 - 8x^2 - 30x^2 + 10$$
$$= 24x^4 - 38x^2 + 10$$

Example 6.4.18 Multiply and simplify the formula for Avery's jam company's revenue, R (in dollars), from Example 6.4.2 where $R = (13 + 0.25x)(1500 - 50x)$ and x represents the number of 25-cent price increases to the selling price of a jar of jam.

Explanation. To multiply this, we'll use FOIL:

$$R = (13 + 0.25x)(1500 - 50x)$$
$$= (13 \cdot 1500) + (13 \cdot (-50x)) + (0.25x \cdot 1500) + (0.25x \cdot (-50x))$$
$$= 19500 - 650x + 375x - 12.5x^2$$
$$= -12.5x^2 - 275x + 19500$$

Example 6.4.19 Tyrone is an artist and he sells each of his paintings for $200. Currently, he can sell 100 paintings per year. Thus, his annual income from paintings is $200 \cdot 100 = 20000$ dollars. He plans to raise the price. However, for each $20 price increase per painting, his customers would buy 5 fewer paintings annually.

Assume Tyrone would raise the price of his paintings x times, each time by $20. Use an expanded polynomial to represent his new income per year.

Explanation. Currently, each painting costs $200. After raising the price x times, each time by $20, each painting's new price would be $200 + 20x$ dollars.

Currently, Tyrone sells 100 paintings per year. After raising the price x times, each time selling 5 fewer paintings, he would sell $100 - 5x$ paintings per year.

His annual income can be calculated by multiplying each painting's price by the number of paintings he would sell:

$$\text{annual income} = (200 + 20x)(100 - 5x)$$
$$= 200(100) + 200(-5x) + 20x(100) + 20x(-5x)$$
$$= 20000 - 1000x + 2000x - 100x^2$$
$$= -100x^2 + 1000x + 20000$$

After raising the price x times, each time by $20, Tyrone's annual income from paintings would be $-100x^2 + 1000x + 20000$ dollars.

6.4.4 Multiplying Polynomials Larger Than Binomials

The foundation for multiplying any pair of polynomials is distribution and monomial multiplication. Whether we are working with binomials, trinomials, or larger polynomials, the process is fundamentally the same.

Example 6.4.20 Multiply $(x + 5)\left(x^2 - 4x + 6\right)$.

We can approach this product using either distribution generic rectangles. We cannot directly use the FOIL method, although it can be helpful to draw arrows to the six pairs of products that will occur.

Using the distributive property, we begin by distributing across $\left(x^2 - 4x + 6\right)$, perform a second step of distribution, and then combine like terms.

$$\begin{aligned}(x + 5)\left(x^2 - 4x + 6\right) &= x\left(x^2 - 4x + 6\right) + 5\left(x^2 - 4x + 6\right) \\ &= x \cdot x^2 - x \cdot 4x + x \cdot 6 + 5 \cdot x^2 - 5 \cdot 4x + 5 \cdot 6 \\ &= x^3 - 4x^2 + 6x + 5x^2 - 20x + 30 \\ &= x^3 + x^2 - 14x + 30\end{aligned}$$

With the foundation of monomial multiplication and understanding how distribution applies in this context, we are able to find the product of any two polynomials.

Checkpoint 6.4.21. Multiply the polynomials.

$(a - 3b)(a^2 + 7ab + 9b^2) = $ _____

Explanation. We multiply the polynomials by using the terms from $a - 3b$ successively.

$$\begin{aligned}(a - 3b)\left(a^2 + 7ab + 9b^2\right) &= aa^2 + a \cdot 7ab + a \cdot 9b^2 - 3ba^2 - 3b \cdot 7ab - 3b \cdot 9b^2 \\ &= a^3 + 4a^2b - 12ab^2 - 27b^3\end{aligned}$$

Exercises

Review and Warmup   Use the properties of exponents to simplify the expression.

1. $r^{14} \cdot r^7$

2. $x^{16} \cdot x^{19}$

3. $(-5x^{19}) \cdot (-4x^{13})$

4. $(9y^2) \cdot (5y^6)$

5. $\left(-8r^3\right)^3$

6. $\left(-4y^4\right)^3$

7. Count the number of terms in each expression.

  a. $-2y - 5y$

  b. $7y - x$

  c. $7s$

  d. $4z - s + 3y^2$

8. Count the number of terms in each expression.

  a. $3y - 8t - 3s$

  b. $s^2 + 6s - 5x + 5z$

  c. $6s - s^2$

  d. $-4y^2 - 5x + 8 + 7x$

9. List the terms in each expression.

  a. $-5.5y^2 - 0.5s + 8.2x$

  b. $1.1y + 3.1t$

  c. $2.2s^2 - 0.5z - 7.3y + 8.5s$

  d. $8.3x^2 - 5.2y - 5y^2$

10. List the terms in each expression.

   a. $6y + 8.6 + 8.6x^2$

   b. $-2.5z^2$

   c. $0.9y - 0.1t$

   d. $-5.5x^2 + 2.1s + 4.5y$

11. List the terms in each expression.

   a. $-8.9y^2 + 4.2z - 8.2 + 3.4z$

   b. $5.9y^2 - 7y^2 - 7.8y$

   c. $-7.5z + 1.4y - 3t$

   d. $1.8z - 2s + 8.7z + 6.6x^2$

12. List the terms in each expression.

   a. $-8.9y + 6.9z^2 + 7.3 - 3.8x$

   b. $7.5y + 6.5y$

   c. $-4.6s^2$

   d. $-6.2y^2 - 5.3y - 6s^2$

13. Simplify each expression, if possible, by combining like terms.

   a. $-9y^2 - 8y^2$

   b. $-5t^2 + 4t^2$

   c. $-9s^2 + 7t^2 - 4t$

   d. $-7x - z + 9z$

14. Simplify each expression, if possible, by combining like terms.

   a. $-6y + 5$

   b. $-6t^2 - 2x^2 - 2x^2 - 7x^2$

   c. $-t^2 + 2$

   d. $-y + 4y^2$

15. Simplify each expression, if possible, by combining like terms.

   a. $y - 9y$

   b. $-y^2 + \frac{9}{5}z^2 - \frac{7}{9}t^2$

   c. $-\frac{4}{9}t^2 + \frac{2}{3}t + t$

   d. $\frac{1}{4}x^2 - 9z^2 + 4x^2 + \frac{2}{3}z$

16. Simplify each expression, if possible, by combining like terms.

   a. $\frac{1}{3}x - \frac{7}{9}s$

   b. $-\frac{2}{9}z + \frac{9}{4}y$

   c. $-\frac{3}{5}z + \frac{5}{2}y - \frac{9}{8}s - \frac{1}{4}z$

   d. $\frac{1}{3}z^2 + \frac{1}{8}z^2 + \frac{6}{5}z^2$

Multiplying Monomials with Binomials Multiply the polynomials.

17. $-x(x - 7) = \boxed{}$

18. $x(x + 5) = \boxed{}$

19. $-7x(3x - 3) = \boxed{}$

20. $8x(-6x + 3) = \boxed{}$

21. $8x^2(x + 5) = \boxed{}$

22. $10x^2(x - 9) = \boxed{}$

23. $7x^2(-7x^2 + 2x) = \boxed{}$

24. $-4x^2(-5x^2 - 6x) = \boxed{}$

25. $10y^2(-2y^2 + 8y - 9) = \boxed{}$

26. $-7y^2(-9y^2 - 6y + 2) = \boxed{}$

27. $(-6x^8y^{15})(-2x^{17} + 5y^{16}) = \boxed{}$ **28.** $(7x^{10}y^{19})(-9x^{13} + 2y^5) = \boxed{}$ **29.** $(8a^{12}b^8)(4a^{19}b^{13} - 2a^4b^{15}) =$
$\boxed{}$

30. $(-9a^{13}b^{16})(8a^6b^3 + 2a^{15}b^{19}) =$ **31.** $(10a^7)(-3a^6 - 10a^5b^8 - 8b^9) =$ **32.** $(-2a^9)(-6a^{10} + 10a^{10}b^8 + 9b^8) =$
$\boxed{}$ $\boxed{}$ $\boxed{}$

Applications of Multiplying Monomials with Binomials

33. A rectangle's length is 3 feet shorter than twice its width. If we use w to represent the rectangle's width, use a polynomial to represent the rectangle's area in expanded form.

area = $\boxed{}$ square feet

34. A rectangle's length is 4 feet shorter than 5 times its width. If we use w to represent the rectangle's width, use a polynomial to represent the rectangle's area in expanded form.

area = $\boxed{}$ square feet

35. A triangle's height is 6 feet longer than 4 times its base. If we use b to represent the triangle's base, use a polynomial to represent the triangle's area in expanded form. A triangle's area can be calculated by $A = \frac{1}{2}bh$, where b stands for base, and h stands for height.

area = $\boxed{}$ square feet

36. A triangle's height is 6 feet longer than twice its base. If we use b to represent the triangle's base, use a polynomial to represent the triangle's area in expanded form. A triangle's area can be calculated by $A = \frac{1}{2}bh$, where b stands for base, and h stands for height.

area = $\boxed{}$ square feet

37. A trapezoid's top base is 2 feet longer than its height, and its bottom base is 6 feet longer than its height. If we use h to represent the trapezoid's height, use a polynomial to represent the trapezoid's area in expanded form. A trapezoid's area can be calculated by $A = \frac{1}{2}(a + b)h$, where a stands for the top base, b stands for the bottom base, and h stands for height.

area = $\boxed{}$ square feet

38. A trapezoid's top base is 10 feet longer than its height, and its bottom base is 6 feet longer than its height. If we use h to represent the trapezoid's height, use a polynomial to represent the trapezoid's area in expanded form. A trapezoid's area can be calculated by $A = \frac{1}{2}(a + b)h$, where a stands for the top base, b stands for the bottom base, and h stands for height.

area = $\boxed{}$ square feet

Multiplying Binomials   Multiply the polynomials.

39. $(t + 1)(t + 5) = \boxed{}$ **40.** $(t + 8)(t + 9) = \boxed{}$ **41.** $(5x + 3)(x + 7) = \boxed{}$

42. $(2x + 8)(x + 4) = \boxed{}$ **43.** $(y + 7)(y - 9) = \boxed{}$ **44.** $(y + 4)(y - 5) = \boxed{}$

45. $(r-1)(r-5) = \boxed{}$

46. $(r-4)(r-7) = \boxed{}$

47. $(3t+8)(t+8) = \boxed{}$

48. $(6t+3)(5t+8) = \boxed{}$

49. $(4t-6)(5t-1) = \boxed{}$

50. $(3x-10)(2x-3) = \boxed{}$

51. $(10x-6)(x-1) = \boxed{}$

52. $(6y-2)(y-4) = \boxed{}$

53. $(3y-8)(y+4) = \boxed{}$

54. $(9r-3)(r+1) = \boxed{}$

55. $(4r-9)(5r^2-4) = \boxed{}$

56. $(2t-5)(4t^2-6) = \boxed{}$

57. $(9t^3+10)(t^2+3) = \boxed{}$

58. $(6t^3+4)(t^2+10) = \boxed{}$

59. $(2x^2-3)(5x^2-4) = \boxed{}$

60. $(5x^2-8)(3x^2-4) = \boxed{}$

61. $(a-4b)(a-7b) = \boxed{}$

62. $(a-5b)(a+2b) = \boxed{}$

63. $(a-8b)(6a+7b) = \boxed{}$

64. $(a+5b)(7a+4b) = \boxed{}$

65. $(8a-4b)(2a+6b) = \boxed{}$

66. $(9a+8b)(8a-6b) = \boxed{}$

67. $(10ab-3)(4ab+6) = \boxed{}$

68. $(2ab-6)(10ab-6) = \boxed{}$

69. $4(x+10)(x+9) = \boxed{}$

70. $-2(y-3)(y+3) = \boxed{}$

71. $y(y+9)(y+6) = \boxed{}$

72. $-3r(r-9)(r-1) = \boxed{}$

73. $-3(4r+3)(r-1) = \boxed{}$

74. $4(2t-1)(t-1) = \boxed{}$

Applications of Multiplying Binomials

75. An artist sells his paintings at \$19.00 per piece. Currently, he can sell 130 paintings per year. Thus, his annual income from paintings is $19 \cdot 130 = 2470$ dollars. He plans to raise the price. However, for each \$3.00 of price increase per painting, his customers would buy 6 fewer paintings annually.

Assume the artist would raise the price of his painting x times, each time by \$3.00. Use an expanded polynomial to represent his new income per year.

new annual income = $\boxed{}$

76. An artist sells his paintings at \$20.00 per piece. Currently, he can sell 110 paintings per year. Thus, his annual income from paintings is $20 \cdot 110 = 2200$ dollars. He plans to raise the price. However, for each \$5.00 of price increase per painting, his customers would buy 10 fewer paintings annually.

Assume the artist would raise the price of his painting x times, each time by \$5.00. Use an expanded polynomial to represent his new income per year.

new annual income = $\boxed{}$

77. A rectangle's base can be modeled by $x + 2$ meters, and its height can be modeled by $x + 4$ meters. Use a polynomial to represent the rectangle's area in expanded form.

area = [   ]
square meters

78. A rectangle's base can be modeled by $x + 3$ meters, and its height can be modeled by $x - 8$ meters. Use a polynomial to represent the rectangle's area in expanded form.

area = [   ]
square meters

Multiplying Larger Polynomials   Multiply the polynomials.

79. $(-3x - 2)(x^2 + 3x + 5) =$ [   ]

80. $(3x + 4)(x^2 - 3x - 2) =$ [   ]

81. $(-4x + 5)(3x^3 - 2x^2 - 4x + 5) =$ [   ]

82. $(-4x - 3)(2x^3 + 5x^2 + 3x - 2) =$ [   ]

83. $(x^2 + 5x - 5)(x^2 + 3x - 3) =$ [   ]

84. $(x^2 - 5x + 2)(x^2 - 3x + 3) =$ [   ]

85. $(a - 10b)(a^2 - 7ab + 4b^2) =$ [   ]

86. $(a + 2b)(a^2 - 2ab - 4b^2) =$ [   ]

87. $(a + b - 3)(a + b + 3) =$ [   ]

88. $(a + b - 4)(a + b + 4) =$ [   ]

Challenge

89. Fill in the blanks with algebraic expressions that make the equation true. You may not use 0 or 1 in any of the blank spaces. An example is ? + ? = 8x, where one possible answer is $3x + 5x = 8x$. There are infinitely many correct answers to this problem. *Be creative.* After finding a correct answer, see if you can come up with a different answer that is also correct.

a. [   ] + [   ] $= -13xy$

b. [   ] + [   ] $= -19x^{15}y^9$

c. [   ] · [   ] · [   ] · [   ] · [   ] $= 17x^{50}y^{75}$

6.5 Special Cases of Multiplying Polynomials

Since we are now able to multiply polynomials together, we will look at a few special cases of polynomial multiplication.

6.5.1 Squaring a Binomial

Example 6.5.2 To "square a binomial" is to take a binomial and multiply it by itself. We know that exponent notation means that $4^2 = 4 \cdot 4$. Applying this to a binomial, we'll see that $(x+4)^2 = (x+4)(x+4)$. To expand this expression, we'll simply distribute $(x + 4)$ across $(x + 4)$:

$$(x + 4)^2 = (x + 4)(x + 4)$$
$$= x^2 + 4x + 4x + 16$$
$$= x^2 + 8x + 16$$

Similarly, to expand $(y - 7)^2$, we'll have:

$$\left(y - 7\right)^2 = \left(y - 7\right)\left(y - 7\right)$$
$$= y^2 - 7y - 7y + 49$$
$$= y^2 - 14y + 49$$

These two examples might look like any other example of multiplying binomials, but looking closely we can see that something very specific (or *special*) happened. Focusing on the original expression and the simplified one, we can see that a specific pattern occurred in each:

$$(x + 4)^2 = x^2 + 4x + 4x + 4 \cdot 4$$
$$(x + 4)^2 = x^2 + 2(4x) + 4^2$$

And:

$$\left(y - 7\right)^2 = y^2 - 7y - 7y + 7 \cdot 7$$
$$\left(y - 7\right)^2 = y^2 - 2(7y) + 7^2$$

Notice that the two middle terms are not only the same, they are also exactly the product of the two terms in the binomial. Furthermore, the last term is the square of the second term in each original binomial.

What we're seeing is a pattern that relates to two important phrases: The process is called **squaring a binomial**, and the result is called a **perfect square trinomial**. The first phrase is a description of what we're doing, we are literally squaring a binomial. The second phrase is a description of what you end up with. This second name will become important in a future chapter.

Example 6.5.3 The general way this pattern is presented is by squaring the two most general binomials possible, $(a + b)$ and $(a - b)$. We will establish the pattern for $(a + b)^2$ and $(a - b)^2$. Once we have done so, we will be able to substitute anything in place of a and b and rely upon the general pattern to simplify squared binomials.

We first must expand $(a + b)^2$ as $(a + b)(a + b)$ and then we can multiply those binomials:

$$(a + b)^2 = (a + b)(a + b)$$
$$= a^2 + ab + ba + b^2$$
$$= a^2 + 2ab + b^2$$

Notice the final simplification step was to add $ab + ba$. Since these are like terms, we can combine them into $2ab$.

Similarly, we can find a general formula for $(a - b)^2$:

$$(a - b)^2 = (a - b)(a - b)$$
$$= a^2 - ab - ba + b^2$$
$$= a^2 - 2ab + b^2$$

Fact 6.5.4 Squaring a Binomial Formulas. *If a and b are real numbers or variable expressions, then we have the following formulas:*

$$(a + b)^2 = a^2 + 2ab + b^2$$
$$(a - b)^2 = a^2 - 2ab + b^2$$

These formulas will allow us to multiply this type of special product more quickly.

Remark 6.5.5. Notice that when both $(a + b)^2$ and $(a - b)^2$ are expanded in Example 6.5.3, the last term was a *positive* b^2 in both. This is because any number or expression, regardless of its sign, is positive after it is squared.

6.5.2 Further Examples of Squaring Binomials

Example 6.5.6 Expand $(2x - 3)^2$ using the squaring a binomial formula.

For this example we need to recognize that to apply the formula $(a - b)^2 = a^2 - 2ab + b^2$ in this situation, $a = 2x$ and $b = 3$. Expanding this, we have:

$$(a - b)^2 = a^2 \quad - 2ab \quad + b^2$$
$$(2x - 3)^2 = (2x)^2 - 2(2x)(3) + (3)^2$$
$$= 4x^2 \quad - 12x \quad + 9$$

Remark 6.5.7. While we rely on the formula for squaring a binomial in Example 6.5.6, we will often omit the step of formally writing the formula and jump to the simplification, in this way:

$$(2x - 3)^2 = 4x^2 - 12x + 9$$

Example 6.5.8 Multiply the following using the squaring a binomial formula:

a. $(5xy + 1)^2$

b. $4(3x - 7)^2$

Explanation.

a. $(5xy + 1)^2 = (5xy)^2 + 2(5xy)(1) + 1^2$
$$= 25x^2y^2 + 10xy + 1$$

b. With this expression, we will first note that the factor of 4 is *outside* the portion of the expression that is squared. Using the order of operations, we will first expand $(3x - 7)^2$ and then multiply that expression by 4:

$$4(3x - 7)^2 = 4\left((3x)^2 - 2(3x)(7) + 7^2\right)$$
$$= 4\left(9x^2 - 42x + 49\right)$$
$$= 36x^2 - 168x + 196$$

Example 6.5.9 A circle's area can be calculated by the formula

$$A = \pi r^2$$

where A stands for area, and r stands for radius. If a certain circle's radius can be modeled by $x - 5$ feet, use an expanded polynomial to model the circle's area.

Explanation. The circle's area would be:

$$A = \pi r^2$$
$$= \pi(x - 5)^2$$
$$= \pi\left[(x)^2 - 2(x)(5) + (5)^2\right]$$
$$= \pi\left[x^2 - 10x + 25\right]$$
$$= \pi x^2 - 10\pi x + 25\pi$$

The circle's area can be modeled by $\pi x^2 - 10\pi x + 25\pi$ square feet.

Checkpoint 6.5.10. Expand the square of a *binomial*.

$\left(y^3 - 12\right)^2 = $ [   ]

Explanation. We use the FOIL technique: *First Outside Inside Last*

$$\left(y^3 - 12\right)^2 = \left(y^3 - 12\right)\left(y^3 - 12\right)$$
$$= y^6 - 12y^3 - 12y^3 + 144$$
$$= y^6 - 24y^3 + 144$$

Alternatively, we might observe that this is the square of *the difference of two terms*, in which case we may use the formula

$$(a - b)^2 = a^2 - 2ab + b^2$$

and write

$$\left(y^3 - 12\right)\left(y^3 - 12\right) = \left(y^3\right)^2 - 2 \cdot y^3 \cdot 12 + 12^2$$
$$= y^6 - 24y^3 + 144$$

which is the same result we obtained using the FOIL method.

6.5.3 The Product of the Sum and Difference of Two Terms

To identify the next "special case" for multiplying polynomials, we'll look at a couple of examples.

Example 6.5.11 Multiply the following binomials:

a. $(x + 5)(x - 5)$
b. $(y - 8)(y + 8)$

Explanation. We can approach these as using distribution, FOIL, or generic rectangles, and obtain the following:

a. $(x + 5)(x - 5) = x^2 - 5x + 5x - 25$
$$= x^2 - 25$$

b. $(y + 8)(y - 8) = y^2 - 8y + 8y - 4$
$$= y^2 - 64$$

Notice that for each of these products, we multiplied the sum of two terms by the difference of the *same* two terms. Notice also in these three examples that once these expressions were multiplied, the two middle terms were opposites and thus canceled to zero.

These pairs, generally written as $(a + b)$ and $(a - b)$, are known as **conjugates**. If we multiply $(a + b)(a - b)$, we can see this general pattern more clearly:

$$(a + b)(a - b) = a^2 - ab + ab - b^2$$
$$= a^2 - b^2$$

As with the previous special case, this one also has two names. This can be called the **product of the sum and difference of two terms**, because this pattern is built on multiplying two binomials that have the same two terms, except one binomial is a sum and the other binomial is a difference. The second name is a **difference of squares**, because the end result of the multiplication is a binomial that is the difference of two perfect squares. As before, the second name will become useful in a future chapter when using exactly the technique described in this section will be pertinent.

Fact 6.5.12 The Product of the Sum and Difference of Two Terms Formula. *If a and b are real numbers or variable expressions, then we have the following formula:*

$$(a + b)(a - b) = a^2 - b^2$$

⬛ Checkpoint 6.5.13. Multiply the polynomials.

$(4x + 2)(4x - 2) =$ [   ]

Explanation. We use the FOIL technique: *First Outside Inside Last*

$$(4x + 2)(4x - 2) = 16x^2 - 8x + 8x - 4$$
$$= 16x^2 - 4$$

Alternatively, we might observe that this is the product of *the sum and difference of two terms*, in which case we may use the formula

$$(a - b)(a + b) = a^2 - b^2$$

and write

$$(4x + 2)(4x - 2) = (4x)^2 - 2^2$$
$$= 16x^2 - 4$$

which is the same result we obtained using the FOIL method.

Example 6.5.14 Multiply the following using Fact 6.5.12.

a. $(4x - 7y)(4x + 7y)$

b. $-2(3x + 1)(3x - 1)$

Explanation. The first step to using this method is to identify the values of a and b.

a. In this instance, $a = 4x$ and $b = 7y$. Using the formula,

$$(4x - 7y)(4x + 7y) = (4x)^2 - (7y)^2$$
$$= 16x^2 - 49y^2$$

b. In this instance, we have a constant factor as well as a product in the form $(a + b)(a - b)$. We will first expand $(3x + 1)(3x - 1)$ by identifying $a = 3x$ and $b = 1$ and using the formula. Then we will multiply the factor of -2 through this expression. So,

$$-2(3x + 1)(3x - 1) = -2\left((3x)^2 - 1^2\right)$$
$$= -2\left(9x^2 - 1\right)$$
$$= -18x^2 + 2$$

Checkpoint 6.5.15. Multiply the polynomials.

$$\left(x^7 - 2\right)\left(x^7 + 2\right) = \boxed{}$$

Explanation. We use the FOIL technique: *First Outside Inside Last*

$$\left(x^7 - 2\right)\left(x^7 + 2\right) = x^{14} + 2x^7 - 2x^7 - 4$$
$$= x^{14} - 4$$

Alternatively, we might observe that this is the product of *the sum and difference of two terms,* in which case we may use the formula

$$(a - b)(a + b) = a^2 - b^2$$

and write

$$\left(x^7 - 2\right)\left(x^7 + 2\right) = \left(x^7\right)^2 - 2^2$$
$$= x^{14} - 4$$

which is the same result we obtained using the FOIL method.

If a and b are real numbers or variable expressions, then we have the following formulas:

Squaring a Binomial (Sum) $(a + b)^2 = a^2 + 2ab + b^2$

Squaring a Binomial (Difference) $(a - b)^2 = a^2 - 2ab + b^2$

Product of the Sum and Difference of Two Terms $(a + b)(a - b) = a^2 - b^2$

List 6.5.16: Special Cases of Multiplication Formulas

Warning 6.5.17 Common Mistakes. We've found that

$$(a + b)(a - b) = a^2 - b^2$$

However,

$$(a - b)^2 \neq a^2 - b^2 \text{ because } (a - b)^2 = a^2 - 2ab + b^2$$

Similarly,

$$(a + b)^2 \neq a^2 + b^2 \text{ because } (a + b)^2 = a^2 + 2ab + b^2$$

6.5.4 Binomials Raised to Other Powers

Example 6.5.18 Simplify the expression $(x + 5)^3$ into an expanded polynomial.

Before we start expanding this expression, it is important to recognize that $(x + 5)^3 \neq x^3 + 5^3$. We can see that this doesn't work by inputting 1 for x and applying the order of operations:

$$(1 + 5)^3 = 6^3 \qquad\qquad 1^3 + 5^3 = 1 + 125$$
$$= 216 \qquad\qquad\qquad = 126$$

With this in mind, we will need to rely on distribution to expand this expression. The first step in expanding $(x + 5)^3$ is to remember that the exponent of 3 indicates that

$$(x + 5)^3 = \overbrace{(x + 5)(x + 5)(x + 5)}^{3 \text{ times}}$$

Once we rewrite this in an expanded form, we next multiply the two binomials on the left and then finish by multiplying that result by the remaining binomial:

$$(x + 5)^3 = [(x + 5)(x + 5)](x + 5)$$
$$= \left[x^2 + 10x + 25\right](x + 5)$$
$$= x^3 + 5x^2 + 10x^2 + 50x + 25x + 125$$
$$= x^3 + 15x^2 + 75x + 125$$

Checkpoint 6.5.19. Simplify the given expression into an expanded polynomial.

$(2y - 6)^3 = $ [_____]

Explanation. The main thing to notice on this problem is that we can write $(2y - 6)^3$ as

$$(2y - 6)^3 = (2y - 6)(2y - 6)^2$$

This means that we can use the FOIL technique on the second binomial multiplication, and then multiply the first factor $2y - 6$ by the result.

$$(2y - 6)^3 = (2y - 6)\left[(2y - 6)(2y - 6)\right]$$
$$= (2y - 6)\left[4y^2 - 12y - 12y + 36\right]$$
$$= (2y - 6)\left[4y^2 - 24y + 36\right]$$
$$= 8y^3 - 48y^2 + 72y - 24y^2 + 144y - 216$$
$$= 8y^3 - 72y^2 + 216y - 216$$

You might like to know that the formula for the cube of the *difference* of two terms is

$$(a - b)^3 = a^3 - 3a^2b + 3ab^2 - b^3$$

If you have the time, you can verify that this formula works in this problem as an exercise.

If we wanted to expand a binomial raised to any power, we always start by rewriting the expression without an exponent.

To multiply $(x - 3)^4$, we'd start by rewriting $(x - 3)^4$ in expanded form as:

$$\overbrace{(x - 3)^4 = (x - 3)(x - 3)(x - 3)(x - 3)}^{4 \text{ times}}$$

We will then multiply pairs of polynomials from the left to the right.

$$\begin{aligned}
(x - 3)^4 &= [(x - 3)(x - 3)](x - 3)(x - 3) \\
&= \left[(x^2 - 6x + 9)(x - 3)\right](x - 3) \\
&= \left[x^3 - 9x^2 + 27x - 27\right](x - 3) \\
&= x^2 - 9x^3 + 27x^2 - 27x - 3x^3 + 27x^2 - 81x + 81 \\
&= x^4 - 12x^3 + 54x^2 - 108x + 81
\end{aligned}$$

Exercises

Review and Warmup   Use the properties of exponents to simplify the expression.

1. $\left(3t^{12}\right)^4$ **2.** $\left(5r^2\right)^3$ **3.** $(2x)^2$ **4.** $(5r)^2$

5. $\left(-10t^6\right)^3$ **6.** $\left(-6r^7\right)^2$ **7.** $-4\left(-3x^8\right)^3$ **8.** $-5\left(-8x^{10}\right)^3$

Simplify each expression, if possible, by combining like terms.

9. a. $-9x + 3s$

   b. $-4x^2 + 8x^2$

   c. $5x^2 + 8x^2$

   d. $2y^2 - 5y^2$

10. a. $-8x^2 - 7$

   b. $-3t^2 - 3s^2$

   c. $-9t - 6t^2$

   d. $-3z^2 + 6y^2$

11. a. $-6x + t$

   b. $-2y^2 - 9y^2 + 7z^2$

   c. $4x^2 - 2 - 7t$

   d. $3z^2 + 9y^2$

12. a. $-4x^2 - 9x^2$

   b. $-s + 2s^2$

   c. $3x + 6t - 3x + 5s$

   d. $-7x - 8x + 5t + 9s$

13. Determine if the following statements are true or false.

  a. $(a - b)^2 = a^2 - b^2$

    ($\square$ True $\square$ False)

  b. $(a + b)^2 = a^2 + b^2$

    ($\square$ True $\square$ False)

  c. $(a + b)(a - b) = a^2 - b^2$

    ($\square$ True $\square$ False)

14. Determine if the following statements are true or false.

  a. $(2(a - b))^2 = 4(a - b)^2$

    ($\square$ True $\square$ False)

  b. $2(a + b)^2 = 2a^2 + 2b^2$

    ($\square$ True $\square$ False)

  c. $2(a + b)(a - b) = 2a^2 - 2b^2$

    ($\square$ True $\square$ False)

Perfect Square Trinomial Formula Expand the square of a *binomial*.

15. $(r + 1)^2 = $ [____]

16. $(r + 8)^2 = $ [____]

17. $(5r + 1)^2 = $ [____]

18. $(2t + 5)^2 = $ [____]

19. $(t - 4)^2 = $ [____]

20. $(x - 7)^2 = $ [____]

21. $(10x - 3)^2 = $ [____]

22. $(7y - 9)^2 = $ [____]

23. $(4y^2 - 5)^2 = $ [____]

24. $(10r^2 - 1)^2 = $ [____]

25. $(r^6 + 3)^2 = $ [____]

26. $(r^9 - 7)^2 = $ [____]

27. $(9a - 10b)^2 = $ [____]

28. $(10a + 6b)^2 = $ [____]

29. $(2ab - 3)^2 = $ [____]

30. $(3ab - 9)^2 = $ [____]

31. $(x^2 + 4y^2)^2 = $ [____]

32. $(x^2 - 5y^2)^2 = $ [____]

Difference of Squares Formula Multiply the polynomials.

33. $(r + 9)(r - 9) = $ [____]

34. $(r - 1)(r + 1) = $ [____]

35. $(2r + 1)(2r - 1) = $ [____]

36. $(5t + 10)(5t - 10) = $ [____]

37. $(5 - 5t)(5 + 5t) = $ [____]

38. $(1 + 9x)(1 - 9x) = $ [____]

39. $(x^4 + 6)(x^4 - 6) = $ [____]

40. $(y^8 - 3)(y^8 + 3) = $ [____]

41. $(2y^{10} - 12)(2y^{10} + 12) = $ [____]

42. $(5r^7 - 1)(5r^7 + 1) = $ [____]

43. $(1 + 7r^5)(1 - 7r^5) = $ [____]

44. $(1 - 13r^3)(1 + 13r^3) = $ [____]

45. $(9x + 7y)(9x - 7y) =$ ◻

46. $(10x - 4y)(10x + 4y) =$ ◻

47. $(ab + 2)(ab - 2) =$ ◻

48. $(ab + 3)(ab - 3) =$ ◻

49. $3(y - 7)(y + 7) =$ ◻

50. $5(y + 11)(y - 11) =$ ◻

51. $2(4r - 6)(4r + 6) =$ ◻

52. $6(3r + 3)(3r - 3) =$ ◻

53. $2(r + 9)^2 =$ ◻

54. $4(t + 6)^2 =$ ◻

55. $2(3t + 9)^2 =$ ◻

56. $6(9x + 4)^2 =$ ◻

57. $(x^2 - 3y^2)(x^2 + 3y^2) =$ ◻

58. $(x^2 - 4y^2)(x^2 + 4y^2) =$ ◻

59. $(5x^8 + 9y^4)(5x^8 - 9y^4) =$ ◻

60. $(6x^6 + 5y^4)(6x^6 - 5y^4) =$ ◻

61. $(7x^4y^3 - 2y^4)(7x^4y^3 + 2y^4) =$

62. $(8x^2y^6 + 8y^4)(8x^2y^6 - 8y^4) =$

Binomials Raised to Other Powers Simplify the given expression into an expanded polynomial.

63. $(t + 3)^3 =$ ◻

64. $(t + 2)^3 =$ ◻

65. $(x - 5)^3 =$ ◻

66. $(x - 3)^3 =$ ◻

67. $(6y + 5)^3 =$ ◻

68. $(5y + 3)^3 =$ ◻

69. $(3r - 6)^3 =$ ◻

70. $(6r - 2)^3 =$ ◻

6.6 Dividing by a Monomial

Now that we know how to add, subtract, and multiply polynomials, we will learn how to divide a polynomial by a monomial.

6.6.1 Dividing a Polynomial by a Monomial

One example of dividing a polynomial is something we already studied in Section 4.7, when we rewrote an equation in standard form in slope-intercept form. We'll briefly review this process.

> **Example 6.6.2** Rewrite $4x - 2y = 10$ in slope-intercept form.
>
> In being asked to rewrite this equation in slope-intercept form, we're really being asked to solve the equation $4x - 2y = 10$ for y.
>
> $$7x - 2y = 10$$
> $$7x - 2y - 7x = 10 - 7x$$
> $$-2y = -7x + 10$$
> $$\frac{-2y}{-2} = \frac{-7x + 10}{-2}$$
> $$y = -\frac{7}{2}x - 5$$
>
> This is an example of polynomial division that we have already done. We'll extend it to more complicated examples, many of which involve dividing polynomials by variables (instead of just numbers).
>
> In the final step of work, we divided each term on the right side of the equation by -2.

Like polynomial multiplication, polynomial division will rely upon distribution.

It's important to remember that dividing by a number c is the same as multiplying by the reciprocal $\frac{1}{c}$:

$$\frac{8}{2} = \frac{1}{2} \cdot 8 \quad \text{and} \quad \frac{9}{3} = \frac{1}{3} \cdot 9$$

If we apply this idea to a situation involving polynomials, say $\frac{a+b}{c}$, we can show that distribution works for division as well:

$$\frac{a + b}{c} = \frac{1}{c} \cdot (a + b)$$
$$= \frac{1}{c} \cdot a + \frac{1}{2} \cdot b$$
$$= \frac{a}{c} + \frac{b}{c}$$

Once we recognize that the division distributes just as multiplication distributed, we are left with individual monomial pairs that we will divide.

> **Example 6.6.3** Simplify $\dfrac{2x^3 + 4x^2 - 10x}{2}$.
>
> The first step will be to recognize that the 2 we're dividing by will be divided into every term of the

numerator. Once we recognize that, we will simply perform that division.

$$\frac{2x^3 + 4x^2 - 10x}{2} = \frac{2x^3}{2} + \frac{4x^2}{2} + \frac{-10x}{2}$$
$$= x^3 + 2x^2 - 5x$$

Example 6.6.4 Simplify $\dfrac{15x^4 - 9x^3 + 12x^2}{3x^2}$

Explanation. The key to simplifying $\frac{15x^4-9x^3+12x^2}{3x^2}$ is to recognize that each term in the numerator will be divided by $3x^2$. In doing this, each coefficient and exponent will change. Performing this division by distributing, we get:

$$\frac{15x^4 - 9x^3 + 12x^2}{3x^2} = \frac{15x^4}{3x^2} + \frac{-9x^3}{3x^2} + \frac{12x^2}{3x^2}$$
$$= 5x^2 - 3x + 4$$

Remark 6.6.5. Once you become comfortable with this process, you will often leave out the step where we wrote out the distribution. You will do the distribution in your head and this will often become a one-step problem. Here's how Example 6.6.4 would be visualized:

$$\frac{15x^4 - 9x^3 + 12x^2}{3x^2} = \square x^{\square} - \square x^{\square} + \square x^{\square}$$

And when calculated, we'd get:

$$\frac{15x^4 - 9x^3 + 12x^2}{3x^2} = 5x^2 - 3x + 4$$

(Note that $\frac{x^2}{x^2}$ is technically x^0, which is equivalent to 1.)

Example 6.6.6 Simplify $\dfrac{20x^3y^4 + 30x^2y^3 - 5x^2y^2}{-5xy^2}$

Explanation.

$$\frac{20x^3y^4 + 30x^2y^3 - 5x^2y^2}{-5xy^2} = \frac{20x^3y^4}{-5xy^2} + \frac{30x^2y^3}{-5xy^2} + \frac{-5x^2y^2}{-5xy^2}$$
$$= -4x^2y^2 - 6xy + x$$

Checkpoint 6.6.7. Simplify the following expression

$$\frac{18r^{20} + 18r^{16} - 54r^{14}}{-6r^2} = \boxed{}$$

Explanation. We divide each term by $-6r^2$ as follows.

$$\frac{18r^{20} + 18r^{16} - 54r^{14}}{-6r^2} = \frac{18r^{20}}{-6r^2} + \frac{18r^{16}}{-6r^2} + \frac{-54r^{14}}{-6r^2}$$
$$= -\frac{18}{6}r^{18} - \frac{18}{6}r^{14} + \frac{54}{6}r^{12}$$
$$= -3r^{18} - 3r^{14} + 9r^{12}$$

Example 6.6.8 A rectangular prism's volume can be calculated by the formula

$$V = Bh$$

where V stands for volume, B stands for base area, and h stands for height. A certain rectangular prism's volume can be modeled by $4x^3 - 6x^2 + 8x$ cubic units. If its height is $2x$ units, find the prism's base area.

Explanation. Since $V = Bh$, we can use $B = \frac{V}{h}$ to calculate the base area. After substitution, we have:

$$\begin{aligned} B &= \frac{V}{h} \\ &= \frac{4x^3 - 6x^2 + 8x}{2x} \\ &= \frac{4x^3}{2x} - \frac{6x^2}{2x} + \frac{8x}{2x} \\ &= 2x^2 - 3x + 4 \end{aligned}$$

The prism's base area can be modeled by $2x^2 - 3x + 4$ square units.

Exercises

Review and Warmup Use the properties of exponents to simplify the expression.

1. $\dfrac{t^{17}}{t^5} =$ [   ]

2. $\dfrac{y^{19}}{y^{17}} =$ [   ]

3. $\dfrac{8r^{16}}{2r^4} =$ [   ]

4. $\dfrac{-12x^{11}}{4x^3} =$ [   ]

5. $\dfrac{5x^{14}}{30x^3} =$ [   ]

6. $\dfrac{7x^4}{28x^3} =$ [   ]

Rewrite the expression simplified and using only positive exponents.

7. $\dfrac{y^6}{y^{11}} =$ [   ]

8. $\dfrac{r^{37}}{r^{42}} =$ [   ]

9. $\dfrac{-3r^9}{11r^{25}} =$ [   ]

10. $\dfrac{-13t^7}{3t^9} =$ [   ]

Dividing Polynomials by Monomials Simplify the following expression

11. $\dfrac{56t^5 + 7t^3}{7} =$ [   ]

12. $\dfrac{-6x^{13} - 18x^4}{3} =$ [   ]

13. $\dfrac{-66x^{10} + 78x^8 + 78x^7}{6x^3} =$ [   ]

14. $\dfrac{-54y^{10} + 27y^7 - 54y^6}{-9y^3} =$ [   ]

15. $\dfrac{-21y^{18} - 14y^4}{7y} = $ ⬚

16. $\dfrac{-24y^8 - 18y^4}{2y} = $ ⬚

17. $\dfrac{25r^{20} - 60r^{18} - 60r^9 - 60r^8}{5r^4} = $ ⬚

18. $\dfrac{-40r^{11} - 100r^{10} - 10r^9 - 40r^8}{10r^4} = $ ⬚

19. $\dfrac{18x^2y^2 - 26xy - 10xy^2}{2xy} = $ ⬚

20. $\dfrac{24x^2y^2 + 8xy + 22xy^2}{2xy} = $ ⬚

21. $\dfrac{60x^{25}y^{18} + 25x^{19}y^9 - 35x^{15}y^7}{-5x^5y^2} = $ ⬚

22. $\dfrac{72x^{18}y^{14} - 96x^{17}y^6 - 48x^{10}y^9}{-8x^5y^2} = $ ⬚

23. $\dfrac{18y^8 + 12y^5 + 12y^4}{6y^2} = $ ⬚

24. $\dfrac{60y^{15} + 110y^{14} - 40y^{13}}{-10y^2} = $ ⬚

Application Problems

25. A rectangular prism's volume can be calculated by the formula $V = Bh$, where V stands for volume, B stands for base area, and h stands for height. A certain rectangular prism's volume can be modeled by $16x^5 - 24x^3 - 40x$ cubic units. If its height is $4x$ units, find the prism's base area.

$B = $ ⬚ square units

26. A rectangular prism's volume can be calculated by the formula $V = Bh$, where V stands for volume, B stands for base area, and h stands for height. A certain rectangular prism's volume can be modeled by $28x^6 - 20x^4 - 40x^2$ cubic units. If its height is $4x$ units, find the prism's base area.

$B = $ ⬚ square units

27. A cylinder's volume can be calculated by the formula $V = Bh$, where V stands for volume, B stands for base area, and h stands for height. A certain cynlinder's volume can be modeled by $12\pi x^5 - 36\pi x^4 + 40\pi x^3$ cubic units. If its base area is $4\pi x^2$ square units, find the prism's height.

$h = $ ⬚ units

28. A cylinder's volume can be calculated by the formula $V = Bh$, where V stands for volume, B stands for base area, and h stands for height. A certain cynlinder's volume can be modeled by $45\pi x^6 + 20\pi x^4 - 50\pi x^2$ cubic units. If its base area is $5\pi x^2$ square units, find the prism's height.

$h = $ ⬚ units

6.7 Exponents and Polynomials Chapter Review

6.7.1 Exponent Rules

In Section 6.1 we covered the exponent rules and how to use them.

Example 6.7.1 Quotients and Exponents. Let t and q be real numbers, where $q \neq 0$ and $t \neq 0$. Find another way to write $\left(\frac{q^9}{t \cdot q^3}\right)^2$.

Explanation. We first use the Quotient Rule, then the Quotient to a Power Rule, then the Power to a Power Rule.

$$\left(\frac{q^9}{t \cdot q^3}\right)^2 = \left(\frac{q^{9-3}}{t}\right)^2$$
$$= \left(\frac{q^6}{t}\right)^2$$
$$= \frac{q^{6 \cdot 2}}{t^2}$$
$$= \frac{q^{12}}{t^2}$$

Example 6.7.2 The Zero Exponent. Recall that the Zero Exponent Rule says that any real number raised to the 0-power is 1. Using this, and the other exponent rules, find another way to write -9^0.

Explanation. Remember that in expressions like -9^0, the exponent only applies to what it is directly next to! In this case, the 0 only applies to the 9 and not the negative sign. So,

$$-9^0 = -1$$

Example 6.7.3 Negative Exponents. Write $5x^{-3}$ without any negative exponents.

Explanation. Recall that the Negative Exponent Rule says that a factor in the numerator with a negative exponent can be flipped into the denominator. So

$$5x^{-3} = \frac{5}{x^3}$$

Note that the 5 does not move to the denominator because the -3 exponent *only applies* to the x to which it is directly attached.

Example 6.7.4 Summary of Exponent Rules. Use the exponent rules in List 6.1.15 to write the expressions in a different way. Reduce and simplify when possible. Always find a way to write your final simplification without any negative exponents.

a. $\dfrac{24p^3}{20p^{12}}$ b. $\left(\dfrac{2v^5}{4g^{-2}}\right)^4$ c. $12n^7 \left(m^0 \cdot n^2\right)^2$ d. $\dfrac{k^5}{k^{-4}}$

Explanation.

a. $\dfrac{24p^3}{20p^{12}} = \dfrac{24}{20} \cdot \dfrac{p^3}{p^{12}}$

$= \dfrac{6}{5} \cdot p^{3-12}$

$= \dfrac{6}{5} \cdot p^{-9}$

$= \dfrac{6}{5} \cdot \dfrac{1}{p^9}$

$= \dfrac{6}{5p^9}$

b. $\left(\dfrac{2v^5}{4g^{-2}}\right)^4 = \left(\dfrac{v^5}{2g^{-2}}\right)^4$

$= \left(\dfrac{v^5 g^2}{2}\right)^4$

$= \dfrac{v^{5\cdot4} g^{2\cdot4}}{2^4}$

$= \dfrac{v^{20} g^8}{16}$

c. $12n^7 \left(m^0 \cdot n^2\right)^2$

$= 12n^7 \left(1 \cdot n^2\right)^2$

$= 12n^7 \left(n^2\right)^2$

$= 12n^7 n^{2\cdot2}$

$= 12n^7 n^4$

$= 12n^{7+4}$

$= 12n^{11}$

d. $\dfrac{k^5}{k^{-4}} = k^5 \cdot k^4$

$= k^{5+4}$

$= k^9$

6.7.2 Scientific Notation

In Section 6.2 we covered the definition of scientific notation, how to convert to and from scientific notation, and how to do some calculations in scientific notation.

Example 6.7.5 Scientific Notation for Large Numbers.

a. The distance to the star Betelgeuse is about 3,780,000,000,000,000 miles. Write this number in scientific notation.

b. The gross domestic product (GDP) of California in the year 2017 was about $\$2.746 \times 10^{13}$. Write this number in standard notation.

Explanation.

a. $3{,}780{,}000{,}000{,}000{,}000 = 3.78 \times 10^{15}$.

b. $\$2.746 \times 10^{13} = \$2{,}746{,}000{,}000{,}000$.

Example 6.7.6 Scientific Notation for Small Numbers.

a. Human DNA forms a double helix with diameter 2×10^{-9} meters. Write this number in standard notation.

b. A single grain of Forget-me-not (Myosotis) pollen is about 0.00024 inches in diameter. Write this number in scientific notation.

Explanation.

a. $2 \times 10^{-9} = 0.000000002$.

b. $0.00024 = 2.4 \times 10^{-4}$.

Example 6.7.7 Multiplying and Dividing Using Scientific Notation. The fastest spacecraft so far have traveled about 5×10^6 miles per day.

a. If that spacecraft traveled at that same speed for 2×10^4 days (which is about 55 years), how far would it have gone? Write your answer in scientific notation.

b. The nearest star to Earth, besides the Sun, is Proxima Centauri, about 2.5×10^{13} miles from Earth. How many days would you have to fly in that spacecraft at top speed to reach Proxima Centauri

Explanation.

a. Remember that you can find the distance traveled by multiplying the rate of travel times the time traveled: $d = r \cdot t$. So this problem turns into

$$d = r \cdot t$$
$$d = \left(5 \times 10^6\right) \cdot \left(2 \times 10^4\right)$$

Multiply coefficient with coefficient and power of 10 with power of 10.

$$= (5 \cdot 2)\left(10^6 \times 10^4\right)$$
$$= 10 \times 10^{10}$$

Remember that this still isn't in scientific notation. So we convert like this:

$$= 1.0 \times 10^1 \times 10^{10}$$
$$= 1.0 \times 10^{11}$$

So, after traveling for 2×10^4 days (55 years), we will have traveled about 1.0×10^{11} miles. That's one-hundred million miles. I hope someone remembered the snacks.

b. Since we are looking for time, let's solve the equation $d = r \cdot t$ for t by dividing by r on both sides: $t = \frac{d}{r}$. So we have:

$$t = \frac{d}{r}$$
$$t = \frac{2.5 \times 10^{13}}{5 \times 10^6}$$

Now we can divide coefficient by coefficient and power of 10 with power of 10.

$$t = \frac{2.5}{5} \times \frac{10^{13}}{10^6}$$
$$t = 0.5 \times 10^7$$
$$t = 5 \times 10^{-1} \times 10^7$$
$$t = 5 \times 10^6$$

This means that to get to Proxima Centauri, even in our fastest spacecraft, would take 5×10^6 years. Converting to standard form, this is 5,000,000 years. I think we're going to need a faster ship.

6.7.3 Adding and Subtracting Polynomials

In Section 6.3 we covered the definitions of a polynomial, a term of a polynomial, a coefficient of a term, the degree of a term, the degree of a polynomial, theleading term of a polynomial, a constant term, monomials, binomials, and trinomials, and how to write a polynomial in standard form.

Example 6.7.8 Polynomial Vocabulary. Decide if the following statements are true or false.

a. The expression $\frac{3}{5}x^2 - \frac{1}{5}x^7 + \frac{x}{2} - 4$ is a polynomial.

b. The expression $4x^6 - 3x^{-2} - x + 1$ is a polynomial.

c. The degree of the polynomial $\frac{3}{5}x^2 - \frac{1}{5}x^7 + \frac{x}{2} - 4$ is 10.

d. The degree of the term $5x^2y^4$ is 6.

e. The leading coefficient of $\frac{3}{5}x^2 - \frac{1}{5}x^7 + \frac{x}{2} - 4$ is $\frac{3}{5}$.

f. There are 4 terms in the polynomial $\frac{3}{5}x^2 - \frac{1}{5}x^7 + \frac{x}{2} - 4$.

g. The polynomial $\frac{3}{5}x^2 - \frac{1}{5}x^7 + \frac{x}{2} - 4$ is in standard form.

Explanation.

a. True. The expression $\frac{3}{5}x^2 - \frac{1}{5}x^7 + \frac{x}{2} - 4$ is a polynomial.

b. False. The expression $4x^6 - 3x^{-2} - x + 1$ is *not* a polynomial. Variables are only allowed to have whole number exponents in polynomials and the second term has a -2 exponent.

c. False. The degree of the polynomial $\frac{3}{5}x^2 - \frac{1}{5}x^7 + \frac{x}{2} - 4$ is *not* 10. It is 7, which is the highest power of any variable in the expression.

d. True. The degree of the term $5x^2y^4$ is 6.

e. False. The leading coefficient of $\frac{3}{5}x^2 - \frac{1}{5}x^7 + \frac{x}{2} - 4$ is *not* $\frac{3}{5}$. The leading coefficient comes from the degree 7 term which is $-\frac{1}{5}$.

f. True. There are 4 terms in the polynomial $\frac{3}{5}x^2 - \frac{1}{5}x^7 + \frac{x}{2} - 4$.

g. False. The polynomial $\frac{3}{5}x^2 - \frac{1}{5}x^7 + \frac{x}{2} - 4$ is *not* in standard form. The exponents have to be written from highest to lowest, i.e. $-\frac{1}{5}x^7 + \frac{3}{5}x^2 + \frac{x}{2} - 4$.

Example 6.7.9 Adding and Subtracting Polynomials. Simplify the expression $\left(\frac{2}{9}x - 4x^2 - 5\right) + \left(6x^2 - \frac{1}{6}x - 3\right)$.

Explanation. First identify like terms and group them either physically or mentally. Then we will look for common denominators for these like terms and combine appropriately.

$$\left(\frac{2}{9}x - 4x^2 - 5\right) + \left(6x^2 - \frac{1}{6}x - 3\right)$$
$$= \frac{2}{9}x - 4x^2 - 5 + 6x^2 - \frac{1}{6}x - 3$$
$$= \left(-4x^2 + 6x^2\right) + \left(\frac{2}{9}x - \frac{1}{6}x\right) + (-3 - 5)$$
$$= 2x^2 + \left(\frac{4}{18}x - \frac{3}{18}x\right) - 8$$
$$= 2x^2 + \frac{1}{18}x - 8$$

6.7.4 Multiplying Polynomials

In Section 6.4 we covered how to multiply two polynomials together using distribution, FOIL, and generic rectangles.

Example 6.7.10 Multiplying Binomials. Expand the expression $(5x - 6)(3 + 2x)$ using the binomial multiplication method of your choice: distribution, FOIL, or generic rectangles.

Explanation. We will show work using the FOIL method.

$$(5x - 6)(3 - 2x) = (5x \cdot 3) + (5x \cdot (-2x)) + (-6 \cdot 3) + (-6 \cdot (-2x))$$
$$= 15x - 10x^2 - 18 + 12x$$
$$= -10x^2 + 27x - 18$$

Example 6.7.11 Multiplying Polynomials Larger than Binomials. Expand the expression $(3x-2)\left(4x^2 - 2x + 5\right)$ by multiplying every term in the first factor with every term in the second factor.

Explanation. $(3x - 2)\left(4x^2 - 2x + 5\right)$
$$= 3x \cdot 4x^2 + 3x \cdot (-2x) + 3x \cdot 5 + (-2) \cdot 4x^2 + (-2) \cdot (-2x) + (-2) \cdot 5$$
$$= 12x^3 - 6x^2 + 15x - 8x^2 + 4x - 10$$
$$= 12x^3 - 14x^2 + 19x - 10$$

6.7.5 Special Cases of Multiplying Polynomials

In Section 6.5 we covered how to square a binomial and how to find the product of the sum or difference of two terms.

Example 6.7.12 Squaring a Binomial. Recall that Fact 6.5.4 gives formulas that help square a binomial. Simplify the expression $(2x + 3)^2$.

Explanation. Remember that you *can* use FOIL to do these problems, but in the interest of understanding concepts at a higher level for use in later chapters, we will use the relevant formula from Fact 6.5.4. In this case, since we have a sum of two terms being squared, we will use $(a + b)^2 = a^2 + 2ab + b^2$.

First identify a and b. In this case, $a = 2x$ and $b = 3$. So, we have:

$$(a + b)^2 = (a)^2 + 2(a)(b) + (b)^2$$
$$(2x + 3)^2 = (2x)^2 + 2(2x)(3) + (3)^2$$
$$= 4x^2 + 12x + 9$$

Example 6.7.13 The Product of the Sum and Difference of Two Terms. Recall that Fact 6.5.12 gives a formula to help multiply things that look like $(a + b)(a - b)$.

Simplify the expression $(7x + 4)(7x - 4)$.

Explanation. Remember that you *can* use FOIL to do these problems, but in the interest of understanding concepts at a higher level for use in later chapters, we will use the formula from Fact 6.5.12. In this case, that means we will use $(a + b)(a - b) = a^2 - b^2$.

First identify a and b. In this case, $a = 7x$ and $b = 4$. So, we have:

$$(a + b)(a - b) = (a)^2 - (b)^2$$
$$(7x + 4)(7x - 4) = (7x)^2 - (4)^2$$
$$= 49x^2 - 16$$

Example 6.7.14 Binomials Raised to Other Powers. To raise binomials to powers higher than 2, we start by expanding the expression and multiplying all factors together from left to right.

Expand the expression $(2x - 5)^3$.

Explanation.

$$
\begin{aligned}
(2x &- 5)^3 \\
&= (2x - 5)(2x - 5)(2x - 5) \\
&= \left[(2x)^2 - 2(2x)(5) + 5^2\right](2x - 5) \\
&= \left[4x^2 - 20x + 25\right](2x - 5) \\
&= \left[4x^2\right](2x) + \left[4x^2\right](-5) + [-20x](2x) + [-20x](-5) + [25](2x) + [25](-5) \\
&= 8x^3 - 20x^2 - 40x^2 + 100x + 50x - 125 \\
&= 8x^3 - 60x^2 + 150x - 125
\end{aligned}
$$

6.7.6 Dividing by a Monomial

In Section 6.6 we covered how you can split a fraction up into multiple terms if there is a sum or difference in the numerator. Mathematically, this happens using the rule $\frac{a+b}{c} = \frac{a}{c} + \frac{b}{c}$. This formula can be used for any number of terms in the numerator, and for both sums and differences.

Example 6.7.15 Expand the expression $\frac{12x^5 + 2x^3 - 4x^2}{4x^2}$.

Explanation.

$$
\begin{aligned}
\frac{12x^5 + 2x^3 - 4x^2}{4x^2} &= \frac{12x^5}{4x^2} + \frac{2x^3}{4x^2} - \frac{4x^2}{4x^2} \\
&= 3x^3 + \frac{x}{2} - 1
\end{aligned}
$$

Exercises

Exponent Rules   Use the properties of exponents to simplify the expression.

1. $\left(4r^{12}\right)^3$

2. $\left(2y^2\right)^2$

3. $(6y^4) \cdot (9y^{17})$

4. $(10y^7) \cdot (-8y^{11})$

5. $\left(-\dfrac{y^9}{4}\right) \cdot \left(\dfrac{y^4}{9}\right)$

6. $\left(-\dfrac{r^{11}}{7}\right) \cdot \left(-\dfrac{r^{16}}{8}\right)$

7. $(-21)^0 = \boxed{}$

8. $(-16)^0 = \boxed{}$

9. $-42^0 = \boxed{}$

10. $-47^0 = \boxed{}$

11. $\left(\dfrac{-3}{2x^9}\right)^3 = \boxed{}$

12. $\left(\dfrac{-3}{8x^4}\right)^3 = \boxed{}$

13. $\dfrac{5r^{14}}{15r^3} = \boxed{}$
14. $\dfrac{7r^4}{14r^3} = \boxed{}$
15. $\left(\dfrac{x^7}{2y^6z^8}\right)^3 = \boxed{}$
16. $\left(\dfrac{x^4}{2y^{10}z^5}\right)^3 = \boxed{}$

Rewrite the expression simplified and using only positive exponents.

17. $\left(\dfrac{1}{8}\right)^{-3} = \boxed{}$
18. $\left(\dfrac{1}{9}\right)^{-3} = \boxed{}$
19. $9x^{-12} = \boxed{}$

20. $19x^{-3} = \boxed{}$
21. $\dfrac{14}{x^{-4}} = \boxed{}$
22. $\dfrac{8}{x^{-5}} = \boxed{}$

23. $\dfrac{18x^{-9}}{x^{-26}} = \boxed{}$
24. $\dfrac{8x^{-11}}{x^{-17}} = \boxed{}$
25. $\dfrac{r^{-3}}{\left(r^4\right)^{10}} = \boxed{}$

26. $\dfrac{r^{-2}}{\left(r^{11}\right)^7} = \boxed{}$
27. $t^{-11} \cdot t^6 = \boxed{}$
28. $t^{-5} \cdot t^4 = \boxed{}$

29. $\left(9x^{-17}\right) \cdot \left(6x^6\right) = \boxed{}$
30. $\left(6x^{-10}\right) \cdot \left(10x^7\right) = \boxed{}$
31. $\left(-5y^{-4}\right)^{-2} \boxed{}$

32. $\left(-2y^{-16}\right)^{-3} \boxed{}$
33. $\left(3y^8\right)^4 \cdot y^{-22} = \boxed{}$
34. $\left(3r^3\right)^3 \cdot r^{-4} = \boxed{}$

35. $\left(r^4t^8\right)^{-3} = \boxed{}$
36. $\left(t^6y^{14}\right)^{-3} = \boxed{}$
37. $\left(t^{-11}x^{10}\right)^{-3} = \boxed{}$

38. $\left(x^{-4}r^6\right)^{-3} = \boxed{}$
39. $\left(\dfrac{x^6}{2}\right)^{-3} = \boxed{}$
40. $\left(\dfrac{y^{15}}{4}\right)^{-4} = \boxed{}$

Scientific Notation   Write the following number in scientific notation.

41. $42000 = \boxed{}$
42. $5200 = \boxed{}$

43. $0.062 = \boxed{}$
44. $0.0072 = \boxed{}$

Write the following number in decimal notation without using exponents.

45. $8.24 \times 10^2 = \boxed{}$
46. $9.24 \times 10^5 = \boxed{}$
47. $1.23 \times 10^0 = \boxed{}$

48. $2.23 \times 10^0 =$ [_____]   **49.** $3.23 \times 10^{-2} =$ [_____]   **50.** $4.23 \times 10^{-3} =$ [_____]

Multiply the following numbers, writing your answer in scientific notation.

51. $(5 \times 10^2)(7 \times 10^2) =$ [_____]   **52.** $(6 \times 10^5)(9 \times 10^3) =$ [_____]

Divide the following numbers, writing your answer in scientific notation.

53. $\dfrac{3.5 \times 10^3}{7 \times 10^{-5}} =$ [_____]   **54.** $\dfrac{1.6 \times 10^4}{8 \times 10^{-2}} =$ [_____]

Adding and Subtracting Polynomials   Is the following expression a monomial, binomial, or trinomial?

55. $10t^{12} - 14t^{10}$ is a   (□ monomial □ binomial   **56.** $-4x^7 + 3x^6$ is a   (□ monomial □ binomial
□ trinomial) of degree [_____]   □ trinomial) of degree [_____]

57. $-18x^6 - 19x^5 + 20x^3$ is a   (□ monomial □ bi-   **58.** $8x^6 + 11x^5 - 3x$ is a   (□ monomial □ bino-
nomial □ trinomial) of degree [_____]   mial □ trinomial) of degree [_____]

Find the degree of the following polynomial.

59. $-6x^5 y^7 - 6x^2 y^3 + 14x^2 - 1$ **60.** $-x^5 y^9 - 20x^2 y^4 - 10x^2 - 12$

The degree of this polynomial is [_____].   The degree of this polynomial is [_____].

Add the polynomials.

61. $\left(7x^2 - 2x - 1\right) + \left(4x^2 - 2x + 4\right)$ **62.** $\left(-8x^2 + 2x - 9\right) + \left(10x^2 - 8x + 9\right)$

63. $\left(-10t^6 - 5t^4 - 2t^2\right) + \left(-7t^6 + 6t^4 + t^2\right)$ **64.** $\left(7t^6 - 2t^4 + 8t^2\right) + \left(8t^6 - 9t^4 - 7t^2\right)$

65. $\left(-2x^3 + 5x^2 - 2x + \dfrac{7}{10}\right) + \left(3x^3 + 9x^2 + 6x + \dfrac{1}{8}\right)$

66. $\left(3x^3 + 8x^2 + 2x + \dfrac{9}{4}\right) + \left(-5x^3 + 6x^2 + 8x + \dfrac{1}{6}\right)$

Subtract the polynomials.

67. $\left(-5x^2 + 2x\right) - \left(-7x^2 - 5x\right)$ **68.** $\left(-3x^2 - 6x\right) - \left(2x^2 + 10x\right)$

69. $\left(6x^2 + 2x + 10\right) - \left(-5x^2 - 7x + 4\right)$

70. $\left(7x^2 - 5x + 10\right) - \left(-9x^2 + 10x - 8\right)$

71. $\left(3r^6 - 5r^4 - 2r^2\right) - \left(10r^6 - 7r^4 + 7r^2\right)$

72. $\left(-9t^6 + 9t^4 + 8t^2\right) - \left(5t^6 - 4t^4 - 6t^2\right)$

73. Add or subtract the given polynomials as indicated.

$\left(-10x^4 + 8xy - 9y^2\right) - \left(-6x^4 - 6xy + 10y^2\right)$

74. Add or subtract the given polynomials as indicated.

$\left(-2x^8 - 2xy + 9y^9\right) - \left(10x^8 + 7xy - 6y^9\right)$

75. A handyman is building two pig pens sharing the same side. Assume the length of the shared side is x meters. The cost of building one pen would be $44x^2 + 9.5x - 13.5$ dollars, and the cost of building the other pen would be $23.5x^2 - 9.5x + 41.5$ dollars. What's the total cost of building those two pens?

A polynomial representing the total cost of building those two pens is _____ dollars.

76. A handyman is building two pig pens sharing the same side. Assume the length of the shared side is x meters. The cost of building one pen would be $33.5x^2 - 8.5x + 40.5$ dollars, and the cost of building the other pen would be $27x^2 + 8.5x + 27$ dollars. What's the total cost of building those two pens?

A polynomial representing the total cost of building those two pens is _____ dollars.

Multiplying Polynomials   Multiply the polynomials.

77. $-3x\left(x - 2\right) = $ _____

78. $-x\left(x + 9\right) = $ _____

79. $-5r^2\left(10r^2 - 5r - 7\right) = $ _____

80. $2r^2\left(7r^2 - 9r + 2\right) = $ _____

81. $\left(8t + 1\right)\left(t + 4\right) = $ _____

82. $\left(5t + 2\right)\left(t + 9\right) = $ _____

83. $\left(x + 1\right)\left(x - 2\right) = $ _____

84. $\left(x + 7\right)\left(x - 8\right) = $ _____

85. $\left(3x - 4\right)\left(2x - 3\right) = $ _____

86. $\left(6y - 10\right)\left(5y - 2\right) = $ _____

87. $3(x + 2)(x + 3) = $ _____

88. $-3(x + 2)(x + 3) = $ _____

89. $x(x - 2)(x + 2) = $ _____

90. $-x(x + 2)(x + 3) = $ _____

91. $(a - 10b)(a^2 + 3ab - 8b^2) = $ _____

92. $(a + 2b)(a^2 - 7ab + 8b^2) = $ _____

93. A rectangle's length is 2 feet shorter than 4 times its width. If we use w to represent the rectangle's width, use a polynomial to represent the rectangle's area in expanded form.

area = [   ] square feet

94. A rectangle's length is 3 feet shorter than twice its width. If we use w to represent the rectangle's width, use a polynomial to represent the rectangle's area in expanded form.

area = [   ] square feet

Special Cases of Multiplying Polynomials Expand the square of a *binomial*.

95. $(9y + 6)^2 = $ [   ]

96. $(6y + 3)^2 = $ [   ]

97. $(r - 9)^2 = $ [   ]

98. $(r - 2)^2 = $ [   ]

99. $(9a + 6b)^2 = $ [   ]

100. $(10a + 2b)^2 = $ [   ]

Multiply the polynomials.

101. $(x + 7)(x - 7) = $ [   ]

102. $(x - 2)(x + 2) = $ [   ]

103. $(1 - 3x)(1 + 3x) = $ [   ]

104. $(8 - 7y)(8 + 7y) = $ [   ]

105. $(3y^8 - 13)(3y^8 + 13) = $ [   ]

106. $(2r^6 - 2)(2r^6 + 2) = $ [   ]

Simplify the given expression into an expanded polynomial.

107. $(r + 5)^3 = $ [   ]

108. $(t + 3)^3 = $ [   ]

Dividing by a Monomial Simplify the following expression

109. $\dfrac{52t^{15} + 12t^8}{4} = $ [   ]

110. $\dfrac{24x^4 - 32x^3}{8} = $ [   ]

111. $\dfrac{-12x^{12} - 24x^{11} - 2x^9}{2x^3} = $ [   ]

112. $\dfrac{-24x^{15} - 24x^{13} - 16x^{12}}{-2x^3} = $ [   ]

113. $\dfrac{4y^{10} - 4y^5}{2y} = $ [   ]

114. $\dfrac{-56y^{17} - 56y^9}{8y} = $ [   ]

CHAPTER 7

Factoring

7.1 Factoring out the Common Factor

In Chapter 6, we learned how to multiply polynomials, such as when you start with $(x+2)(x+3)$ and obtain $x^2 + 5x + 6$. This chapter, starting with this section, is about the *opposite* process—factoring. For example, starting with $x^2 + 5x + 6$ and obtaining $(x + 2)(x + 3)$. We will start with the simplest kind of factoring: for example starting with $x^2 + 2x$ and obtaining $x(x + 2)$.

7.1.1 Motivation for Factoring

When you write $x^2 + 2x$, you have an algebraic expression built with two terms—two parts that are *added* together. When you write $x(x + 2)$, you have an algebraic expression built with two factors—two parts that are *multiplied* together. Factoring is useful, because sometimes (but not always) having your expression written as parts that are *multiplied* together makes it easy to simplify the expression.

You've seen this with fractions. To simplify $\frac{15}{35}$, breaking down the numerator and denominator into factors is useful: $\frac{3 \cdot 5}{7 \cdot 5}$. Now you can see that the factors of 5 cancel.

There are a few other reasons to appreciate the value of factoring that will float to the surface in this chapter and beyond.

7.1.2 Identifying the Greatest Common Factor

The most basic technique for factoring involves recognizing the **greatest common factor** between two expressions, which is the largest factor that goes in evenly to both expressions. For example, the greatest common factor between 6 and 8 is 2 since 2 goes in nicely into both 6 and 8 and no larger number would divide both 6 and 8 nicely.

Similarly, the greatest common factor between $4x$ and $3x^2$ is x. If you write $4x$ as a product of its factors, you have $2 \cdot 2 \cdot x$. And if you fully factor $3x^2$, you have $3 \cdot x \cdot x$. The only factor they have in common is x, so that is the greatest common factor. No larger expression goes in nicely to both expressions.

> **Example 7.1.2 Finding the Greatest Common Factor.** What is the common factor between $6x^2$ and $70x$? Break down each of these into its factors:
>
> $$6x^2 = 2 \cdot 3 \cdot x \cdot x \qquad\qquad 70x = 2 \cdot 5 \cdot 7 \cdot x$$

And identify the common factors:

$$6x^2 = \overset{\downarrow}{2} \cdot 3 \cdot \overset{\downarrow}{x} \cdot x \qquad\qquad 70x = \overset{\downarrow}{2} \cdot 5 \cdot 7 \cdot \overset{\downarrow}{x}$$

With 2 and x in common, the greatest common factor is $2x$.

Let's try a few more examples.

Checkpoint 7.1.3.

    a. The greatest common factor between $6x$ and $8x$ is [   ].

    b. The greatest common factor between $14x^2$ and $10x$ is [   ].

    c. The greatest common factor between $6y^2$ and $7y^2$ is [   ].

    d. The greatest common factor between $12xy^2$ and $9xy$ is [   ].

    e. The greatest common factor between $6x^3$, $2x^2$, and $8x$ is [   ].

Explanation.

    a. Since $6x$ completely factors as $\overset{\downarrow}{2} \cdot 3 \cdot \overset{\downarrow}{x}$ ...

      ... and $8x$ completely factors as $\overset{\downarrow}{2} \cdot 2 \cdot 2 \cdot \overset{\downarrow}{x}$, ...

      ... the greatest common factor is $2x$.

    b. Since $14x^2$ completely factors as $\overset{\downarrow}{2} \cdot 7 \cdot \overset{\downarrow}{x} \cdot x$ ...

      ... and $10x$ completely factors as $\overset{\downarrow}{2} \cdot 5 \cdot \overset{\downarrow}{x}$, ...

      ... the greatest common factor is $2x$.

    c. Since $6y^2$ completely factors as $2 \cdot 3 \cdot \overset{\downarrow}{y} \cdot \overset{\downarrow}{y}$ ...

      ... and $7y^2$ completely factors as $7 \cdot \overset{\downarrow}{y} \cdot \overset{\downarrow}{y}$, ...

      ... the greatest common factor is y^2.

    d. Since $12xy^2$ completely factors as $2 \cdot 2 \cdot \overset{\downarrow}{3} \cdot \overset{\downarrow}{x} \cdot \overset{\downarrow}{y} \cdot y$ ...

      ... and $9xy$ completely factors as $\overset{\downarrow}{3} \cdot 3 \cdot \overset{\downarrow}{x} \cdot \overset{\downarrow}{y}$, ...

      ... the greatest common factor is $3xy$.

    e. Since $6x^3$ completely factors as $\overset{\downarrow}{2} \cdot 3 \cdot \overset{\downarrow}{x} \cdot x \cdot x$ ...

      ..., $2x^2$ completely factors as $\overset{\downarrow}{2} \cdot \overset{\downarrow}{x} \cdot x$, ...

      ... and $8x$ completely factors as $\overset{\downarrow}{2} \cdot 2 \cdot 2 \cdot \overset{\downarrow}{x}$, ...

      ... the greatest common factor is $2x$.

7.1.3 Factoring Out the Greatest Common Factor

We have learned the distributive property: $a(b + c) = ab + ac$. Perhaps you have thought of this as a way to "distribute" the number a to each of b and c. In this section, we will use the distributive property in the opposite way. If you have an expression $ab + ac$, it is equal to $a(b + c)$. In that example, we factored out a, which is the common factor between ab and ac.

The following steps use the distributive property to factor out the greatest common factor between two or more terms.

Factoring Out the Greatest Common Factor by Filling in the Blank

Process 7.1.4.

1. *Identify the common factor in all terms.*

2. *Write the common factor outside a pair of parentheses with the appropriate addition or subtraction signs inside.*

3. *For each term from the original expression, what would you multiply the greatest common factor by to result in that term? Write your answer in the parentheses.*

Example 7.1.5 To factor $12x^2 + 15x$:

1. The common factor between $12x^2$ and $15x$ is $3x$.

2. $3x(\quad + \quad)$

3. $3x(4x + 5)$

Let's look at a few examples.

Example 7.1.6 Factor the polynomial $3x^3 + 3x^2 - 9$.

a. We identify the common factor as 3, because 3 is the only common factor between $3x^3$, $3x^2$ and 9.

b. We write:
$$3x^3 + 3x^2 - 9 = 3(\quad + \quad - \quad).$$

c. We ask the question "3 times what gives $3x^3$?" The answer is x^3. Now we have:
$$3x^3 + 3x^2 - 9 = 3(x^3 + \quad - \quad).$$

We ask the question "3 times what gives $3x^2$?" The answer is x^2. Now we have:
$$3x^3 + 3x^2 - 9 = 3(x^3 + x^2 - \quad).$$

We ask the question "3 times what gives 9?" The answer is 3. Now we have:
$$3x^3 + 3x^2 - 9 = 3(x^3 + x^2 - 3).$$

To check that this is correct, multiplying through $3(x^3 + x^2 - 3)$ should give the original expression $3x^3 + 3x^2 - 9$. We check this, and it does.

Checkpoint 7.1.7. Factor the polynomial $4x^3 + 12x^2 - 12x$.

Explanation. In this exercise, $4x$ is the greatest common factor. We find

$$4x^3 + 12x^2 - 12x = 4x(\quad + \quad - \quad)$$
$$= 4x(x^2 + \quad - \quad)$$
$$= 4x(x^2 + 3x - \quad)$$
$$= 4x(x^2 + 3x - 3)$$

Note that you might fail to recognize that $4x$ is the greatest common factor. At first you might only find that, say, 4 is a common factor. This is OK—you can factor out the 4 and continue from there:

$$4x^3 + 12x^2 - 12x = 4(\quad + \quad - \quad)$$
$$= 4(x^3 + \quad - \quad)$$
$$= 4(x^3 + 3x^2 - \quad)$$
$$= 4(x^3 + 3x^2 - 3x)$$

Now examine the second factor here and there is *still* a common factor, x. Factoring this out too.

$$= 4x(x^2 + 3x - 3)$$

So there is more than one way to find the answer here.

7.1.4 Visualizing With Rectangles

In Section 6.4, we learned one way to multiply polynomials using rectangle diagrams. Similarly, we can factor a polynomial with a rectangle diagram.

Factoring Out the Greatest Common Factor Using Rectangles

Process 7.1.8 Factoring Out the Greatest Common Factor Using Rectangles.

1. *Put the terms into adjacent rectangles. Think of these as labeling the areas of each rectangle.*

2. *Identify the common factor, and mark the height of the overall rectangle with it.*

3. *Mark the base of each rectangle based on each rectangle's area and height.*

4. *Since the overall rectangle's area equals its base times its height, the height is one factor, and the sum of the widths is another factor.*

Example 7.1.9 We will factor $12x^2 + 15x$, the same polynomial from the example in Algorithm 7.1.4, so that you may compare the two styles.

So $12x^2 + 15x$ factors as $3x(4x + 5)$.

7.1.5 More Examples of Factoring out the Common Factor

Previous examples did not cover every nuance with factoring out the greatest common factor. Here are a few more factoring examples that attempt to do so.

Example 7.1.10 Factor $-35m^5 + 5m^4 - 10m^3$.

First, we identify the common factor. The number 5 is the greatest common factor of the three coefficients (which were $-35, 5$, and -10) and also m^3 is the largest expression that divides m^5, m^4, and m^3. Therefore the greatest common factor is $5m^3$.

In this example, the leading term is a negative number. When this happens, we will make it common practice to take that negative as part of the greatest common factor. So we will proceed by factoring out $-5m^3$. Note the sign changes.

$$-35m^5 + 5m^4 - 10m^3 = -5m^3(\quad - \quad + \quad)$$
$$= -5m^3(7m^2 - \quad + \quad)$$
$$= -5m^3(7m^2 - m + \quad)$$
$$= -5m^3(7m^2 - m + 2)$$

Example 7.1.11 Factor $14 - 7n^2 + 28n^4 - 21n$.

Notice that the terms are not in a standard order, with powers of n decreasing as you read left to right. It is usually a best practice to rearrange the terms into the standard order first.

$$14 - 7n^2 + 28n^4 - 21n = 28n^4 - 7n^2 - 21n + 14.$$

The number 7 divides all of the numerical coefficients. Separately, no power of n is part of the greatest common factor because the 14 term has no n factors. So the greatest common factor is just 7. We proceed by factoring that out:

$$14 - 7n^2 + 28n^4 - 21n = 28n^4 - 7n^2 - 21n + 14$$
$$= 7(4n^4 - n^2 - 3n + 2)$$

Example 7.1.12 Factor $24ab^2 + 16a^2b^3 - 12a^3b^2$.

There are two variables in this polynomial, but that does not change the factoring strategy. The greatest numerical factor between the three terms is 4. The variable a divides all three terms, and b^2 divides all three terms. So we have:

$$24ab^2 + 16a^2b^3 - 12a^3b^2 = 4ab^2(6 + 4ab - 3a^2)$$

Example 7.1.13 Factor $4m^2n - 3xy$.

There are no common factors in those two terms (unless you want to count 1 or -1, but we do not count these for the purposes of identifying a greatest common factor). In this situation we can say the polynomial is **prime** or **irreducible**, and leave it as it is.

Example 7.1.14 Factor $-x^3 + 2x + 18$.

There are no common factors in those three terms, and it would be correct to state that this polynomial is prime or irreducible. However, since its leading coefficient is negative, it may be wise to factor out a negative sign. So, it could be factored as $-\left(x^3 - 2x - 18\right)$. Note that *every* term is negated as the leading negative sign is extracted.

Exercises

Review and Warmup   Multiply the polynomials.

1. $6x(x - 3) =$

2. $8x(x - 8) =$

3. $10x(9x - 5) =$

4. $-2x(4x - 8) =$

5. $-6x^2(x + 9) =$

6. $-4x^2(x - 5) =$

7. $2y^2\left(-8y^2 + 4y\right) =$

8. $8r^2\left(-6r^2 - 8r\right) =$

Identifying Common Factors   Find the greatest common factor of the following terms.

9. 5 and $20r$

10. 2 and $18t$

11. $8t$ and $32t^2$

12. $5t$ and $35t^2$

13. $2x^3$ and $-20x^4$

14. $8x^3$ and $-56x^4$

15. $4y^{19}$ and $-24y^{10}$

16. $10y^{12}$ and $-80y^9$

17. $7r^{17}, -14r^{10}, 21r^8$

18. $4r^{11}, -16r^9, 32r^8$

19. $5x^{11}y^9, -45x^9y^{17}, 35x^8y^{18}$

20. $3x^{11}y^{13}, -12x^9y^{14}, 27x^6y^{19}$

Factoring out the Common Factor   Factor the given polynomial.

21. $4t + 4 =$

22. $10x + 10 =$

23. $6x - 6 =$

24. $3y - 3 =$

25. $-9y - 9 = $

26. $-6r - 6 = $

27. $3r - 27 = $

28. $9t + 36 = $

29. $18t^2 + 48 = $

30. $21t^2 - 6 = $

31. $14x^2 - 20x + 8 = $

32. $50x^2 + 90x + 70 = $

33. $16y^4 + 24y^3 + 72y^2 = $

34. $12y^4 + 6y^3 + 16y^2 = $

35. $50r^5 - 70r^4 + 100r^3 = $

36. $40r^5 + 12r^4 + 4r^3 = $

37. $24t - 64t^2 + 80t^3 = $

38. $8t + 6t^2 + 18t^3 = $

39. $27t^2 + 5 = $

40. $20x^2 + 9 = $

41. $3xy + 3y = $

42. $4xy + 4y = $

43. $5x^{13}y^9 + 35y^9 = $

44. $6x^7y^9 + 24y^9 = $

45. $20x^5y^6 - 32x^4y^6 + 40x^3y^6 = $

46. $80x^5y^9 + 10x^4y^9 + 60x^3y^9 = $

47. $24x^5y^7z^6 + 20x^4y^7z^5 + 32x^3y^7z^4 = $

48. $50x^5y^8z^8 + 35x^4y^8z^7 + 10x^3y^8z^6 = $

7.2 Factoring by Grouping

This section covers a technique for factoring polynomials like x^3+3x^2+2x+6, which factors as $(x^2+2)(x+3)$. If there are four terms, the technique in this section *might* help you to factor the polynomial. Additionally, this technique is a stepping stone to a factoring technique in Section 7.3 and Section 7.4.

7.2.1 Factoring out Common Polynomials

Recall that to factor $3x + 6$, we factor out the common factor 3:

$$3x + 6 = 3x + 3 \cdot 2$$
$$= 3(x + 2)$$

The "3" here could have been something more abstract, and it still would be valid to factor it out:

$$xA + 2A = x\overset{\downarrow}{A} + 2\overset{\downarrow}{A} \qquad\qquad x🍎 + 2🍎 = x\overset{\downarrow}{🍎} + 2\overset{\downarrow}{🍎}$$
$$= A(x + 2) \qquad\qquad\qquad = 🍎(x + 2)$$

In fact, even "larger" things can be factored out, as in this example:

$$x(a + b) + 2(a + b) = x\overbrace{(a + b)}^{\downarrow} + 2\overbrace{(a + b)}^{\downarrow}$$
$$= (a + b)(x + 2)$$

In this last example, we factored out the binomial factor $(a + b)$. Factoring out binomials is the essence of this section, so let's see that a few more times:

$$x(x + 2) + 3(x + 2) = x\overbrace{(x + 2)}^{\downarrow} + 3\overbrace{(x + 2)}^{\downarrow}$$
$$= (x + 2)(x + 3)$$

$$z^2(2y + 5) + 3(2y + 5) = z^2\overbrace{(2y + 5)}^{\downarrow} + 3\overbrace{(2y + 5)}^{\downarrow}$$
$$= (2y + 5)(z^2 + 3)$$

And even with an expression like $Q^2(Q - 3) + Q - 3$, if we re-write it in the right way using a 1 and some parentheses, then it too can be factored:

$$Q^2(Q - 3) + Q - 3 = Q^2(Q - 3) + 1(Q - 3)$$
$$= Q^2\overbrace{(Q - 3)}^{\downarrow} + 1\overbrace{(Q - 3)}^{\downarrow}$$
$$= (Q - 3)(Q^2 + 1)$$

The truth is you are unlikely to come upon an expression like $x(x + 2) + 3(x + 2)$, as in these examples. Why wouldn't someone have multiplied that out already? Or factored it all the way? So far in this section, we have only been looking at a stepping stone to a real factoring technique called **factoring by grouping**.

7.2.2 Factoring by Grouping

Factoring by grouping is a factoring technique that *sometimes* works on polynomials with four terms. Here is an example.

Example 7.2.2 Suppose we must factor $x^3 - 3x^2 + 5x - 15$. Note that there are four terms, and they are written in descending order of the powers of x. "Grouping" means to group the first two terms and the last two terms together:

$$x^3 - 3x^2 + 5x - 15 = (x^3 - 3x^2) + (5x - 15)$$

Now, each of these two groups has its own greatest common factor we can factor out:

$$= x^2(x - 3) + 5(x - 3)$$

In a sense, we are "lucky" because we now see matching binomials that can themselves be factored out:

$$= x^2\overbrace{(x - 3)} + 5\overbrace{(x - 3)}$$
$$= (x - 3)(x^2 + 5)$$

And so we have factored $x^3 - 3x^2 + 5x - 15$ as $(x - 3)(x^2 + 5)$. But to be sure, if we multiply this back out, it should recover the original $x^3 - 3x^2 + 5x - 15$. To confirm your answers are correct, you should always make checks like this.

Checkpoint 7.2.3. Factor $x^3 + 4x^2 + 2x + 8$.

Explanation. We will break the polynomial into two groups: $x^3 + 4x^2$ and $2x + 8$.

$$x^3 + 4x^2 + 2x + 8 = (x^3 + 4x^2) + (2x + 8)$$

and now each group has its own greatest common factor to factor out:

$$= x^2(x + 4) + 2(x + 4)$$

and now the binomial $(x + 4)$ appears twice and can be factored out:

$$= (x + 4)(x^2 + 2)$$

Example 7.2.4 Factor $t^3 - 5t^2 - 3t + 15$. This example has a complication with negative signs. If we try to break up this polynomial into two groups as $(t^3 - 5t^2) - (3t + 15)$, then we've made an error! In that last expression, we are *subtracting* a group with the term 15, so overall it subtracts 15. The original polynomial *added* 15, so we are off course.

One way to handle this is to treat subtraction as addition of a negative:

$$t^3 - 5t^2 - 3t + 15 = t^3 - 5t^2 + (-3t) + 15$$
$$= (t^3 - 5t^2) + (-3t + 15)$$

Now we can proceed to factor out common factors from each group. Since the second group leads with a negative coefficient, we'll factor out -3. This will result in the " $+ 15$" becoming " $- 5$."

$$= t^2(t - 5) + (-3)(t - 5)$$

$$\downarrow \qquad \downarrow$$
$$= t^2\overbrace{(t-5)} - 3\overbrace{(t-5)}$$
$$= (t-5)(t^2-3)$$

And remember that we can confirm this is correct by multiplying it out. If we made no mistakes, it should result in the original $t^3 - 5t^2 - 3t + 15$.

Checkpoint 7.2.5. Factor $6q^3 - 9q^2 - 4q + 6$.

Explanation. We will break the polynomial into two groups: $6q^3 - 9q^2$ and $-4q + 6$.

$$6q^3 - 9q^2 - 4q + 6 = \left(6q^3 - 9q^2\right) + (-4q + 6)$$

and now each group has its own greatest common factor to factor out:

$$= 3q^2(2q - 3) - 2(2q - 3)$$

and now the binomial $(2q - 3)$ appears twice and can be factored out:

$$= (2q - 3)(3q^2 - 2)$$

Example 7.2.6 Factor $x^3 - 3x^2 + x - 3$. To succeed with this example, we will need to "factor out" a trivial number 1 that isn't apparent until we make it so.

$$x^3 - 3x^2 + x - 3 = \left(x^3 - 3x^2\right) + (x - 3)$$
$$= x^2(x - 3) + 1(x - 3)$$
$$\downarrow \qquad \downarrow$$
$$= x^2\overbrace{(x - 3)} + 1\overbrace{(x - 3)}$$
$$= (x - 3)(x^2 + 1)$$

Notice how we changed $x - 3$ to $+1(x - 3)$, so we wouldn't forget the $+1$ in the final factored form. As always, we should check this is correct by multiplying it out.

Checkpoint 7.2.7. Factor $6t^6 + 9t^4 + 2t^2 + 3$.

Explanation. We will break the polynomial into two groups: $6t^6 + 9t^4$ and $2t^2 + 3$.

$$6t^6 + 9t^4 + 2t^2 + 3 = \left(6t^6 + 9t^4\right) + \left(2t^2 + 3\right)$$

the first group has its own greatest common factor to factor out, and with the second group we can write a 1:

$$= 3t^4\left(2t^2 + 3\right) + 1\left(2t^2 + 3\right)$$

and now the binomial $\left(2t^2 + 3\right)$ appears twice and can be factored out:

$$= \left(2t^2 + 3\right)\left(3t^4 + 1\right)$$

Example 7.2.8 Factor $xy^2 - 10y^2 - 2x + 20$. The technique can work when there are multiple variables too.

$$xy^2 - 10y^2 - 2x + 20 = \left(xy^2 - 10y^2\right) + (-2x + 20)$$

$$= y^2(x - 10) + (-2)(x - 10)$$

$$= y^2\overbrace{(x - 10)}-2\overbrace{(x - 10)}$$

$$= (x - 10)(y^2 - 2).$$

Unfortunately, this technique is not guaranteed to work on every polynomial with four terms. In fact, *most* randomly selected four-term polynomials will not factor using this method and those selected here should be considered "nice." Here is an example that will not factor with grouping:

$$x^3 + 6x^2 + 11x + 6 = (x^3 + 6x^2) + (11x + 6)$$

$$= x^2 \underbrace{(x + 6)}_{?} + 1 \underbrace{(11x + 6)}_{?}$$

In this example, at the step where we hope to see the same binomial appearing twice, we see two different binomials. It doesn't mean that this kind of polynomial can't be factored, but it does mean that "factoring by grouping" is not going to help. This polynomial actually factors as $(x + 1)(x + 2)(x + 3)$. So the fact that grouping fails to factor the polynomial doesn't tell us whether or not it is prime.

Exercises

Review and Warmup   Factor the given polynomial.

1. $-6y - 6 = $

2. $-3r - 3 = $

3. $8r + 16 = $

4. $5t - 30 = $

5. $14t^2 - 20 = $

6. $56t^2 + 40 = $

Factoring out Common Polynomials   Factor the given polynomial.

7. $x(x + 5) + 7(x + 5) = $

8. $x(x - 2) + 5(x - 2) = $

9. $x(y + 4) - 7(y + 4) = $

10. $x(y - 5) + 2(y - 5) = $

11. $6x(x + y) + 7(x + y) = $

12. $7x(x + y) + 4(x + y) = $

13. $4t^7(4t - 7) + 4t - 7 = $

14. $10t^{10}(8t + 5) + 8t + 5 = $

15. $42t^4(t + 18) + 21t^3(t + 18) + 56t^2(t + 18) = $

16. $14x^4(x+2) + 21x^3(x+2) + 14x^2(x+2) =$ [ ]

Factoring by Grouping   Factor the given polynomial.

17. $x^2 + 10x - 9x - 90 =$ [ ]

18. $y^2 - 6y - 7y + 42 =$ [ ]

19. $y^2 + 3y - 4y - 12 =$ [ ]

20. $r^2 - 9r - 2r + 18 =$ [ ]

21. $r^3 + 6r^2 - 8r - 48 =$ [ ]

22. $r^3 - 3r^2 - 6r + 18 =$ [ ]

23. $t^3 - 8t^2 - 3t + 24 =$ [ ]

24. $t^3 + 6t^2 - 10t - 60 =$ [ ]

25. $xy - 2x + 4y - 8 =$ [ ]

26. $xy + 3x - 8y - 24 =$ [ ]

27. $xy - 4x - 3y + 12 =$ [ ]

28. $xy - 5x + 6y - 30 =$ [ ]

29. $6x^2 + 30xy + 5xy + 25y^2 =$ [ ]

30. $7x^2 + 14xy + 4xy + 8y^2 =$ [ ]

31. $8x^2 - 80xy + 7xy - 70y^2 =$ [ ]

32. $9x^2 - 36xy + 8xy - 32y^2 =$ [ ]

33. $x^3 - 10 - 7x^3y + 70y =$ [ ]

34. $x^3 + 2 - 4x^3y - 8y =$ [ ]

35. $x^3 + 3 + 6x^3y + 18y =$ [ ]

36. $x^3 - 4 + 9x^3y - 36y =$ [ ]

37. $18y^5 + 36y^4 - 9y^4 - 18y^3 + 18y^3 + 36y^2 =$ [ ]

38. $8r^5 + 24r^4 - 8r^4 - 24r^3 + 12r^3 + 36r^2 =$ [ ]

7.3 Factoring Trinomials with Leading Coefficient One

In Chapter 6, we learned how to multiply binomials like $(x + 2)(x + 3)$ and obtain the trinomial $x^2 + 5x + 6$. In this section, we will learn how to undo that. So we'll be starting with a trinomial like $x^2 + 5x + 6$ and obtaining its factored form $(x + 2)(x + 3)$. The trinomials that we'll factor in this section all have leading coefficient 1, but Section 7.4 will cover some more general trinomials.

7.3.1 Factoring Trinomials by Listing Factor Pairs

Consider the example $x^2 + 5x + 6 = (x + 2)(x + 3)$. There are at least three things that are important to notice:

- The leading coefficient of $x^2 + 5x + 6$ is 1.

- The two factors on the right use the numbers 2 and 3, and when you *multiply* these you get the 6.

- The two factors on the right use the numbers 2 and 3, and when you *add* these you get the 5.

So the idea is that if you need to factor $x^2 + 5x + 6$ and you somehow discover that 2 and 3 are special numbers (because $2 \cdot 3 = 6$ and $2 + 3 = 5$), then you can conclude that $(x + 2)(x + 3)$ is the factored form of the given polynomial.

Example 7.3.2 Factor $x^2 + 13x + 40$. Since the leading coefficient is 1, we are looking to write this polynomial as $(x + ?)(x + ?)$ where the question marks are two possibly different, possibly negative, numbers. We need these two numbers to multiply to 40 and add to 13. How can you track these two numbers down? Since the numbers need to multiply to 40, one method is to list all **factor pairs** of 40 in a table just to see what your options are. We'll write every *pair of factors* that multiply to 40.

$1 \cdot 40$	$-1 \cdot (-40)$
$2 \cdot 20$	$-2 \cdot (-20)$
$4 \cdot 10$	$-4 \cdot (-10)$
$5 \cdot 8$	$-5 \cdot (-8)$

We wanted to find *all* factor pairs. To avoid missing any, we started using 1 as a factor, and then slowly increased that first factor. The table skips over using 3 as a factor, because 3 is not a factor of 40. Similarly the table skips using 6 and 7 as a factor. And there would be no need to continue with 8 and beyond, because we already found "large" factors like 8 as the partners of "small" factors like 5.

There is an entire second column where the signs are reversed, since these are also ways to multiply two numbers to get 40. In the end, there are eight factor pairs.

We need a pair of numbers that *also* adds to 13. So we check what each of our factor pairs add up to:

Factor Pair	Sum of the Pair		Factor Pair	Sum of the Pair
$1 \cdot 40$	41		$-1 \cdot (-40)$	(no need to go this far)
$2 \cdot 20$	22		$-2 \cdot (-20)$	(no need to go this far)
$4 \cdot 10$	14		$-4 \cdot (-10)$	(no need to go this far)
$5 \cdot 8$	13 (what we wanted)		$-5 \cdot (-8)$	(no need to go this far)

The winning pair of numbers is 5 and 8. Again, what matters is that $5 \cdot 8 = 40$, and $5 + 8 = 13$. So we can conclude that $x^2 + 13x + 40 = (x + 5)(x + 8)$.

To ensure that we made no mistakes, here are some possible checks.

Multiply it Out   Multiplying out our answer $(x + 5)(x + 8)$ should give us $x^2 + 13x + 40$.

We could also use a rectangular area diagram to verify the factorization is correct:

$$(x + 5)(x + 8) = (x + 5) \cdot x + (x + 5) \cdot 8$$
$$= x^2 + 5x + 8x + 40$$
$$\overset{\checkmark}{=} x^2 + 13x + 40$$

	x	5
x	x^2	$5x$
8	$8x$	40

Evaluating   If the answer really is $(x + 5)(x + 8)$, then notice how evaluating at -5 would result in 0. So the original expression should also result in 0 if we evaluate at -5. And similarly, if we evaluate it at -8, $x^2 + 13x + 40$ should be 0.

$$(-5)^2 + 13(-5) + 40 \overset{?}{=} 0 \qquad\qquad (-8)^2 + 13(-8) + 40 \overset{?}{=} 0$$
$$25 - 65 + 40 \overset{?}{=} 0 \qquad\qquad 64 - 104 + 40 \overset{?}{=} 0$$
$$0 \overset{\checkmark}{=} 0 \qquad\qquad 0 \overset{\checkmark}{=} 0.$$

This also gives us evidence that the factoring was correct.

Example 7.3.3 Factor $y^2 - 11y + 24$. The negative coefficient is a small complication from Example 7.3.2, but the process is actually still the same.

Explanation.   We need a pair of numbers that multiply to 24 and add to -11. Note that we *do* care to keep track that they sum to a negative total.

Factor Pair	Sum of the Pair
$1 \cdot 24$	25
$2 \cdot 12$	14
$3 \cdot 8$	11 (close; wrong sign)
$4 \cdot 6$	10

Factor Pair	Sum of the Pair
$-1 \cdot (-24)$	-25
$-2 \cdot (-12)$	-14
$-3 \cdot (-8)$	-11 (what we wanted)
$-4 \cdot (-6)$	(no need to go this far)

So $y^2 - 11y + 24 = (y - 3)(y - 8)$. To confirm that this is correct, we should check. Either by multiplying out the factored form:

$$(y - 3)(y - 8) = (y - 3) \cdot y - (y - 3) \cdot 8$$
$$= y^2 - 3y - 8y + 24$$
$$\overset{\checkmark}{=} y^2 - 11y + 24$$

	y	-3
y	y^2	$-3y$
-8	$-8y$	24

Or by evaluating the original expression at 3 and 8:

$$3^2 - 11(3) + 24 \overset{?}{=} 0 \qquad\qquad 8^2 - 11(8) + 24 \overset{?}{=} 0$$

$$9 - 33 + 24 \stackrel{?}{=} 0 \qquad\qquad 64 - 88 + 24 \stackrel{?}{=} 0$$

$$0 \stackrel{\checkmark}{=} 0 \qquad\qquad 0 \stackrel{\checkmark}{=} 0.$$

Our factorization passes the tests.

Example 7.3.4 Factor $z^2 + 5z - 6$. The negative coefficient is again a small complication from Example 7.3.2, but the process is actually still the same.

Explanation. We need a pair of numbers that multiply to -6 and add to 5. Note that we *do* care to keep track that they multiply to a negative product.

Factor Pair	Sum of the Pair		Factor Pair	Sum of the Pair
$1 \cdot (-6)$	-5 (close; wrong sign)		$-1 \cdot 6$	5 (what we wanted)
$2 \cdot (-3)$	14		$-2 \cdot 3$	(no need to go this far)

So $z^2 + 5z - 6 = (z - 1)(z + 6)$. To confirm that this is correct, we should check. Either by multiplying out the factored form:

$$
\begin{aligned}
(z - 1)(z + 6) &= (z - 1) \cdot z + (z - 1) \cdot 6 \\
&= z^2 - z + 6z - 6 \\
&\stackrel{\checkmark}{=} z^2 + 5z - 6
\end{aligned}
$$

	z	-1
z	z^2	$-z$
6	$6z$	-6

Or by evaluating the original expression at 1 and -6:

$$1^2 + 5(1) - 6 \stackrel{?}{=} 0 \qquad\qquad (-6)^2 + 5(-6) - 6 \stackrel{?}{=} 0$$

$$1 + 5 - 6 \stackrel{?}{=} 0 \qquad\qquad 36 - 30 - 6 \stackrel{?}{=} 0$$

$$0 \stackrel{\checkmark}{=} 0 \qquad\qquad 0 \stackrel{\checkmark}{=} 0.$$

Our factorization passes the tests.

Checkpoint 7.3.5. Factor $m^2 - 6m - 40$.

Explanation. We need a pair of numbers that multiply to -40 and add to -6. Note that we *do* care to keep track that they multiply to a negative product and sum to a negative total.

Factor Pair	Sum of the Pair
$1 \cdot (-40)$	-39
$2 \cdot (-20)$	-18
$4 \cdot (-10)$	-6 (what we wanted)
(no need to continue)	...

So $m^2 - 6m - 40 = (m + 4)(m - 10)$.

7.3.2 Connection to Grouping

The factoring method we just learned takes a bit of a shortcut. To prepare yourself for a more complicated factoring technique in Section 7.4, you may want to try taking the "scenic route" instead of that shortcut.

Example 7.3.6 Let's factor $x^2 + 13x + 40$ again (the polynomial from Example 7.3.2). As before, it is important to discover that 5 and 8 are important numbers, because they multiply to 40 and add to 13. As before, listing out all of the factor pairs is one way to discover the 5 and the 8.

Instead of jumping to the factored answer, we can show how $x^2 + 13x + 40$ factors in a more step-by-step fashion using 5 and 8. Since they add up to 13, we can write:

$$x^2 + \overset{\downarrow}{13}x + 40 = x^2 + 5x + \overbrace{8x}^{\downarrow} + 40$$

We have intentionally split up the trinomial into an unsimplified polynomial with four terms. In Section 7.2, we handled such four-term polynomials by grouping:

$$= \left(x^2 + 5x\right) + (8x + 40)$$

Now we can factor out each group's greatest common factor:

$$= x(x + 5) + 8(x + 5)$$
$$= x\overbrace{(x + 5)}^{\downarrow} + 8\overbrace{(x + 5)}^{\downarrow}$$
$$= (x + 5)(x + 8)$$

And we have found that $x^2 + 13x + 40$ factors as $(x + 5)(x + 8)$ without memorizing the shortcut.

This approach takes more time, and ultimately you may not use it much. However, if you try a few examples this way, it may make you more comfortable with the more complicated technique in Section 7.4.

7.3.3 Trinomials with Higher Powers

So far we have only factored examples of **quadratic** trinomials: trinomials whose highest power of the variable is 2. However, this technique can also be used to factor trinomials where there is a larger highest power of the variable. It only requires that the highest power is even, that the next highest power is half of the highest power, and that the third term is a constant term.

In the four examples below, check:

1. if the highest power is even

2. if the next highest power is half of the highest power

3. if the last term is constant

Factor pairs *will* help with...

- $y^6 - 23y^3 - 50$

- $h^{16} + 22h^8 + 105$

Factor pairs *won't* help with...

- $y^5 - 23y^3 - 50$

- $h^{16} + 22h^8 + 105h^2$

Example 7.3.7 Factor $h^{16} + 22h^8 + 105$. This polynomial is one of the examples above where using factor pairs will help. We find that $7 \cdot 15 = 105$, and $7 + 15 = 22$, so the numbers 7 and 15 can be used:

$$h^{16} + 22h^8 + 105 = h^{16} + \overbrace{7h^8 + 15h^8} + 105$$
$$= \left(h^{16} + 7h^8\right) + \left(15h^8 + 105\right)$$
$$= h^8 \left(h^8 + 7\right) + 15 \left(h^8 + 7\right)$$
$$= \left(h^8 + 7\right) \left(h^8 + 15\right)$$

Actually, once we settled on using 7 and 15, we could have concluded that $h^{16} + 22h^8 + 105$ factors as $\left(h^8 + 7\right) \left(h^8 + 15\right)$, if we know which power of h to use. We'll always use half the highest power in these factorizations.

In any case, to confirm that this is correct, we should check by multiplying out the factored form:

$$(h^8 + 7)(h^8 + 15) = (h^8 + 7) \cdot h^8 + (h^8 + 7) \cdot 15$$
$$= h^{16} + 7h^8 + 15h^8 + 105$$
$$\overset{\checkmark}{=} h^{16} + 22h^8 + 15$$

	h^8	7
h^8	h^{16}	$7h^8$
15	$15h^8$	105

Our factorization passes the tests.

Checkpoint 7.3.8. Factor $y^6 - 23y^3 - 50$.

Explanation. We need a pair of numbers that multiply to -50 and add to -23. Note that we *do* care to keep track that they multiply to a negative product and sum to a negative total.

Factor Pair	Sum of the Pair
$1 \cdot (-50)$	-49
$2 \cdot (-25)$	-23 (what we wanted)
(no need to continue)	...

So $y^6 - 23y^3 - 50 = \left(y^3 - 25\right) \left(y^3 + 2\right)$.

7.3.4 Factoring in Stages

Sometimes factoring a polynomial will take two or more "stages." Always begin factoring a polynomial by factoring out its greatest common factor, and *then* apply a second stage where you use a technique from this section. The process of factoring a polynomial is not complete until each of the factors cannot be factored further.

Example 7.3.9 Factor $2z^2 - 6z - 80$.

Explanation. We will first factor out the common factor, 2:

$$2z^2 - 6z - 80 = 2\left(z^2 - 3z - 40\right)$$

Now we are left with a factored expression that might factor more. Looking inside the parentheses, we ask ourselves, "what two numbers multiply to be -40 and add to be -3?" Since 5 and -8 do the job the full factorization is:

$$2z^2 - 6z - 80 = 2\left(z^2 - 3z - 40\right)$$
$$= 2(z + 5)(z - 8)$$

Example 7.3.10 Factor $-r^2 + 2r + 24$.

Explanation. The three terms don't exactly have a common factor, but as discussed in Section 7.1, when the leading term has a negative sign, it is often helpful to factor out that negative sign:

$$-r^2 + 2r + 24 = -\left(r^2 - 2r - 24\right).$$

Looking inside the parentheses, we ask ourselves, "what two numbers multiply to be -24 and add to be -2?" Since -6 and 4 work here and the full factorization is shown:

$$-r^2 + 2r + 24 = -\left(r^2 - 2r - 24\right)$$
$$= -(r - 6)(r + 4)$$

Example 7.3.11 Factor $p^2q^3 + 4p^2q^2 - 60p^2q$.

Explanation. First, always look for the greatest common factor: in this trinomial it is p^2q. After factoring this out, we have

$$p^2q^3 + 4p^2q^2 - 60p^2q = p^2q\left(q^2 + 4q - 60\right).$$

Looking inside the parentheses, we ask ourselves, "what two numbers multiply to be -60 and add to be 4?" Since 10 and -6 fit the bill, the full factorization can be shown below:

$$p^2q^3 + 4p^2q^2 - 60p^2q = p^2q\left(q^2 + 4q - 60\right)$$
$$= p^2q(q + 10)(q - 6)$$

7.3.5 More Trinomials with Two Variables

You might encounter a trinomial with two variables that can be factored using the methods we've discussed in this section. It can be tricky though: $x^2 + 5xy + 6y^2$ has two variables and it *can* factor using the methods from this section, but $x^2 + 5x + 6y^2$ also has two variables and it *cannot* be factored. So in examples of this nature, it is even more important to check that factorizations you find actually work.

Example 7.3.12 Factor $x^2 + 5xy + 6y^2$. This is a trinomial, and the coefficient of x is 1, so maybe we can factor it. We want to write $(x + ?)(x + ?)$ where the question marks will be *something* that makes it all multiply out to $x^2 + 5xy + 6y^2$.

Since the last term in the polynomial has a factor of y^2, it is natural to wonder if there is a factor of y in each of the two question marks. If there were, these two factors of y would multiply to y^2. So it is natural to wonder if we are looking for $(x + ?y)(x + ?y)$ where now the question marks are just numbers.

At this point we can think like we have throughout this section. Are there some numbers that multiply to 6 and add to 5? Yes, specifically 2 and 3. So we suspect that $(x+2y)(x+3y)$ might be the factorization. To confirm that this is correct, we should check by multiplying out the factored form:

$$(x+2y)(x+3y) = (x+2y) \cdot x + (x+2y) \cdot 3y$$
$$= x^2 + 2xy + 3xy + 6y^2$$
$$\overset{\checkmark}{=} x^2 + 5xy + 6y^2$$

	x	$2y$
x	x^2	$2xy$
$3y$	$3xy$	$6y^2$

Our factorization passes the tests.

In Section 7.4, there is a more definitive method for factoring polynomials of this form.

Exercises

Review and Warmup   Multiply the polynomials.

1. $(t+6)(t+10) = \boxed{}$ **2.** $(t+3)(t+4) = \boxed{}$ **3.** $(x+9)(x-3) = \boxed{}$

4. $(x+6)(x-8) = \boxed{}$ **5.** $(y-9)(y-4) = \boxed{}$ **6.** $(y-2)(y-10) = \boxed{}$

7. $3(x+2)(x+3) = \boxed{}$ **8.** $-4(y-1)(y-9) = \boxed{}$ **9.** $2(y-10)(y-3) = \boxed{}$

10. $-2(r+7)(r+6) = \boxed{}$

Factoring Trinomials with Leading Coefficient One   Factor the given polynomial.

11. $t^2 + 3t + 2 = \boxed{}$ **12.** $x^2 + 14x + 48 = \boxed{}$ **13.** $x^2 + 14x + 40 = \boxed{}$

14. $y^2 + 5y + 4 = \boxed{}$ **15.** $y^2 + 3y - 28 = \boxed{}$ **16.** $r^2 - 3r - 4 = \boxed{}$

17. $r^2 + 2r - 80 = \boxed{}$ **18.** $r^2 - 2r - 35 = \boxed{}$ **19.** $t^2 - 9t + 20 = \boxed{}$

20. $t^2 - 19t + 90 = \boxed{}$ **21.** $x^2 - 10x + 21 = \boxed{}$ **22.** $x^2 - 11x + 24 = \boxed{}$

23. $y^2 + 12y + 20 = \boxed{}$ **24.** $y^2 + 15y + 54 = \boxed{}$ **25.** $r^2 + 13r + 30 = \boxed{}$

26. $r^2 + 13r + 36 = $ ☐

27. $r^2 - 4r - 60 = $ ☐

28. $t^2 + 6t - 16 = $ ☐

29. $t^2 + 4t - 45 = $ ☐

30. $x^2 - 3x - 10 = $ ☐

31. $x^2 - 7x + 10 = $ ☐

32. $y^2 - 17y + 72 = $ ☐

33. $y^2 - 8y + 15 = $ ☐

34. $r^2 - 9r + 8 = $ ☐

35. $r^2 + 2r + 7 = $ ☐

36. $r^2 + 5 = $ ☐

37. $t^2 - 2t + 5 = $ ☐

38. $t^2 + 9 = $ ☐

39. $x^2 + 10x + 25 = $ ☐

40. $x^2 + 24x + 144 = $ ☐

41. $y^2 + 16y + 64 = $ ☐

42. $y^2 + 8y + 16 = $ ☐

43. $r^2 - 24r + 144 = $ ☐

44. $r^2 - 16r + 64 = $ ☐

45. $r^2 - 8r + 16 = $ ☐

46. $t^2 - 22t + 121 = $ ☐

47. $9t^2 - 9t - 18 = $ ☐

48. $7x^2 + 7x - 42 = $ ☐

49. $6x^2 - 6x - 12 = $ ☐

50. $4y^2 - 16 = $ ☐

51. $2y^2 - 8y + 6 = $ ☐

52. $2y^2 - 14y + 20 = $ ☐

53. $4r^2 - 20r + 16 = $ ☐

54. $2r^2 - 12r + 10 = $ ☐

55. $9t^4 + 27t^3 + 18t^2 = $ ☐

56. $2t^4 + 22t^3 + 20t^2 = $ ☐

57. $5x^6 + 20x^5 + 15x^4 = $ ☐

58. $2x^6 + 18x^5 + 16x^4 = $ ☐

59. $5y^6 - 10y^5 - 40y^4 = $ ☐

60. $6y^4 - 18y^3 - 24y^2 = $ ☐

61. $4y^{10} - 12y^9 - 16y^8 = $ ☐

62. $2r^8 + 12r^7 - 32r^6 = $ ☐

63. $4r^6 - 12r^5 + 8r^4 = $ ☐

64. $7t^9 - 28t^8 + 21t^7 = $ ☐

65. $4t^{10} - 20t^9 + 16t^8 = $ ☐

66. $9x^6 - 27x^5 + 18x^4 = $ ☐

67. $-x^2 + 16 = $ ☐

68. $-y^2 + 4y + 21 = $ ☐

69. $-y^2 + 6y + 40 = $ ☐

70. $-y^2 - 5y + 6 = $ ☐

71. $r^2 + 10rx + 16x^2 = \boxed{}$ **72.** $r^2 + 9rt + 18t^2 = \boxed{}$ **73.** $t^2 - 7tr - 30r^2 = \boxed{}$

74. $t^2 + tx - 12x^2 = \boxed{}$ **75.** $x^2 - 6xt + 8t^2 = \boxed{}$ **76.** $x^2 - 11xy + 24y^2 = \boxed{}$

77. $y^2 + 16yx + 64x^2 = \boxed{}$ **78.** $y^2 + 2yt + t^2 = \boxed{}$ **79.** $y^2 - 4yr + 4r^2 = \boxed{}$

80. $r^2 - 22rx + 121x^2 = \boxed{}$ **81.** $4r^2 + 20r + 16 = \boxed{}$ **82.** $2t^2 + 14t + 12 = \boxed{}$

83. $2x^2y + 10xy + 8y = \boxed{}$ **84.** $8x^2y + 24xy + 16y = \boxed{}$ **85.** $3a^2b - 21ab - 24b = \boxed{}$

86. $7a^2b - 7b = \boxed{}$ **87.** $2x^2y - 6xy + 4y = \boxed{}$ **88.** $2x^2y - 16xy + 14y = \boxed{}$

89. $2x^3y + 12x^2y + 16xy = \boxed{}$ **90.** $2x^3y + 14x^2y + 20xy = \boxed{}$ **91.** $x^2y^2 - 9x^2z^2 = \boxed{}$

92. $x^2y^2 - 5x^2yz - 14x^2z^2 = \boxed{}$ **93.** $x^2 + x + 0.24 = \boxed{}$ **94.** $x^2 + x + 0.21 = \boxed{}$

95. $y^2t^2 + 12yt + 20 = \boxed{}$ **96.** $y^2r^2 + 11yr + 28 = \boxed{}$ **97.** $y^2x^2 + 7yx - 30 = \boxed{}$

98. $r^2t^2 + 8rt - 20 = \boxed{}$ **99.** $r^2y^2 - 12ry + 20 = \boxed{}$ **100.** $t^2x^2 - 11tx + 24 = \boxed{}$

101. $2t^2r^2 + 22tr + 20 = \boxed{}$ **102.** $2x^2y^2 + 20xy + 18 = \boxed{}$ **103.** $10x^2y^2 - 10 = \boxed{}$

104. $5y^2r^2 - 10yr - 15 = \boxed{}$ **105.** $2x^2y^3 - 16xy^2 + 30y = \boxed{}$ **106.** $2x^2y^3 - 6xy^2 + 4y = \boxed{}$

107. $(a + b)r^2 + 10(a + b)r + 16(a + b) = \boxed{}$

108. $(a + b)r^2 + 10(a + b)r + 24(a + b) = \boxed{}$

Challenge

109. What integers can go in the place of b so that the quadratic expression $x^2 + bx - 16$ is factorable?

7.4 Factoring Trinomials with a Nontrivial Leading Coefficient

In Section 7.3, we learned how to factor $ax^2 + bx + c$ when $a = 1$. In this section, we will examine the situation when $a \neq 1$. The techniques are similar to those in the last section, but there are a few important differences that will make-or-break your success in factoring these.

7.4.1 The AC Method

The AC Method is a technique for factoring trinomials like $4x^2 + 5x - 6$, where there is no greatest common factor, and the leading coefficient is not 1.

Please note at this point that if we try the method in the previous section and ask ourselves the question "what two numbers multiply to be -6 and add to be 5?," we might come to the *erroneous* conclusion that $4x^2 + 5x - 6$ factors as $(x + 6)(x - 1)$. If we expand $(x + 6)(x - 1)$, we get

$$(x + 6)(x - 1) = x^2 + 5x - 6$$

This expression is *almost* correct, except for the missing leading coefficient, 4. Dealing with this missing coefficient requires starting over with the AC method. If you are only interested in the steps for using the technique, skip ahead to Algorithm 7.4.3.

The example below explains *why* the AC Method works. Understanding all of the details might take a few rereads, and coming back to this example after mastering the algorithm may be the best course of action.

> **Example 7.4.2** Expand the expression $(px + q)(rx + s)$ and analyze the result to gain an insight into *why* the AC method works. Then use this information to factor $4x^2 + 5x - 6$.
>
> **Explanation.** Factoring is the opposite process from multiplying polynomials together. We can gain some insight into how to factor complicated polynomials by taking a closer look at what happens when two generic polynomials are multiplied together:
>
> $$\begin{aligned} (px + q)(rx + s) &= (px + q)(rx) + (px + q)s \\ &= (px)(rx) + q(rx) + (px)s + qs \\ &= (pr)x^2 + qrx + psx + qs \\ &= (pr)x^2 + (qr + ps)x + qs \end{aligned} \quad (7.4.1)$$
>
> When you encounter a trinomial like $4x^2 + 5x - 6$ and you wish to factor it, the leading coefficient, 4, is the (pr) from Equation (7.4.1). Similarly, the -6 is the qs, and the 5 is the $(qr + ps)$.
>
> Now, if you multiply the leading coefficient and constant term from Equation (7.4.1), you have $(pr)(qs)$, which equals $pqrs$. Notice that if we factor this number in just the right way, $(qr)(ps)$, then we have two factors that add to the middle coefficient from Equation (7.4.1), $(qr + ps)$.
>
> Can we do all this with the example $4x^2 + 5x - 6$? Multiplying 4 and -6 makes -24. Is there some way to factor -24 into two factors which add to 5? We make a table of factor pairs for -24 to see:

Factor Pair	Sum of the Pair		Factor Pair	Sum of the Pair
$-1 \cdot 24$	23		$1 \cdot (-24)$	(no need to go this far)
$-2 \cdot 12$	10		$2 \cdot (-12)$	(no need to go this far)
$-3 \cdot 8$	5 (what we wanted)		$3 \cdot (-8)$	(no need to go this far)
$-4 \cdot 6$	(no need to go this far)		$4 \cdot (-6)$	(no need to go this far)

So that 5 in $4x^2 + 5x - 6$, which is equal to the abstract $(qr + ps)$ from Equation (7.4.1), breaks down as $-3 + 8$. We can take -3 to be the qr and 8 to be the ps. Once we intentionally break up the 5 this way, factoring by grouping (see Section 7.2) can take over and is guaranteed to give us a factorization.

$$4x^2 \overbrace{+ 5x} -6 = 4x^2 \overbrace{-3x + 8x} -6$$

Now that there are four terms, group them and factor out each group's greatest common factor.

$$= \left(4x^2 - 3x\right) + (8x - 6)$$
$$= x(4x - 3) + 2(4x - 3)$$
$$= (4x - 3)(x + 2)$$

And this is the factorization of $4x^2 + 5x - 6$. This whole process is known as the "AC method," since it begins by multiplying a and c from the generic $ax^2 + bx + c$.

The AC Method   Here is a summary of the algorithm:

Process 7.4.3 The AC Method. *To factor $ax^2 + bx + c$:*

1. *Multiply $a \cdot c$.*

2. *Make a table of factor pairs for ac. Look for a pair that adds to b. If you cannot find one, the polynomial is irreducible.*

3. *If you did find a factor pair summing to b, replace b with an explicit sum, and distribute x. With the four terms you have at this point, use factoring by grouping to continue. You are guaranteed to find a factorization.*

Example 7.4.4 Factor $10x^2 + 23x + 6$.

1. $10 \cdot 6 = 60$

2. Use a list of factor pairs for 60 to find that 3 and 20 are a pair that sums to 23.

3. Intentionally break up the 23 as $3 + 20$:

$$10x^2 \overbrace{+ 23x} +6$$

$$= 10x^2 \overbrace{+ 3x + 20x} +6$$
$$= \left(10x^2 + 3x\right) + (20x + 6)$$
$$= x(10x + 3) + 2(10x + 3)$$
$$= (10x + 3)(x + 2)$$

Example 7.4.5 Factor $2x^2 - 5x - 3$.

Explanation. Always start the factoring process by examining if there is a greatest common factor. Here there is not one. Next, note that this is a trinomial with a leading coefficient that is not 1. So the AC Method may be of help.

1. Multiply $2 \cdot (-3) = -6$.

2. Examine factor pairs that multiply to -6, looking for a pair that sums to -5:

Factor Pair	Sum of the Pair		Factor Pair	Sum of the Pair
$1 \cdot -6$	-5 (what we wanted)		$-1 \cdot 6$	(no need to go this far)
$2 \cdot -3$	(no need to go this far)		$-2 \cdot 3$	(no need to go this far)

3. Intentionally break up the -5 as $1 + (-6)$:

$$2x^2 \overbrace{- 5x} - 3 = 2x^2 \overbrace{+ x - 6x} - 3$$
$$= \left(2x^2 + x\right) + (-6x - 3)$$
$$= x(2x + 1) - 3(2x + 1)$$
$$= (2x + 1)(x - 3)$$

So we believe that $2x^2 - 5x - 3$ factors as $(2x + 1)(x - 3)$, and we should check by multiplying out the factored form:

$$(2x + 1)(x - 3) = (2x + 1) \cdot x + (2x + 1) \cdot (-3)$$
$$= 2x^2 + x - 6x - 3$$
$$\overset{\checkmark}{=} 2x^2 - 5x - 3$$

	$2x$	1
x	$2x^2$	x
-3	$-6x$	-3

Our factorization passes the tests.

Example 7.4.6 Factor $6p^2 + 5pq - 6q^2$. Note that this example has two variables, but that does not really change our approach.

Explanation. There is no greatest common factor. Since this is a trinomial, we try the AC Method.

1. Multiply $6 \cdot (-6) = -36$.

2. Examine factor pairs that multiply to -36, looking for a pair that sums to 5:

Factor Pair	Sum of the Pair
$1 \cdot -36$	-35
$2 \cdot -18$	-16
$3 \cdot -12$	-9
$4 \cdot -9$	-5 (close; wrong sign)
$6 \cdot -6$	0

Factor Pair	Sum of the Pair
$-1 \cdot 36$	35
$-2 \cdot 18$	16
$-3 \cdot 12$	9
$-4 \cdot 9$	5 (what we wanted)

3. Intentionally break up the 5 as $-4 + 9$:

$$6p^2 + \overbrace{5pq} - 6q^2 = 6p^2 \overbrace{- 4pq + 9pq} - 6q^2$$
$$= (6p^2 - 4pq) + (9pq - 6q^2)$$
$$= 2p(3p - 2q) + 3q(3p - 2q)$$
$$= (3p - 2q)(2p + 3q)$$

So we believe that $6p^2 + 5pq - 6q^2$ factors as $(3p - 2q)(2p + 3q)$, and we should check by multiplying out the factored form:

$$(3p - 2q)(2p + 3q) = (3p - 2q) \cdot 2p + (3p - 2q) \cdot 3q$$
$$= 6p^2 - 4pq + 9pq - 6q^2$$
$$\overset{\checkmark}{=} 6p^2 + 5pq - 6q^2$$

	$3p$	$-2q$
$2p$	$6p^2$	$-4pq$
$3q$	$9pq$	$-6q^2$

Our factorization passes the tests.

7.4.2 Factoring in Stages

Sometimes factoring a polynomial will take two or more "stages." For instance you may need to begin factoring a polynomial by factoring out its greatest common factor, and *then* apply a second stage where you use a technique from this section. The process of factoring a polynomial is not complete until each of the factors cannot be factored further.

Example 7.4.7 Factor $18n^2 - 21n - 60$.

Explanation. Notice that 3 is a common factor in this trinomial. We should factor it out first:

$$18n^2 - 21n - 60 = 3\left(6n^2 - 7n - 20\right)$$

Now we are left with two factors, one of which is $6n^2 - 7n - 20$, which might factor further. Using the AC Method:

1. $6 \cdot -20 = -120$

2. Examine factor pairs that multiply to -120, looking for a pair that sums to -7:

Factor Pair	Sum of the Pair	Factor Pair	Sum of the Pair
$1 \cdot -120$	-119	$-1 \cdot 120$	(no need to go this far)
$2 \cdot -60$	-58	$-2 \cdot 60$	(no need to go this far)
$3 \cdot -40$	-37	$-3 \cdot 40$	(no need to go this far)
$4 \cdot -30$	-26	$-4 \cdot 30$	(no need to go this far)
$5 \cdot -24$	-19	$-5 \cdot 24$	(no need to go this far)
$6 \cdot -20$	-14	$-6 \cdot 20$	(no need to go this far)
$8 \cdot -15$	-7 (what we wanted)	$-8 \cdot 15$	(no need to go this far)
$10 \cdot -12$	(no need to go this far)	$-10 \cdot 12$	(no need to go this far)

3. Intentionally break up the -7 as $8 + (-15)$:

$$18n^2 - 21n - 60 = 3\left(6n^2 \overbrace{-7n} -20\right)$$

$$= 3\left(6n^2 \overbrace{+8n - 15n} -20\right)$$

$$= 3\left((6n^2 + 8n) + (-15n - 20)\right)$$

$$= 3\left(2n(3n + 4) - 5(3n + 4)\right)$$

$$= 3(3n + 4)(2n - 5)$$

So we believe that $18n^2 - 21n - 60$ factors as $3(3n + 4)(2n - 5)$, and you should check by multiplying out the factored form.

Example 7.4.8 Factor $-16x^3y - 12x^2y + 18xy$.

Explanation. Notice that $2xy$ is a common factor in this trinomial. Also the leading coefficient is negative, and as discussed in Section 7.1, it is wise to factor that out as well. So we find:

$$-16x^3y - 12x^2y + 18xy = -2xy\left(8x^2 + 6x - 9\right)$$

Now we are left with one factor being $8x^2 + 6x - 9$, which might factor further. Using the AC Method:

1. $8 \cdot -9 = -72$

2. Examine factor pairs that multiply to -72, looking for a pair that sums to 6:

Factor Pair	Sum of the Pair	Factor Pair	Sum of the Pair
$1 \cdot -72$	-71	$-1 \cdot 72$	71
$2 \cdot -36$	-34	$-2 \cdot 36$	34
$3 \cdot -24$	-21	$-3 \cdot 24$	21
$4 \cdot -18$	-14	$-4 \cdot 18$	14
$6 \cdot -12$	-6 (close; wrong sign)	$-6 \cdot 12$	6 (what we wanted)
$8 \cdot -9$	-1	$-8 \cdot 9$	(no need to go this far)

3. Intentionally break up the 6 as $-6 + 12$:

$$-16x^3y - 12x^2y + 18xy = -2xy\left(8x^2 \overbrace{+ 6x} - 9\right)$$

$$= -2xy\left(8x^2 \overbrace{- 6x + 12x} - 9\right)$$

$$= -2xy\left((8x^2 - 6x) + (12x - 9)\right)$$

$$= -2xy\left(2x(4x - 3) + 3(4x - 3)\right)$$

$$= -2xy(4x - 3)(2x + 3)$$

So we believe that $-16x^3y - 12x^2y + 18xy$ factors as $-2xy(4x - 3)(2x + 3)$, and you should check by multiplying out the factored form.

Exercises

Review and Warmup Multiply the polynomials.

1. $(5t + 4)(6t + 8) =$ ⬚

2. $(3x + 9)(5x + 8) =$ ⬚

3. $(6x - 8)(3x - 4) =$ ⬚

4. $(5y - 4)(y - 3) =$ ⬚

5. $(4y - 10)(y + 9) =$ ⬚

6. $(10y - 6)(y + 7) =$ ⬚

7. $(7r^3 + 9)(r^2 + 4) =$ ⬚

8. $(4r^3 + 1)(r^2 + 7) =$ ⬚

Factoring Trinomials with a Nontrivial Leading Coefficient Factor the given polynomial.

9. $3t^2 + 22t + 7 =$ ⬚

10. $3t^2 + 17t + 10 =$ ⬚

11. $2x^2 + 5x - 7 =$ ⬚

12. $5x^2 + 3x - 14 =$ ⬚

13. $2y^2 - 15y + 7 =$ ⬚

14. $3y^2 - 7y + 2 =$ ⬚

15. $2y^2 + y + 1 =$ ⬚

16. $3r^2 + 3r + 7 =$ ⬚

17. $4r^2 + 25r + 6 =$ ⬚

18. $6t^2 + 7t + 1 =$ ⬚

19. $6t^2 + t - 7 =$ ⬚

20. $4x^2 + 5x - 9 =$ ⬚

21. $8x^2 - 21x + 10 =$ ⬚

22. $6x^2 - 11x + 5 =$ ⬚

23. $6y^2 + 11y + 3 =$ ⬚

24. $15y^2 + 29y + 8 =$ ☐

25. $9r^2 - 3r - 20 =$ ☐

26. $6r^2 + 7r - 20 =$ ☐

27. $12t^2 - 25t + 12 =$ ☐

28. $10t^2 - 17t + 3 =$ ☐

29. $6x^2 + 8x + 2 =$ ☐

30. $10x^2 + 25x + 15 =$ ☐

31. $15x^2 - 12x - 27 =$ ☐

32. $4y^2 - 14y - 18 =$ ☐

33. $6y^2 - 27y + 12 =$ ☐

34. $6r^2 - 20r + 16 =$ ☐

35. $4r^9 + 14r^8 + 6r^7 =$ ☐

36. $6t^9 + 15t^8 + 6t^7 =$ ☐

37. $4t^6 + 30t^5 - 16t^4 =$ ☐

38. $4x^7 - 14x^6 - 30x^5 =$ ☐

39. $6x^{10} - 27x^9 + 21x^8 =$ ☐

40. $6x^5 - 26x^4 + 24x^3 =$ ☐

41. $2y^2t^2 + 15yt + 18 =$ ☐

42. $2y^2r^2 + 19yr + 9 =$ ☐

43. $5r^2x^2 + rx - 4 =$ ☐

44. $2r^2y^2 - 13ry - 7 =$ ☐

45. $3t^2y^2 - 4ty + 1 =$ ☐

46. $2t^2x^2 - 17tx + 8 =$ ☐

47. $3x^2 + 20xr + 12r^2 =$ ☐

48. $5x^2 + 7xy + 2y^2 =$ ☐

49. $5x^2 - 8xt - 21t^2 =$ ☐

50. $5y^2 - 14yr - 3r^2 =$ ☐

51. $3y^2 - 10yx + 8x^2 =$ ☐

52. $2r^2 - 13rt + 18t^2 =$ ☐

53. $8r^2 + 17rx + 9x^2 =$ ☐

54. $6t^2 + 7tx + x^2 =$ ☐

55. $4t^2 - 11tx - 3x^2 =$ ☐

56. $4x^2 - 3xr - r^2 =$ ☐

57. $4x^2 - 17xr + 4r^2 =$ ☐

58. $4x^2 - 9xt + 2t^2 =$ ☐

59. $8y^2 + 14yr + 3r^2 =$ ☐

60. $12y^2 + 17yx + 6x^2 =$ ☐

61. $15r^2 + 2rt - 24t^2 =$ ☐

62. $25r^2 - y^2 =$ ☐

63. $8t^2 - 22tx + 15x^2 =$ ☐

64. $8t^2 - 14tr + 5r^2 =$ ☐

65. $10x^2y^2 + 18xy + 8 =$ ☐

66. $10x^2y^2 + 34xy + 12 =$ ☐

67. $6x^2r^2 + 3xr - 18 =$ ☐

68. $6y^2t^2 - 9yt - 6 =$ ☐

69. $21y^{10}x^2 - 35y^9x + 14y^8 = \boxed{}$ **70.** $25r^6x^2 - 35r^5x + 10r^4 = \boxed{}$ **71.** $10x^2 + 25xy + 15y^2 = \boxed{}$

72. $9x^2 + 24xy + 12y^2 = \boxed{}$ **73.** $4a^2 + 6ab - 4b^2 = \boxed{}$ **74.** $6a^2 - 9ab - 15b^2 = \boxed{}$

75. $10x^2 - 15xy + 5y^2 = \boxed{}$ **76.** $4x^2 - 18xy + 20y^2 = \boxed{}$ **77.** $6x^2y + 26xy^2 + 20y^3 = \boxed{}$

78. $6x^2y + 38xy^2 + 12y^3 = \boxed{}$

79. $12x^2(y - 3) + 20x(y - 3) + 8(y - 3) = \boxed{}$

80. $10x^2(y + 3) + 34x(y + 3) + 12(y + 3) = \boxed{}$

81. $6x^2(y - 6) + 15x(y - 6) + 6(y - 6) = \boxed{}$

82. $15x^2(y + 7) + 18x(y + 7) + 3(y + 7) = \boxed{}$

83.a. Factor the given polynomial.

$2x^2 + 17x + 21 = \boxed{}$

b. Use your previous answer to factor

$2(y - 1)^2 + 17(y - 1) + 21 = \boxed{}$

84.a. Factor the given polynomial.

$5x^2 + 11x + 6 = \boxed{}$

b. Use your previous answer to factor

$5(y + 7)^2 + 11(y + 7) + 6 = \boxed{}$

Challenge

85. What integers can go in the place of b so that the quadratic expression $3x^2 + bx + 4$ is factorable?

7.5 Factoring Special Polynomials

Certain polynomials have patterns that you can train yourself to recognize. And when they have these patterns, there are formulas you can use to factor them, much more quickly than using the techniques from Section 7.3 and Section 7.4.

7.5.1 Difference of Squares

If b is some positive integer, then when you multiply $(x - b)(x + b)$:

$$(x - b)(x + b) = x^2 - bx + bx - b^2$$
$$= x^2 - b^2.$$

The $-bx$ and the $+bx$ cancel each other out. So this is telling us that

$$x^2 - b^2 = (x - b)(x + b).$$

And so if we ever encounter a polynomial of the form $x^2 - b^2$ (a "difference of squares") then we have a quick formula for factoring it. Just identify what "b" is, and use that in $(x - b)(x + b)$.

To use this formula, it's important to recognize which numbers are perfect squares, as in Table 1.3.7.

Example 7.5.2 Factor $x^2 - 16$.

Explanation. The "16" being subtracted here is a perfect square. It is the same as 4^2. So we can take $b = 4$ and write:

$$x^2 - 16 = (x - b)(x + b)$$
$$= (x - 4)(x + 4)$$

⭐ **Checkpoint 7.5.3.** Try to factor one yourself:

Factor $x^2 - 49$.

Explanation. The "49" being subtracted here is a perfect square. It is the same as 7^2. So we can take $b = 7$ and write:

$$x^2 - 49 = (x - b)(x + b)$$
$$= (x - 7)(x + 7)$$

We can do a little better. There is nothing special about starting with "x^2" in these examples. In full generality:

Fact 7.5.4 The Difference of Squares Formula. *If A and B are any algebraic expressions, then:*

$$A^2 - B^2 = (A - B)(A + B).$$

Example 7.5.5 Factor $1 - p^2$.

Explanation. The "1" at the beginning of this expression is a perfect square; it's the same as 1^2. The "p^2" being subtracted here is also perfect square. We can take $A = 1$ and $B = p$, and use The Difference

---

of Squares Formula:

$$1 - p^2 = (A - B)(A + B)$$
$$= (1 - p)(1 + p)$$

Example 7.5.6 Factor $m^2 n^2 - 4$.

Explanation. Is the "$m^2 n^2$" at the beginning of this expression a perfect square? By the rules for exponents, it is the same as $(mn)^2$, so yes, it is a perfect square and we may take $A = mn$. The "4" being subtracted here is also perfect square. We can take $B = 2$. The Difference of Squares Formula tells us:

$$m^2 n^2 - 4 = (A - B)(A + B)$$
$$= (mn - 2)(mn + 2)$$

Checkpoint 7.5.7. Try to factor one yourself:

Factor $4z^2 - 9$.

Explanation. The "$4z^2$" at the beginning here is a perfect square. It is the same as $(2z)^2$. So we can take $A = 2z$. The "9" being subtracted is also a perfect square, so we can take $B = 3$:

$$4z^2 - 9 = (A - B)(A + B)$$
$$= (2z - 3)(2z + 3)$$

Example 7.5.8 Factor $x^6 - 9$.

Explanation. Is the "x^6" at the beginning of this expression is a perfect square? It may appear to be a *sixth* power, but it is *also* a perfect square because we can write $x^6 = \left(x^3\right)^2$. So we may take $A = x^3$. The "9" being subtracted here is also perfect square. We can take $B = 3$. The Difference of Squares Formula tells us:

$$x^6 - 9 = (A - B)(A + B)$$
$$= (x^3 - 3)(x^3 + 3)$$

Warning 7.5.9. It's a common mistake to write something like $x^2 + 16 = (x + 4)(x - 4)$. This is not what The Difference of Squares Formula allows you to do, and this is in fact incorrect. The issue is that $x^2 + 16$ is a *sum* of squares, not a *difference*. And it happens that $x^2 + 16$ is actually prime. In fact, any sum of squares without a common factor will always be prime.

7.5.2 Perfect Square Trinomials

If we expand $(A + B)^2$:

$$(A + B)^2 = (A + B)(A + B)$$
$$= A^2 + BA + AB + B^2$$
$$= A^2 + 2AB + B^2.$$

The BA and the AB equal each other and double up when added together. So this is telling us that

$$A^2 + 2AB + B^2 = (A + B)^2.$$

And so if we ever encounter a polynomial of the form $A^2 + 2AB + B^2$ (a "perfect square trinomial") then we have a quick formula for factoring it.

The tricky part is recognizing when a trinomial you have encountered is in this special form. Ask yourself:

1. Is the first term a perfect square? If so, jot down what A would be.

2. Is the second term a perfect square? If so, jot down what B would be.

3. When you multiply 2 with what you wrote down for A and B, i.e. $2AB$, do you have the middle term? If you have this middle term exactly, then your polynomial factors as $(A + B)^2$. If the middle term is the negative of $2AB$, then the sign on your B can be reversed, and your polynomial factors as $(A - B)^2$.

Fact 7.5.10 The Perfect Square Trinomial Formula. *If A and B are any algebraic expressions, then:*

$$A^2 + 2AB + B^2 = (A + B)^2$$

and

$$A^2 - 2AB + B^2 = (A - B)^2$$

Example 7.5.11 Factor $x^2 + 6x + 9$.

Explanation. The first term, x^2, is clearly a perfect square. So we could take $A = x$. The last term, 9, is also a perfect square since it is equal to 3^2. So we could take $B = 3$. Now we multiply $2AB = 2 \cdot x \cdot 3$, and the result is $6x$. This is the middle term, which is what we hope to see.

So we can use The Perfect Square Trinomial Formula:

$$x^2 + 6x + 9 = (A + B)^2$$
$$= (x + 3)^2$$

Example 7.5.12 Factor $4x^2 - 20xy + 25y^2$.

Explanation. The first term, $4x^2$, is a perfect square because it equals $(2x)^2$. So we could take $A = 2x$. The last term, $25y^2$, is also a perfect square since it is equal to $(5y)^2$. So we could take $B = 5y$. Now we multiply $2AB = 2 \cdot (2x) \cdot (5y)$, and the result is $20xy$. This is the *negative* of the middle term, which we can work with. The factored form will be $(A - B)^2$ instead of $(A + B)^2$.

So we can use The Perfect Square Trinomial Formula:

$$4x^2 - 20xy + 25y^2 = (A - B)^2$$
$$= (2x - 5y)^2$$

Checkpoint 7.5.13. Try to factor one yourself:

Factor $16q^2 + 56q + 49$.

Explanation. The first term, $16q^2$, is a perfect square because it equals $(4q)^2$. So we could take $A = 4q$. The last term, 49, is also a perfect square since it is equal to 7^2. So we could take $B = 7$. Now we multiply $2AB = 2 \cdot (4q) \cdot 7$, and the result is $56q$. This is the middle term, which is what we hope to see.

So we can use The Perfect Square Trinomial Formula:

$$16q^2 + 56q + 49 = (A + B)^2$$
$$= (4q + 7)^2$$

Warning 7.5.14. It is not enough to just see that the first and last terms are perfect squares. For example, $9x^2 + 10x + 25$ has its first term equal to $(3x)^2$ and its last term equal to 5^2. But when you examine $2 \cdot (3x) \cdot 5$ the result is $30x$, *not* equal to the middle term. So The Perfect Square Trinomial Formula doesn't apply here. In fact, this polynomial doesn't factor at all.

Remark 7.5.15. To factor these perfect square trinomials, we *could* use methods from Section 7.3 and Section 7.4. As an exercise for yourself, try to factor each of the three previous examples using those methods. The advantage to using The Perfect Square Trinomial Formula is that it is much faster. With some practice, all of the work for using it can be done mentally.

7.5.3 Difference/Sum of Cubes Formulas

The following calculation may seem to come from nowhere at first, but see it through. If we multiply $(A - B)\left(A^2 + AB + B^2\right)$:

$$\begin{aligned} (A - B)\left(A^2 + AB + B^2\right) &= A^3 - BA^2 + A^2B - BAB + AB^2 - B^3 \\ &= A^3 - A^2B + A^2B - AB^2 + AB^2 - B^3 \\ &= A^3 \overbrace{- A^2B + A^2B} \overbrace{- AB^2 + AB^2} - B^3 \\ &= A^3 - B^3. \end{aligned}$$

This is telling us that

$$A^3 - B^3 = (A - B)\left(A^2 + AB + B^2\right).$$

And so if we ever encounter a polynomial of the form $A^3 - B^3$ (a "difference of cubes") then we have a quick formula for factoring it.

A similar formula exists for factoring a *sum* of cubes, $A^3 + B^3$. Here are both formulas, followed by some tips on how to memorize them.

Fact 7.5.16 The Difference/Sum of Cubes Formula. *If A and B are any algebraic expressions, then:*

$$A^3 - B^3 = (A - B)\left(A^2 + AB + B^2\right)$$

and

$$A^3 + B^3 = (A + B)\left(A^2 - AB + B^2\right)$$

To memorize this, focus on:

- The factorization is a binomial times a trinomial.

- The sign you start with appears again in the binomial.

- The three terms in the trinomial are all quadratic: A^2, AB, and B^2.

- In the trinomial, it's *always* adding A^2 and B^2. But the sign on AB is always the *opposite* of the sign in the sum/difference of cubes.

> **Differences of cubes *are* sums of cubes.** Technically a difference of cubes, $A^3 - B^3$, is equal to $A^3 + (-B)^3$. So you can treat any difference of cubes as a sum of cubes, $A^3 + B^3$, where B is negative and you only need to memorize the *sum* of cubes formula.

To use these formulas effectively, we need to recognize when numbers are perfect cubes. Perfect cubes become large fast before you can list too many of them. Try to memorize as many of these as you can:

$$1^3 = 1 \qquad\qquad 2^3 = 8 \qquad\qquad 3^3 = 27 \qquad\qquad 4^3 = 64$$

$$5^3 = 125 \qquad 6^3 = 216 \qquad 7^3 = 343 \qquad 8^3 = 512$$
$$9^3 = 729 \qquad 10^3 = 1000 \qquad 11^3 = 1331 \qquad 12^3 = 1728$$

Let's look at a few examples.

Example 7.5.17 Factor $x^3 - 27$.

Explanation. We recognize that x^3 is a perfect cube and $27 = 3^3$, so we can use The Difference/Sum of Cubes Formula to factor the binomial. Note that since we have a *difference* of cubes, the binomial factor from the formula will *subtract* two terms, and the middle term from the trinomial will be $+AB$.

$$
\begin{aligned}
x^3 - 27 &= (A - B)\left(A^2 + AB \quad + B^2\right) \\
&= (x - 3)\left(x^2 \quad + (x)(3) + 3^2\right) \\
&= (x - 3)\left(x^2 \quad + 3x \quad + 9\right)
\end{aligned}
$$

Example 7.5.18 Factor $27m^3 + 64n^3$.

Explanation. We recognize that $27m^3 = (3m)^3$ and $64n^3 = (4n)^3$ are both perfect cubes, so we can use The Difference/Sum of Cubes Formula to factor the binomial. Note that since we have a *sum* of cubes, the binomial factor from the formula will *add* two terms, and the middle term from the trinomial will be $-AB$.

$$
\begin{aligned}
27m^3 + 64n^3 &= (A \ + B)\left(A^2 \quad - AB \quad + B^2\right) \\
&= (3m + 4n)\left((3m)^2 - (3m)(4n) + (4n)^2\right) \\
&= (3m + 4n)\left(9m^2 \quad - 12mn \quad + 16n^2\right)
\end{aligned}
$$

Checkpoint 7.5.19. Try to factor one yourself:

Factor $y^3 + 1000$.

Explanation. We recognize that y^3 is a perfect cube and $1000 = 10^3$, so we can use The Difference/Sum of Cubes Formula to factor the binomial. Note that since we have a *sum* of cubes, the binomial factor from the formula will *add* two terms, and the middle term from the trinomial will be $-AB$.

$$
\begin{aligned}
y^3 + 1000 &= (A + B)\left(A^2 - AB + B^2\right) \\
&= (y + 10)\left(y^2 - (y)(10) + 10^2\right) \\
&= (y + 10)\left(y^2 - 10y + 100\right)
\end{aligned}
$$

7.5.4 Factoring in Stages

Sometimes factoring a polynomial will take two or more "stages." For instance you might use one of the special formulas from this section to factor something into two factors, and *then* those factors might be factor even more. When the task is to *factor* a polynomial, the intention is that you *fully* factor it, breaking down the pieces into even smaller pieces when that is possible.

Example 7.5.20 Factor out any greatest common factor. Factor $12z^3 - 27z$.

Explanation. The two terms of this polynomial have greatest common factor $3z$, so the first step in factoring should be to factor this out:

$$3z\left(4z^2 - 9\right).$$

Now we have two factors. There is nothing for us to do with $3z$, but we should ask if $\left(4z^2 - 9\right)$ can factor further. And in fact, that is a difference of squares. So we can apply The Difference of Squares Formula. The full process would be:

$$
\begin{aligned}
12z^3 - 27z &= 3z\left(4z^2 - 9\right)\\
&= 3z(2z - 3)(2z + 3)
\end{aligned}
$$

Example 7.5.21 Recognize a *second* special pattern. Factor $p^4 - 1$.

Explanation. Since p^4 is the same as $\left(p^2\right)^2$, we have a difference of squares here. We can apply The Difference of Squares Formula:

$$p^4 - 1 = \left(p^2 - 1\right)\left(p^2 + 1\right)$$

It doesn't end here. Of the two factors we found, $\left(p^2 + 1\right)$ cannot be factored further. But the other one, $\left(p^2 - 1\right)$ is *also* a difference of squares. So we should apply The Difference of Squares Formula again:

$$= (p - 1)(p + 1)\left(p^2 + 1\right)$$

Checkpoint 7.5.22. Try to factor one yourself:

Factor $3x^3 - 3$.

Explanation. The two terms of this polynomial have greatest common factor 3, so the first step in factoring should be to factor this out:

$$3\left(x^3 - 1\right).$$

Now we have two factors. There is nothing for us to do with 3, but we should ask if $\left(x^3 - 1\right)$ can factor further. And in fact, that is a difference of cubes. So we can apply The Difference/Sum of Cubes Formula. The full process would be:

$$
\begin{aligned}
3x^3 - 3 &= 3\left(x^3 - 1\right)\\
&= 3(x - 1)(x^2 + x + 1)
\end{aligned}
$$

The trinomial from The Difference/Sum of Cubes Formula will *never* factor further. However, in some cases, the binomial from that formula will factor further, and you should look for this.

Example 7.5.23 Factor $64p^6 - 729$.

Explanation. We recognize that $64p^6$ is a perfect square because it equals $\left(8p^3\right)^2$. And 729 is also a perfect square because $729 = 27^2$. So we can use The Difference of Squares Formula to factor the binomial.

$$
\begin{aligned}
64p^6 - 729 &= (A - B)(A + B)\\
&= \left(8p^3 - 27\right)\left(8p^3 + 27\right)
\end{aligned}
$$

Next, note that in each of the factors we have an $8p^3$ and a 27, both of which can be viewed as perfect cubes: $8p^3 = (2p)^3$ and a $27 = 3^3$. So we will use *both* the difference of cubes and the sum of cubes formulas.

$$= (A - B)\left(A^2 + AB + B^2\right)(A + B)\left(A^2 - AB + B^2\right)$$
$$= (2p - 3)\left((2p)^2 + (2p) \cdot 3 + 3^2\right)(2p + 3)\left((2p)^2 - (2p) \cdot 3 + 3^2\right)$$
$$= (2p - 3)\left(4p^2 + 6p + 9\right)(2p + 3)\left(4p^2 - 6p + 9\right)$$

Example 7.5.24 Factor $32x^6y^2 - 48x^5y + 18x^4$.

Explanation. The first step of factoring any polynomial is to factor out the common factor if possible. For this trinomial, the common factor is $2x^4$, so we write

$$32x^6y^2 - 48x^5y + 18x^4 = 2x^4(16x^2y^2 - 24xy + 9).$$

The square numbers 16 and 9 in $16x^2y^2 - 24xy + 9$ hint that maybe we could use The Perfect Square Trinomial Formula. Taking $A = 4xy$ and $B = 3$, we multiply $2AB = 2 \cdot (4xy) \cdot 3$. The result is $24xy$, which is the negative of our middle term. So the whole process is:

$$32x^6y^2 - 48x^5y + 18x^4 = 2x^4(16x^2y^2 - 24xy + 9)$$
$$= 2x^4(a - b)^2$$
$$= 2x^4(4xy - 3)^2$$

Exercises

Review and Warmup Expand the square of a *binomial*.

1. $(5t + 10)^2 = $ ☐

2. $(2t + 4)^2 = $ ☐

3. $(x - 4)^2 = $ ☐

4. $(x - 7)^2 = $ ☐

5. $\left(y^7 + 13\right)^2 = $ ☐

6. $\left(y^{10} + 4\right)^2 = $ ☐

Multiply the polynomials.

7. $(r - 6)(r + 6) = $ ☐

8. $(r + 12)(r - 12) = $ ☐

9. $(4t - 5)(4t + 5) = $ ☐

10. $(3t + 3)(3t - 3) = $ ☐

11. $\left(6t^5 - 11\right)\left(6t^5 + 11\right) = $ ☐

12. $\left(4x^6 - 13\right)\left(4x^6 + 13\right) = $ ☐

Factoring Factor the given polynomial.

13. $x^2 - 9 = $ ☐

14. $y^2 - 121 = $ ☐

15. $49y^2 - 81 = $ ☐

16. $4r^2 - 121 =$ [   ]

17. $r^2t^2 - 49 =$ [   ]

18. $t^2y^2 - 144 =$ [   ]

19. $36t^2x^2 - 1 =$ [   ]

20. $121t^2y^2 - 64 =$ [   ]

21. $36 - x^2 =$ [   ]

22. $1 - x^2 =$ [   ]

23. $4 - 81y^2 =$ [   ]

24. $49 - 25y^2 =$ [   ]

25. $r^4 - 4 =$ [   ]

26. $r^4 - 25 =$ [   ]

27. $25t^4 - 121 =$ [   ]

28. $144t^4 - 49 =$ [   ]

29. $t^{10} - 64 =$ [   ]

30. $x^6 - 25 =$ [   ]

31. $16x^4 - 9y^4 =$ [   ]

32. $25x^4 - 64y^4 =$ [   ]

33. $x^6 - 36y^{10} =$ [   ]

34. $x^{14} - 49y^{10} =$ [   ]

35. $r^2 + 14r + 49 =$ [   ]

36. $t^2 + 6t + 9 =$ [   ]

37. $t^2 - 22t + 121 =$ [   ]

38. $t^2 - 14t + 49 =$ [   ]

39. $9x^2 + 6x + 1 =$ [   ]

40. $100x^2 + 20x + 1 =$ [   ]

41. $49y^2 - 14y + 1 =$ [   ]

42. $9y^2 - 6y + 1 =$ [   ]

43. $36r^2t^2 - 12rt + 1 =$ [   ]

44. $121r^2y^2 - 22ry + 1 =$ [   ]

45. $t^2 + 8tx + 16x^2 =$ [   ]

46. $t^2 + 18tr + 81r^2 =$ [   ]

47. $t^2 - 4ty + 4y^2 =$ [   ]

48. $x^2 - 2xr + r^2 =$ [   ]

49. $49x^2 + 84xr + 36r^2 =$ [   ]

50. $4y^2 + 20yt + 25t^2 =$ [   ]

51. $16y^2 - 24yt + 9t^2 =$ [   ]

52. $4r^2 - 12rx + 9x^2 =$ [   ]

53. $r^3 + 125 =$ [   ]

54. $t^3 + 1728 =$ [   ]

Hint: $1728 = 12^3$.

55. $t^3 - 512 =$ [   ]

Hint: $-512 = -8^3$.

56. $t^3 - 64 =$ [   ]

57. $1000x^3 + 1 =$ [   ]

Hint: $1000 = 10^3$.

58. $343x^3 + 1 =$ ☐

Hint: $343 = 7^3$.

59. $27y^3 - 1 =$ ☐

60. $729y^3 - 1 =$ ☐

Hint: $729 = 9^3$.

61. $216r^3 + 125 =$ ☐

Hint: $216 = 6^3$.

62. $27r^3 + 125 =$ ☐

63. $729t^3 - 512 =$ ☐

Hint: $729 = 9^3$, $-512 = -8^3$.

64. $216t^3 - 343 =$ ☐

Hint: $216 = 6^3$, $-343 = -7^3$.

65. $x^3y^3 + 1728 =$ ☐

Hint: $1728 = 12^3$.

66. $x^3y^3 + 8 =$ ☐

67. $x^3 - 64y^3 =$ ☐

68. $x^3 - 125y^3 =$ ☐

69. $81y^4 - 16 =$ ☐

70. $16r^4 - 81 =$ ☐

71. $3r^2 - 3 =$ ☐

72. $4r^2 - 64 =$ ☐

73. $9t^3 - 36t =$ ☐

74. $3t^3 - 147t =$ ☐

75. $9x^4r^3 - 81x^2r =$ ☐

76. $7x^4y^4 - 63x^2y^2 =$ ☐

77. $108 - 12y^2 =$ ☐

78. $128 - 2y^2 =$ ☐

79. $80r^2 + 40r + 5 =$ ☐

80. $63r^2 + 42r + 7 =$ ☐

81. $64r^2y^2 + 32ry + 4 =$ ☐

82. $98t^2x^2 + 28tx + 2 =$ ☐

83. $96t^2 - 48t + 6 =$ ☐

84. $24x^2 - 24x + 6 =$ ☐

85. $9x^{10} + 6x^9 + x^8 =$ ☐

86. $121y^6 + 22y^5 + y^4 =$ ☐

87. $49y^9 - 14y^8 + y^7 =$ ☐

88. $9r^5 - 6r^4 + r^3 =$ ☐

89. $18r^8 + 12r^7 + 2r^6 =$ ☐

90. $72r^6 + 24r^5 + 2r^4 =$ ☐

91. $20t^4 - 20t^3 + 5t^2 =$ ☐

92. $36t^9 - 36t^8 + 9t^7 =$ ☐

93. $125x^4 + 64x =$ ☐

94. $8x^4 + 125x =$ ☐

95. $64x^3 + 125y^3 =$ ☐

96. $125x^3 + 216y^3 =$ ☐

Hint: $216 = 6^3$.

97. $3r^4 - 243 = \boxed{}$ **98.** $3r^4 - 48 = \boxed{}$ **99.** $r^{10} + 125r^7 = \boxed{}$

100. $t^6 + 1728t^3 = \boxed{}$ **101.** $t^2 + 64 = \boxed{}$ **102.** $x^2 + 16 = \boxed{}$
    Hint: $1728 = 12^3$.

103. $5x^3 + 20x = \boxed{}$ **104.** $4y^3 + 100y = \boxed{}$ **105.** $0.09y - y^3 = \boxed{}$

106. $0.81r - r^3 = \boxed{}$ **107.** $(r - 8)^2 - 81 = \boxed{}$ **108.** $(r - 7)^2 - 49 = \boxed{}$

109. $x^2 + 14x + 49 - 121y^2 = \boxed{}$ **110.** $x^2 + 20x + 100 - 49y^2 = \boxed{}$

7.6 Factoring Strategies

7.6.1 Factoring Strategies

Deciding which method to use when factoring a random polynomial can seem like a daunting task. Understanding all of the techniques that we have learned and how they fit together can be done using a decision tree.

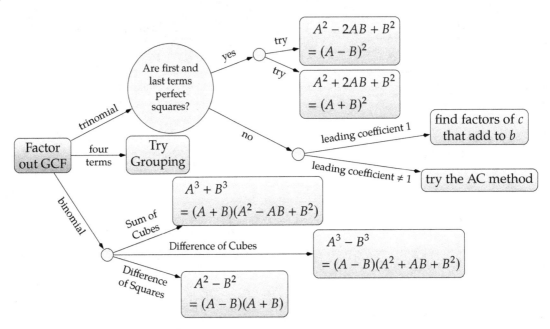

Figure 7.6.2: Factoring Decision Tree

Using the decision tree can guide us when we are given an expression to factor.

Example 7.6.3 Factor the expression $4k^2 + 12k - 40$ completely.

Explanation. Start by noting that the GCF is 4. Factoring this out, we get

$$4k^2 + 12k - 40 = 4\left(k^2 + 3k - 10\right).$$

Following the decision tree, we now have a trinomial where the leading coefficient is 1 and we need to look for factors of -10 that add to 3. We find that -2 and 5 work. So, the full factorization is:

$$4k^2 + 12k - 40 = 4\left(k^2 + 3k - 10\right)$$
$$= 4(k - 2)(k + 5)$$

Example 7.6.4 Factor the expression $64d^2 + 144d + 81$ completely.

Explanation. Start by noting that the GCF is 1, and there is no GCF to factor out. We continue along the decision tree for a trinomial. Notice that both 64 and 81 are perfect squares and that this expression might factor using the pattern $A^2 + 2AB + B^2 = (A + B)^2$. To find A and B, take the square roots of the first and last terms, so $A = 8d$ and $B = 9$. We have to check that the middle term is correct: since

$2AB = 2(8d)(9) = 144d$ matches our middle term, the expression must factor as

$$64d^2 + 144d + 81 = (8d + 9)^2.$$

Example 7.6.5 Factor the expression $10x^2y - 12xy^2$ completely.

Explanation. Start by noting that the GCF is $2xy$. Factoring this out, we get

$$10x^2y - 12xy^2 = 2xy(5x - 6y).$$

Since we have a binomial inside the parentheses, the only options on the decision tree for a binomial involve squares or cubes. Since there are none, we conclude that $2xy(5x - 6y)$ is the complete factorization.

Example 7.6.6 Factor the expression $9b^2 - 25y^2$ completely.

Explanation. Start by noting that the GCF is 1, and there is no GCF to factor out. We continue along the decision tree for a binomial and notice that we now have a difference of squares, $A^2 - B^2 = (A - B)(A + B)$. To find the values for A and B that fit the patterns, just take the square roots. So $A = 3b$ since $(3b)^2 = 9b^2$ and $B = 5y$ since $(5y)^2 = 25y^2$. So, the expression must factor as

$$9b^2 - 25y^2 = (3b - 5y)(3b + 5y).$$

Example 7.6.7 Factor the expression $24w^3 + 6w^2 - 9w$ completely.

Explanation. Start by noting that the GCF is $3w$. Factoring this out, we get

$$24w^3 + 6w^2 - 9w = 3w\left(8w^2 + 2w - 3\right).$$

Following the decision tree, we now have a trinomial inside the parentheses where $a \neq 1$. We should try the AC method because neither 8 nor -3 are perfect squares. In this case, $ac = -24$ and we must find two factors of -24 that add to be 2. The numbers 6 and -4 work in this case. The rest of the factoring process is:

$$\begin{aligned}
24w^3 + 6w^2 - 9w &= 3w\left(8w^2 + \overbrace{2w} - 3\right) \\
&= 3w\left(8w^2 + \overbrace{6w - 4w} - 3\right) \\
&= 3w\left(\left(8w^2 + 6w\right) + (-4w - 3)\right) \\
&= 3w\left(2w(4w + 3) - 1(4w + 3)\right) \\
&= 3w(4w + 3)(2w - 1)
\end{aligned}$$

Example 7.6.8 Factor the expression $q^5 + q^2$ completely.

Explanation. Start by noting that the GCF is q^2. Factoring this out, we find

$$q^5 + q^2 = q^2\left(q^3 + 1\right).$$

Following the decision tree, we now have a binomial with a sum of cubes. (Notice that $1^3 = 1$.) So using the sum of cubes formula, we have the complete factorization:

$$q^5 + q^2 = q^2 \left(q^3 + 1\right)$$
$$= q^2(q + 1)\left(q^2 - q + 1\right).$$

Example 7.6.9 Factor the expression $-6xy + 9y + 2x - 3$ completely.

Explanation. Start by noting that the GCF is 1, and there is no GCF to factor out. We continue along the decision tree. Since we have a four-term polynomial, we should try to factor by grouping. The full process is:

$$-6xy + 9y + 2x - 3 = (-6xy + 9y) + (2x - 3)$$
$$= -3y(2x - 3) + 1(2x - 3)$$
$$= (2x - 3)(-3y + 1)$$

Note that the negative sign in front of the $3y$ can be factored out if you wish. That would look like:

$$= -(2x - 3)(3y - 1)$$

Example 7.6.10 Factor the expression $4w^3 - 20w^2 + 24w$ completely.

Explanation. Start by noting that the GCF is $4w$. Factoring this out, we get

$$4w^3 - 20w^2 + 24w = 4w \left(w^2 - 5w + 6\right).$$

Following the decision tree, we now have a trinomial with $a = 1$ inside the parentheses. So, we can look for factors of 6 that add up to -5. Since -3 and -2 fit the requirements, the full factorization is:

$$4w^3 - 20w^2 + 24w = 4w \left(w^2 - 5w + 6\right)$$
$$= 4w(w - 3)(w - 2)$$

Example 7.6.11 Factor the expression $9 - 24y + 16y^2$ completely.

Explanation. Start by noting that the GCF is 1, and there is no GCF to factor out. Continue along the decision tree. We now have a trinomial where both the first term, 9, and last term, $16y^2$, look like perfect squares. To use the perfect squares difference pattern, $A^2 - 2AB + B^2 = (A - B)^2$, recall that we need to mentally take the square roots of these two terms to find A and B. So, $A = 3$ since $3^2 = 9$, and $B = 4y$ since $(4y)^2 = 16y^2$. Now we have to check that $2AB$ matches $24y$:

$$2AB = 2(3)(4y) = 24y.$$

So the full factorization is:

$$9 - 24y + 16y^2 = (3 - 4y)^2.$$

Example 7.6.12 Factor the expression $9 - 25y + 16y^2$ completely.

Explanation. Start by noting that the GCF is 1, and there is no GCF to factor out. Since we now have a trinomial where both the first term and last term are perfect squares in exactly the same way as in

Example 11. However, we cannot apply the perfect squares method to this problem because it worked when $2AB = 24y$. Since our middle term is $25y$, we can be certain that it won't be a perfect square.

Continuing on with the decision tree, our next option is to use the AC method. You might be tempted to rearrange the order of the terms, but that is unnecessary. In this case, $ac = 144$ and we need to come up with two factors of 144 that add to be -25. After a brief search, we conclude that those values are -16 and -9. The remainder of the factorization is:

$$9 \overbrace{- 25y} + 16y^2 = 9 \overbrace{- 16y - 9y} + 16y^2$$
$$= (9 - 16y) + (-9y + 16y^2)$$
$$= 1(9 - 16y) - y(9 + 16y)$$
$$= (9 - 16y)(1 - y)$$

Example 7.6.13 Factor the expression $20x^4 + 13x^3 - 21x^2$ completely.

Explanation. Start by noting that the GCF is x^2. Factoring this out, we get

$$20x^4 + 13x^3 - 21x^2 = x^2 \left(20x^2 + 13x - 21\right).$$

Following the decision tree, we now have a trinomial inside the parentheses where $a \neq 1$ and we should try the AC method. In this case, $ac = -420$ and we need factors of -420 that add to 13.

Factor Pair	Sum		Factor Pair	Sum		Factor Pair	Sum
$1 \cdot -420$	-419		$5 \cdot -84$	-79		$12 \cdot -35$	-23
$2 \cdot -210$	-208		$6 \cdot -70$	-64		$14 \cdot -30$	-16
$3 \cdot -140$	-137		$7 \cdot -60$	-53		$15 \cdot -28$	-13
$4 \cdot -105$	-101		$10 \cdot -42$	-32		$20 \cdot -21$	-1

In the table of the factor pairs of -420 we find $15 + (-28) = -13$, the opposite of what we want, so we want the opposite numbers: -15 and 28. The rest of the factoring process is shown:

$$20x^4 + 13x^3 - 21x^2 = x^2 \left(20x^2 \overbrace{+ 13x} -21\right)$$
$$= x^2 \left(20x^2 \overbrace{-15x + 28x} -21\right)$$
$$= x^2 \left((20x^2 - 15x) + (28x - 21)\right)$$
$$= x^2 \left(5x(4x - 3) + 7(4x - 3)\right)$$
$$= x^2(4x - 3)(5x + 7)$$

Exercises

Strategies

1. In factoring $7x^3 - 2401y^3$, which factoring techniques/tools will be useful? Check all that apply.

   □ Factoring out a GCF   □ Factoring by grouping   □ Finding two numbers that multiply to c and sum to b □ The AC Method   □ Difference of Squares   □ Difference of Cubes   □ Sum of Cubes   □ Perfect Square Trinomial   □ None of the above

2. In factoring $-7r - 4$, which factoring techniques/tools will be useful? Check all that apply.

   □ Factoring out a GCF   □ Factoring by grouping   □ Finding two numbers that multiply to c and sum to b □ The AC Method   □ Difference of Squares   □ Difference of Cubes   □ Sum of Cubes   □ Perfect Square Trinomial   □ None of the above

3. In factoring $8a^2 - 8$, which factoring techniques/tools will be useful? Check all that apply.

   □ Factoring out a GCF   □ Factoring by grouping   □ Finding two numbers that multiply to c and sum to b □ The AC Method   □ Difference of Squares   □ Difference of Cubes   □ Sum of Cubes   □ Perfect Square Trinomial   □ None of the above

4. In factoring $16b^2 + 8b + 1$, which factoring techniques/tools will be useful? Check all that apply.

   □ Factoring out a GCF   □ Factoring by grouping   □ Finding two numbers that multiply to c and sum to b □ The AC Method   □ Difference of Squares   □ Difference of Cubes   □ Sum of Cubes   □ Perfect Square Trinomial   □ None of the above

5. In factoring $A^2 - 16A + 64$, which factoring techniques/tools will be useful? Check all that apply.

   □ Factoring out a GCF   □ Factoring by grouping   □ Finding two numbers that multiply to c and sum to b □ The AC Method   □ Difference of Squares   □ Difference of Cubes   □ Sum of Cubes   □ Perfect Square Trinomial   □ None of the above

6. In factoring $2560B^3 + 3645$, which factoring techniques/tools will be useful? Check all that apply.

   □ Factoring out a GCF   □ Factoring by grouping   □ Finding two numbers that multiply to c and sum to b □ The AC Method   □ Difference of Squares   □ Difference of Cubes   □ Sum of Cubes   □ Perfect Square Trinomial   □ None of the above

7. In factoring $mx - m - 5x + 5$, which factoring techniques/tools will be useful? Check all that apply.

   □ Factoring out a GCF   □ Factoring by grouping   □ Finding two numbers that multiply to c and sum to b □ The AC Method   □ Difference of Squares   □ Difference of Cubes   □ Sum of Cubes   □ Perfect Square Trinomial   □ None of the above

8. In factoring $200n^2 - 32C^2$, which factoring techniques/tools will be useful? Check all that apply.

   □ Factoring out a GCF   □ Factoring by grouping   □ Finding two numbers that multiply to c and sum to b □ The AC Method   □ Difference of Squares   □ Difference of Cubes   □ Sum of Cubes   □ Perfect Square Trinomial   □ None of the above

9. In factoring $q^2 - 4qa + 4a^2$, which factoring techniques/tools will be useful? Check all that apply.

   □ Factoring out a GCF   □ Factoring by grouping   □ Finding two numbers that multiply to c and sum to b □ The AC Method   □ Difference of Squares   □ Difference of Cubes   □ Sum of Cubes   □ Perfect Square Trinomial   □ None of the above

10. In factoring $-36n - 6$, which factoring techniques/tools will be useful? Check all that apply.

    □ Factoring out a GCF   □ Factoring by grouping   □ Finding two numbers that multiply to c and sum to b □ The AC Method   □ Difference of Squares   □ Difference of Cubes   □ Sum of Cubes   □ Perfect Square Trinomial   □ None of the above

11. In factoring $9r^3 + 9C^3$, which factoring techniques/tools will be useful? Check all that apply.

    □ Factoring out a GCF   □ Factoring by grouping   □ Finding two numbers that multiply to c and sum to b □ The AC Method   □ Difference of Squares   □ Difference of Cubes   □ Sum of Cubes   □ Perfect Square Trinomial   □ None of the above

12. In factoring $56ta + 21t + 8a + 3$, which factoring techniques/tools will be useful? Check all that apply.

    □ Factoring out a GCF   □ Factoring by grouping   □ Finding two numbers that multiply to c and sum to b □ The AC Method   □ Difference of Squares   □ Difference of Cubes   □ Sum of Cubes   □ Perfect Square Trinomial   □ None of the above

13. In factoring $4b^4 - 4b$, which factoring techniques/tools will be useful? Check all that apply.

    □ Factoring out a GCF   □ Factoring by grouping   □ Finding two numbers that multiply to c and sum to b □ The AC Method   □ Difference of Squares   □ Difference of Cubes   □ Sum of Cubes   □ Perfect Square Trinomial   □ None of the above

14. In factoring $16A^2 + 72A + 81$, which factoring techniques/tools will be useful? Check all that apply.

    □ Factoring out a GCF   □ Factoring by grouping   □ Finding two numbers that multiply to c and sum to b □ The AC Method   □ Difference of Squares   □ Difference of Cubes   □ Sum of Cubes   □ Perfect Square Trinomial   □ None of the above

Factoring   Factor the given polynomial.

15. $4r + 4 = $

16. $-10r - 10 = $

17. $14t^2 - 35 = $

18. $30t^4 + 80t^3 + 80t^2 = $

19. $20x - 50x^2 + 90x^3 = $

20. $3xy + 3y = $

21. $64x^5y^8 - 24x^4y^8 + 72x^3y^8 = $

22. $y^2 + 9y + 8y + 72 = $

23. $xy - 6x - 10y + 60 = $

24. $x^3 - 7 - 5x^3y + 35y = $

25. $r^2 - 2r - 80 = $

26. $5t^2 - 23t - 10 = $

27. $3t^2x^2 + tx - 4 = $

28. $3x^2 + x + 8 = $

29. $6x^2 - 17x - 3 = $

30. $4y^2 + 16y + 7 = $

31. $9y^2 - 15y + 4 = $

32. $3y^2 + 5yr + 2r^2 = $

33. $2r^2 - 9rx + 9x^2 = $

34. $8r^2 - 17ry - 21y^2 = $

35. $20t^2 + 13ty + 2y^2 = $

36. $12t^2 - 7ty + y^2 = $

37. $4x^2 + 6x - 28 = $

38. $15x^2y^2 + 27xy - 6 = $

39. $8y^{10} + 28y^9 + 20y^8 = $

40. $10y^5 - 22y^4 + 4y^3 = $

41. $9x^2 + 24xy + 15y^2 = $

42. $9x^2 - 24xy + 12y^2 = $

43. $r^2 + 9r + 18 = $

44. $t^2 - 9t + 14 = $

45. $t^2 + 10tx + 24x^2 = $

46. $x^2y^2 - xy - 30 = $

47. $x^2 - 7xt + 10t^2 = $

48. $2y^2t^2 + 10yt + 8 = $

49. $3y^2 - 12y - 36 = $

50. $10y^7 + 30y^6 + 20y^5 = $

51. $10r^6 - 30r^5 + 20r^4 = $

52. $2x^2y + 10xy + 12y = $

53. $2x^2y - 16xy + 14y = $

54. $3x^2y^3 - 15xy^2 + 12y = $

55. $x^2y^2 + 2x^2yz - 3x^2z^2 = $

56. $x^2 + 0.9x + 0.14 = $

57. $y^2 - 64 = $ [   ]

58. $y^2 r^2 - 100 = $ [   ]

59. $36 - y^2 = $ [   ]

60. $r^4 - 64 = $ [   ]

61. $r^{14} - 100 = $ [   ]

62. $x^{14} - 121y^{10} = $ [   ]

63. $16t^4 - 81 = $ [   ]

64. $2x^3 - 18x = $ [   ]

65. $x^2 + 121 = $ [   ]

66. $40 - 10y^2 = $ [   ]

67. $y^2 + 6y + 9 = $ [   ]

68. $y^2 - 18yt + 81t^2 = $ [   ]

69. $r^2 - 12r + 36 = $ [   ]

70. $9r^2 - 6r + 1 = $ [   ]

71. $t^2 + 6tx + 9x^2 = $ [   ]

72. $16t^2 + 40ty + 25y^2 = $ [   ]

73. $48x^2 t^2 + 24xt + 3 = $ [   ]

74. $100x^5 + 20x^4 + x^3 = $ [   ]

75. $12y^{10} + 12y^9 + 3y^8 = $ [   ]

76. $y^3 + 8 = $ [   ]

77. $512y^3 + 1 = $ [   ]

Hint: $512 = 8^3$.

78. $125r^3 + 729 = $ [   ]

Hint: $729 = 9^3$.

79. $x^3 y^3 + 729 = $ [   ]

Hint: $729 = 9^3$.

80. $343t^4 + 512t = $ [   ]

Hint: $343 = 7^3, 512 = 8^3$.

81. $0.16t - t^3 = $ [   ]

82. $3x^4 - 243 = $ [   ]

83. $x^2 - 14x + 49 - 64y^2 = $ [   ]

84. $30x^2(y + 2) + 36x(y + 2) + 6(y + 2) = $ [   ]

Challenge

85. Select the expression which is equivalent to the following expression: $333^2 - 888^2$ (□ 1221(1221) □ 1221(-555) □ -555(-555) □ none of the above)

7.7 Factoring Chapter Review

7.7.1 Review of Factoring out the GCF

In Section 7.1 we covered how to factor out the greatest common factor. Recall that the **greatest common factor** between two expressions is the largest factor that goes in evenly to both expressions.

Example 7.7.1 Finding the Greatest Common Factor. What is the greatest common factor between $12x^3y$ and $42x^2y^2$?

Explanation. Break down each of these into its factors:

$$12x^3y = (2 \cdot 2) \cdot 3 \cdot (x \cdot x \cdot x) \cdot y \qquad 42x^2y^2 = 2 \cdot 3 \cdot 7 \cdot (x \cdot x) \cdot (y \cdot y)$$

Identify the common factors:

$$12x^3y = \overset{\downarrow}{2} \cdot 2 \cdot \overset{\downarrow}{3} \cdot \overset{\downarrow}{x} \cdot \overset{\downarrow}{x} \cdot x \cdot \overset{\downarrow}{y} \qquad 42x^2y^2 = \overset{\downarrow}{2} \cdot \overset{\downarrow}{3} \cdot 7 \cdot \overset{\downarrow}{x} \cdot \overset{\downarrow}{x} \cdot \overset{\downarrow}{y} \cdot y$$

With $2, 3$, two x's and a y in common, the greatest common factor is $6x^2y$.

Example 7.7.2 What is the greatest common factor between $18c^3y^2$ and $27y^3c$?

Explanation. Break down each into factors. You can definitely do this mentally with practice.

$$18c^3y^2 = 2 \cdot 3 \cdot 3 \cdot c \cdot c \cdot c \cdot y \cdot y \qquad 27y^3c = 3 \cdot 3 \cdot 3 \cdot y \cdot y \cdot y \cdot c$$

And take note of the common factors.

$$18c^3y^2 = 2 \cdot \overset{\downarrow}{3} \cdot \overset{\downarrow}{3} \cdot \overset{\downarrow}{c} \cdot c \cdot c \cdot \overset{\downarrow}{y} \cdot \overset{\downarrow}{y} \qquad 27y^3c = \overset{\downarrow}{3} \cdot \overset{\downarrow}{3} \cdot 3 \cdot \overset{\downarrow}{y} \cdot \overset{\downarrow}{y} \cdot y \cdot c$$

And so the GCF is $9y^2c$

Example 7.7.3 Factoring out the Greatest Common Factor. Factor out the GCF from the expression $32mn^2 - 24m^2n - 12mn$.

Explanation. To factor out the GCF from the expression $32mn^2 - 24m^2n - 12mn$, first note that the GCF to all three terms is $4mn$. Begin by writing that in front of a blank pair of parentheses and fill in the missing pieces.

$$32mn^2 - 24m^2n - 12mn = 4mn(\quad - \quad - \quad)$$
$$= 4mn(8n - 6m - 3)$$

Example 7.7.4 Factor out the GCF from the expression $14x^3 - 35x^2$.

Explanation. First note that the GCF of the terms in $14x^3 - 35x^2$ is $7x^2$. Factoring this out, we have:

$$14x^3 - 35x^2 = 7x^2(\quad - \quad)$$
$$= 7x^2(2x - 5)$$

Example 7.7.5 Factor out the GCF from the expression $36m^3n^2 - 18m^2n^5 + 24mn^3$.

Explanation. First note that the GCF of the terms in $36m^3n^2 - 18m^2n^5 + 24mn^3$ is $6mn^2$. Factoring this out, we have:

$$36m^3n^2 - 18m^2n^5 + 24mn^3 = 6mn^2\,(\quad - \quad + \quad)$$
$$= 6mn^2\left(6m^2 - 3mn^3 + 4n\right)$$

Example 7.7.6 Factor out the GCF from the expression $42f^3w^2 - 8w^2 + 9f^3$.

Explanation. First note that the GCF of the terms in $42f^3w^2 - 8w^2 + 9f^3$ is 1. The only way to factor the GCF out of this expression is:

$$42f^3w^2 - 8w^2 + 9f^3 = 1\left(42f^3w^2 - 8w^2 + 9f^3\right)$$

7.7.2 Review of Factoring by Grouping

In Section 7.2 we covered how to factor by grouping. Recall that factoring using grouping is used on four-term polynomials, and also later in the AC method in Section 7.4. Begin by grouping two pairs of terms and factoring out their respective GCF; if all is well, we should be left with two matching pieces in parentheses that can be factored out in their own right.

Example 7.7.7 Factor the expression $2x^3 + 5x^2 + 6x + 15$ using grouping.

Explanation.

$$2x^3 + 5x^2 + 6x + 15 = \left(2x^3 + 5x^2\right) + (6x + 15)$$
$$= x^2\,(2x + 5) + 3\,(2x + 5)$$
$$= \left(x^2 + 3\right)(2x + 5)$$

Example 7.7.8 Factor the expression $2xy - 3x - 8y + 12$ using grouping.

Explanation.

$$2xy - 3x + 8y - 12 = \left(2xy - 3x\right) + \left(-8y + 12\right)$$
$$= x\,(2y - 3) - 4\,(2y - 3)$$
$$= (x - 4)(2y - 3)$$

Example 7.7.9 Factor the expression $xy - 2 - 2x + y$ using grouping.

Explanation. This is a special example because if we try to simply follow the algorithm without considering the bigger context, we will fail:

$$xy - 2 - 2x + y = \left(xy - 2\right) + \left(-2x + y\right)$$

Note that there is no common factor in either grouping, besides 1, but the groupings themselves don't match. We should now recognize that whatever we are doing isn't working and try something else. It turns out that this polynomial *isn't* prime; all we need to do is rearrange the polynomial into standard form where the degrees decrease from left to right before grouping.

$$xy - 2 - 2x + y = xy - 2x + y - 2$$
$$= (xy - 2x) + (y - 2)$$
$$= x(y - 2) + 1(y - 2)$$
$$= (x + 1)(y - 2)$$

Example 7.7.10 Factor the expression $15m^2 - 3m - 10mn + 2n$ using grouping.

Explanation.

$$15m^2 - 3m - 10mn + 2n = (15m^2 - 3m) + (-10mn + 2n)$$
$$= 3m(5m - 1) - 2n(5m - 1)$$
$$= (3m - 2n)(5m - 1)$$

7.7.3 Review of Factoring Trinomials with Leading Coefficient 1

In Section 7.3 we covered factoring expressions that look like $x^2 + bx + c$. The trick was to look for two numbers whose product was c and whose sum was b. Always remember to look for a greatest common factor first, before looking for factor pairs.

Example 7.7.11 Answer the questions to practice for the factor pairs method.

  a. What two numbers multiply to be 6 and add to be 5?

  b. What two numbers multiply to be −6 and add to be 5?

  c. What two numbers multiply to be −6 and add to be −1?

  d. What two numbers multiply to be 24 and add to be −10?

  e. What two numbers multiply to be −24 and add to be 2?

  f. What two numbers multiply to be −24 and add to be −5?

  g. What two numbers multiply to be 420 and add to be 44?

  h. What two numbers multiply to be −420 and add to be −23?

  i. What two numbers multiply to be 420 and add to be −41?

Explanation.

  a. What two numbers multiply to be 6 and add to be 5? The numbers are 2 and 3.

  b. What two numbers multiply to be −6 and add to be 5? The numbers are 6 and −1.

  c. What two numbers multiply to be −6 and add to be −1? The numbers are −3 and 2.

  d. What two numbers multiply to be 24 and add to be −10? The numbers are −6 and −4.

  e. What two numbers multiply to be −24 and add to be 2? The numbers are 6 and −4.

  f. What two numbers multiply to be −24 and add to be −5? The numbers are −8 and 3.

g. What two numbers multiply to be 420 and add to be 44? The numbers are 30 and 14.

h. What two numbers multiply to be −420 and add to be −23? The numbers are −35 and 12.

i. What two numbers multiply to be 420 and add to be −41? The numbers are −20 and −21.

Note that for parts g–i, the factors of 420 are important. Below is a table of factors of 420 which will make it much clearer how the answers were found. To generate a table like this, we start with 1, and we work our way up the factors of 420.

Factor Pair	Factor Pair	Factor Pair
$1 \cdot 420$	$5 \cdot 84$	$12 \cdot 35$
$2 \cdot 210$	$6 \cdot 70$	$14 \cdot 30$
$3 \cdot 140$	$7 \cdot 60$	$15 \cdot 28$
$4 \cdot 105$	$10 \cdot 42$	$20 \cdot 21$

It is now much easier to see how to find the numbers in question. For example, to find two numbers that multiply to be −420 and add to be −23, simply look in the table for two factors that are 23 apart and assign a negative sign appropriately. As we found earlier, the numbers that are 23 apart are 12 and 35, and making the larger one negative, we have our answer: 12 and −35.

Example 7.7.12 Factor the expression $x^2 - 3x - 28$

Explanation. To factor the expression $x^2 - 3x - 28$, think of two numbers that multiply to be −28 and add to be −3. In the Section 7.3, we created a table of all possibilities of factors, like the one shown, to be sure that we never missed the right numbers; however, we encourage you to try this mentally for most problems.

Factor Pair	Sum of the Pair	Factor Pair	Sum of the Pair
$-1 \cdot 28$	27	$1 \cdot (-28)$	−27
$-2 \cdot 14$	12	$2 \cdot (-14)$	−12
$-4 \cdot 7$	3 (close; wrong sign)	$4 \cdot (-7)$	−3 (what we wanted)

Since the two numbers in question are 4 and −7 that means simply that

$$x^2 - 3x - 28 = (x + 4)(x - 7)$$

Remember that you can always multiply out your factored expression to verify that you have the correct answer. We will use the FOIL expansion.

$$(x + 4)(x - 7) = x^2 - 7x + 4x - 28$$
$$\overset{\checkmark}{=} x^2 - 3x - 28$$

Example 7.7.13 Factoring in Stages. Completely factor the expression $4x^3 - 4x^2 - 120x$.

Explanation. Remember that some expressions require more than one step to completely factor. To factor $4x^3 - 4x^2 - 120x$, first, always look for any GCF; after that is done, consider other options. Since

the GCF is $4x$, we have that
$$4x^3 - 4x^2 - 120x = 4x\left(x^2 - x - 30\right).$$

Now the factor inside parentheses might factor further. The key here is to consider what two numbers multiply to be -30 and add to be -1. In this case, the answer is -6 and 5. So, to completely write the factorization, we have:
$$\begin{aligned} 4x^3 - 4x^2 - 120x &= 4x\left(x^2 - x - 30\right) \\ &= 4x(x-6)(x+5) \end{aligned}$$

Example 7.7.14 Factoring Expressions with Higher Powers. Completely factor the expression $p^{10} - 6p^5 - 72$.

Explanation. If we have a trinomial with an even exponent on the leading term, and the middle term has an exponent that is half the leading term exponent, we can still use the factor pairs method. To factor $p^{10} - 6p^5 - 72$, we note that the middle term exponent 5 is half of the leading term exponent 10, and that two numbers that multiply to be -72 and add to be -6 are -12 and 6. So the factorization of the expression is
$$p^{10} - 6p^5 - 72 = \left(p^5 - 12\right)\left(p^5 + 6\right)$$

Example 7.7.15 Factoring Expressions with Two Variables. Completely factor the expression $x^2 - 3xy - 70y^2$.

Explanation. If an expression has two variables, like $x^2 - 3xy - 70y^2$, we pretend for a moment that the expression is $x^2 - 3x - 70$. To factor this expression we ask ourselves "what two numbers multiply to be -70 and add to be -3?" The two numbers in question are 7 and -10. So $x^2 - 3x - 70$ factors as $(x + 7)(x - 10)$.

To go back to the original problem now, simply make the two numbers $7y$ and $-10y$. So, the full factorization is
$$x^2 - 3xy - 70y^2 = (x + 7y)(x - 10y)$$

With problems like this, it is important to verify the your answer to be sure that all of the variables ended up where they were supposed to. So, to verify, simply FOIL your answer.
$$\begin{aligned} (x + 7y)(x - 10y) &= x^2 - 10xy + 7yx - 70y^2 \\ &= x^2 - 10xy + 7xy - 70y^2 \\ &\overset{\checkmark}{=} x^2 - 3xy - 70y^2 \end{aligned}$$

Example 7.7.16 Completely factor the expressions.

a. $x^2 - 11x + 30$

b. $-s^2 + 3s + 28$

c. $g^2 - 3g - 24$

d. $w^2 - wr - 30r^2$

e. $z^8 + 2z^4 - 63$

Explanation.

a. $x^2 - 11x + 30 = (x - 6)(x - 5)$

b. $-s^2 + 3s + 28 = -(s^2 - 3s - 28)$
$$= -(s - 7)(s + 4)$$

c. $g^2 - 3g - 24$ is prime. No two integers multiply to be -24 and add to be -3.

d. $w^2 - wr - 30r^2 = (w - 6r)(w + 5r)$

e. $z^8 + 2z^4 - 63 = (z^4 - 7)(z^4 + 9)$

7.7.4 Review of Factoring Trinomials with Non-Trivial Leading Coefficient

In Section 7.4 we covered factoring trinomials of the form $ax^2 + bx + c$ when $a \neq 1$ using the AC method.

Example 7.7.17 Using the AC Method. Completely factor the expression $9x^2 - 6x - 8$.

Explanation. To factor the expression $9x^2 - 6x - 8$, we first find ac:

1. $9 \cdot (-8) = -72$.

2. Examine factor pairs that multiply to -72, looking for a pair that sums to -6:

Factor Pair	Sum of the Pair		Factor Pair	Sum of the Pair
$1 \cdot -72$	-71		$-1 \cdot 72$	(no need to go this far)
$2 \cdot -36$	-34		$-2 \cdot 36$	(no need to go this far)
$3 \cdot -24$	-21		$-3 \cdot 24$	(no need to go this far)
$4 \cdot -18$	-14		$-4 \cdot 18$	(no need to go this far)
$6 \cdot -12$	-6		$-6 \cdot 12$	(no need to go this far)
$8 \cdot -9$	(no need to go this far)		$-8 \cdot 9$	(no need to go this far)

3. Intentionally break up the -6 as $6 + (-12)$ and then factor using grouping:

$$9x^2 \overbrace{- 6x} - 8 = 9x^2 \overbrace{+ 6x - 12x} - 8$$
$$= (9x^2 + 6x) + (-12x - 8)$$
$$= 3x(3x + 2) - 4(3x + 2)$$
$$= (3x + 2)(3x - 4)$$

Example 7.7.18 Completely factor the expression $3x^2 + 5x - 6$.

Explanation. First note that there is no GCF besides 1 and that $ac = -18$. To look for two factors of -18 that add up to 5, we will make a factor pair table.

Factor Pair	Sum of the Pair		Factor Pair	Sum of the Pair
$1 \cdot -18$	-17		$-1 \cdot 18$	17
$2 \cdot -9$	-7		$-2 \cdot 9$	7
$3 \cdot -6$	-3		$-3 \cdot 6$	3

Since none of the factor pairs of -18 sum to 5, we must conclude that this trinomial is prime. The only way to factor it is $3x^2 + 5x - 6 = 1\left(3x^2 + 5x - 6\right)$.

Example 7.7.19 Completely factor the expression $3y^2 + 20y - 63$.

Explanation. First note that $ac = -189$. Looking for two factors of -189 that add up to 20, we find 27 and -7. Breaking up the $+20$ into $+27 - 7$, we can factor using grouping.

$$
\begin{aligned}
3y^2 + 20y - 63 &= 3y^2 + \overbrace{27y - 7y} - 63 \\
&= \left(3y^2 + 27y\right) + \left(-7y - 63\right) \\
&= 3y\left(y + 9\right) - 7\left(y + 9\right) \\
&= \left(y + 9\right)\left(3y - 7\right)
\end{aligned}
$$

Example 7.7.20 Factoring in Stages with the AC Method. Completely factor the expression $8y^3 + 54y^2 + 36y$.

Explanation. Recall that some trinomials need to be factored in stages: the first stage is always to factor out the GCF. To factor $8y^3 + 54y^2 + 36y$, first note that the GCF of the three terms in the expression is $2y$. Then apply the AC method:

$$8y^3 + 54y^2 + 36y = 2y\left(4y^2 + 27y + 18\right)$$

Now we find $ac = 4 \cdot 18 = 72$. What two factors of 72 add up to 27? After checking a few numbers, we find that 3 and 24 fit the requirements. So:

$$
\begin{aligned}
&= 2y\left(4y^2 + \overbrace{27y} + 18\right) \\
&= 2y\left(4y^2 + \overbrace{3y + 24y} + 18\right) \\
&= 2y\left(\left(4y^2 + 3y\right) + \left(24y + 18\right)\right) \\
&= 2y\left(y\left(4y + 3\right) + 6\left(4y + 3\right)\right) \\
&= 2y\left(4y + 3\right)\left(y + 6\right)
\end{aligned}
$$

Example 7.7.21 Completely factor the expression $18x^3 + 26x^2 + 4x$.

Explanation. First note that there is a GCF of $2x$ which should be factored out first. Doing this leaves us with $18x^3 + 26x^2 + 8x = 2x\left(9x^2 + 13x + 4\right)$. Now we apply the AC method on the factor in the parentheses. So, $ac = 36$, and we must find two factors of 36 that sum to be 13. These two factors are 9 and 4. Now we can use grouping.

$$
\begin{aligned}
18x^3 + 26x^2 + 8x &= 2x\left(9x^2 + \overbrace{13x} + 4\right) \\
&= 2x\left(9x^2 + \overbrace{9x + 4x} + 4\right)
\end{aligned}
$$

$$= 2x\left((9x^2 + 9x) + (4x + 4)\right)$$
$$= 2x\left(9x(x + 1) + 4(x + 1)\right)$$
$$= 2x(x + 1)(9x + 4)$$

7.7.5 Review of Factoring Special Forms

In Section 7.5 we covered how to factor binomials and trinomials using formulas. Using these formulas, when appropriate, often drastically increased the speed of factoring. Below is a summary of the formulas covered. For each, consider that A and B could be any algebraic expressions.

Difference of Squares $A^2 - B^2 = (A + B)(A - B)$

Perfect Square Sum $A^2 + 2AB + B^2 = (A + B)^2$

Perfect Square Difference $A^2 - 2AB + B^2 = (A + B)^2$

Difference of Cubes $A^3 - B^3 = (A - B)\left(A^2 + AB + B^2\right)$

Sum of Cubes $A^3 + B^3 = (A + B)\left(A^2 - AB + B^2\right)$

Example 7.7.22 Factoring the Form $A^2 - 2AB + B^2$. Completely factor the expression $16y^2 - 24y + 9$.

Explanation. To factor $16y^2 - 24y + 9$ we notice that the expression might be of the form $A^2 - 2AB + B^2$. To find A and B, we mentally take the square root of both the first and last terms of the original expression. The square root of $16y^2$ is $4y$ since $(4y)^2 = 4^2y^2 = 16y^2$. The square root of 9 is 3. So, we conclude that $A = 4y$ and $B = 3$. Recall that we now need to check that the $24y$ matches our $2AB$. Using our values for A and B, we indeed see that $2AB = -2(4y)(3) = 24y$. So, we conclude that

$$16y^2 - 24y + 9 = (4y - 3)^2.$$

Example 7.7.23 Mixed Special Forms Factoring.

   a. Completely factor the expression $v^3 - 27$.

   b. Completely factor the expression $9w^2 + 12w + 4$.

   c. Completely factor the expression $4q^2 - 81$.

   d. Completely factor the expression $9p^2 + 25$.

   e. Completely factor the expression $121b^2 - 36$.

   f. Completely factor the expression $25u^2 - 70u + 49$.

   g. Completely factor the expression $64q^3 - 27y^3$.

Explanation. The first step for each problem is to try to fit the expression to one of the special factoring forms.

   a. To factor $v^3 - 27$ we notice that the expression is of the form $A^3 - B^3$. To find values for A and B, take mentally take the cube root of both terms. So, $A = v$ and $B = 3$. So, using the form $A^3 - B^3 = (A - B)(A^2 + AB + B^2)$, we have that

$$v^3 - 27 = (v - 3)\left(v^2 + (v)(3) + 3^2\right)$$
$$= (v - 3)\left(v^2 + 3v + 9\right)$$

b. To factor $9w^2 + 12w + 4$ we notice that the expression might be of the form $A^2 + 2AB + B^2$ where $A = 3w$ and $B = 2$. With this formula we need to check the value of $2AB$ which in this case is $2AB = 2(3w)(2) = 12w$. Since the value of $2AB$ is correct, the expression must factor as

$$9w^2 + 12w + 4 = (3w + 2)^2$$

c. To factor $4q^2 - 81$ we notice that the expression is of the form $A^2 - B^2$ where $A = 2q$ and $B = 9$. Thus, the expression must factor as

$$4q^2 - 81 = (2q - 9)(2q + 9)$$

d. To factor $9p^2 + 25$ we notice that the expression is of the form $A^2 + B^2$. This is called a sum of squares. If you recall from the section, the sum of squares is always prime. So $9p^2 + 25$ is prime.

e. To completely factor the expression $121b^2 - 36$ first note that the expression is of the form $A^2 - B^2$ where $A = 11b$ and $B = 6$. So, the expression factors as

$$121b^2 - 36 = (11b + 6)(11b - 6).$$

f. To completely factor the expression $25u^2 - 70u + 49$ first note that the expression might be of the form $A^2 - 2AB + B^2$ where $A = 5u$ and $B = 7$. Now, we check that $2AB$ matches the middle term: $2AB = 2(5u)(7) = 70u$. So, the expression factors as

$$25u^2 - 70u + 49 = (5u - 7)^2.$$

g. To completely factor the expression $64q^3 - 27y^3$ first note that the expression is of the form $A^3 - B^3$ where $A = 4q$ and $B = 3y$. So, the expression factors as

$$64q^3 - 27y^3 = (4q - 3y)\left((4q)^2 + (4q)(3y) + (3y)^2\right)$$
$$= (4q - 3y)\left(16q^2 + 12qy + 9y^2\right)$$

7.7.6 Review of Factoring Strategies

In Section 7.6 we covered a factoring decision tree to help us decide what methods to try when factoring a given expression. Remember to always factor out the GCF first.

Example 7.7.24 Factor the expressions using an effective method.

a. $24xy - 20x - 18y + 15$.

c. $8u^2 + 14u - 9$.

b. $12t^2 + 36t + 27$.

d. $18c^2 - 98p^2$.

Explanation.

a. To factor the expression $24xy - 20x - 18y + 15$, we first look for a GCF. Since the GCF is 1, we can move further on the flowchart. Since this is a four-term polynomial, we will try grouping.

$$24xy - 20x - 18y + 15 = 24xy + (-20x) + (-18y) + 15$$
$$= (24xy - 20x) + (-18y + 15)$$

$$= 4x(6y - 5) + (-3)(6y - 5)$$

$$= 4x \overbrace{(6x - 5)} - 3 \overbrace{(6x - 5)}$$

$$= (6x - 5)(4x - 3)$$

b. To factor the expression $12t^2 + 36t + 27$, we first look for a GCF. Since the GCF is 3, first we will factor that out.

$$12t^2 + 36t + 27 = 3\left(4t^2 + 12t + 9\right)$$

Next, we can note that the first and last terms are perfect squares where $A^2 = 4t^2$ and $B = 9$; so $A = 2t$ and $B = 3$. To check the middle term, $2AB = 12t$. So the expression factors as a perfect square.

$$12t^2 + 36t + 27 = 3\left(4t^2 + 12t + 9\right)$$
$$= 3(2t + 3)^2$$

c. To factor the expression $8u^2 + 14u - 9$, we first look for a GCF. Since the GCF is 1, we can move further on the flowchart. Since the expression is a trinomial with leading coefficient other than 1, we should try the AC method. Note that $AC = -72$ and factor pairs of -72 that add up to 14 are 18 and -4.

$$8u^2 + 14u - 9 = 8u^2 + 18u - 4u - 9$$
$$= \left(8u^2 + 18\right) + (-4u - 9)$$
$$= 2u(4u + 9) - 1(4u + 9)$$
$$= (2u - 1)(4u + 9)$$

d. To factor the expression $18c^2 - 98p^2$, we first look for a GCF. Since the GCF is 2, first we will factor that out.

$$18c^2 - 98p^2 = 2\left(9c^2 - 49p^2\right)$$

Now we notice that we have a binomial where both the first and second terms can be written as squares: $9c^2 = (3c)^2$ and $49p^2 = (7p)^2$.

$$18c^2 - 98p^2 = 2\left(9c^2 - 49p^2\right)$$
$$= 2(3c - 7p)(3c + 7p)$$

Exercises

Find the greatest common factor of the following terms.

1. $7y$ and $28y^2$

2. $4r$ and $28r^2$

3. $9r^{10}$ and $-54r^9$

4. $6t^{15}$ and $-54t^9$

5. $10x^{20}y^9, -70x^{18}y^{11}, 80x^7y^{20}$

6. $7x^{20}y^4, -21x^{19}y^9, 21x^{13}y^{10}$

Greatest Common Factor   Factor the given polynomial.

7. $6x - 6 = $ [　　　　]

8. $3y - 3 = $ [　　　　]

9. $9y + 54 = $ [　　　　]

10. $6y + 60 = $ [　　　　]

11. $28r^2 - 16r + 16 = $ [　　　]

12. $9r^2 - 6r + 12 = $ [　　　]

13. $13t^2 + 8 = $ [　　　　]

14. $3t^2 - 20 = $ [　　　　]

15. $x(x - 8) + 3(x - 8) = $ [　　]

16. $x(x - 5) + 9(x - 5) = $ [　　]

17. $4x(x + y) + 7(x + y) = $ [　]

18. $5x(x + y) - 8(x + y) = $ [　]

Factor by Grouping   Factor the given polynomial.

19. $y^2 + 4y + 2y + 8 = $ [　　]

20. $r^2 - 10r + 8r - 80 = $ [　　]

21. $r^3 + 7r^2 + 6r + 42 = $ [　]

22. $t^3 + 3t^2 - 8t - 24 = $ [　　]

23. $xy + 10x + 5y + 50 = $ [　　]

24. $xy + 2x + 9y + 18 = $ [　　]

25. $3x^2 + 18xy + 5xy + 30y^2 = $ [　　　　　　]

26. $4x^2 + 32xy + 9xy + 72y^2 = $ [　　　　　　]

One-variable Polynomials   Factor the given polynomial.

27. $y^2 + 7y + 6 = $ [　　　　]

28. $y^2 + 7y + 10 = $ [　　　　]

29. $r^2 + 6r - 27 = $ [　　　　]

30. $r^2 + 5r - 50 = $ [　　　　]

31. $t^2 - 10t + 16 = $ [　　　　]

32. $t^2 - 11t + 18 = $ [　　　　]

33. $x^2 - x + 1 = $ [　　　　]

34. $x^2 + 5x + 7 = $ [　　　　]

35. $x^2 - 20x + 100 = $ [　　　　]

36. $y^2 - 10y + 25 = $ [　　　　]

37. $7y^2 - 7 = $ [　　　　]

38. $6r^2 + 18r - 24 = $ [　　　　]

39. $3r^2 - 21r + 18 = $ [　　　　]

40. $2t^2 - 16t + 30 = $ [　　　　]

41. $2t^8 - 6t^7 - 36t^6 = $ [　　　　]

42. $6x^4 - 12x^3 - 18x^2 = $ [　　]

43. $-x^2 - 6x + 40 = $ [　　　　]

44. $-x^2 + 5x + 14 = $ [　　　　]

Multivariable Polynomials   Factor the given polynomial.

45. $y^2 - 6yr - 27r^2 = \boxed{}$ **46.** $y^2 + 4yt - 32t^2 = \boxed{}$ **47.** $r^2 - 12ry + 27y^2 = \boxed{}$

48. $r^2 - 3rx + 2x^2 = \boxed{}$ **49.** $2a^2b + 2ab - 24b = \boxed{}$ **50.** $2a^2b + 6ab - 8b = \boxed{}$

51. $2x^3y + 6x^2y + 4xy = \boxed{}$ **52.** $4x^3y + 16x^2y + 12xy = \boxed{}$

Non-trivial Leading Coefficient   Factor the given polynomial.

53. $3x^2 + 2x - 8 = \boxed{}$ **54.** $2y^2 - 3y - 9 = \boxed{}$ **55.** $2y^2 + y + 2 = \boxed{}$

56. $3r^2 - 8r + 6 = \boxed{}$ **57.** $6r^2 + 19r + 10 = \boxed{}$ **58.** $4t^2 + 12t + 5 = \boxed{}$

59. $6t^2 - 19t + 10 = \boxed{}$ **60.** $9x^2 - 21x + 10 = \boxed{}$ **61.** $8x^2 + 20x + 8 = \boxed{}$

62. $15x^2 + 25x + 10 = \boxed{}$ **63.** $4y^2 - 22y + 28 = \boxed{}$ **64.** $6y^2 - 10y + 4 = \boxed{}$

65. $4r^4 - 14r^3 + 12r^2 = \boxed{}$ **66.** $6r^6 - 14r^5 + 8r^4 = \boxed{}$ **67.** $5t^2r^2 + 18tr + 16 = \boxed{}$

68. $2t^2y^2 + 9ty + 10 = \boxed{}$ **69.** $3x^2 - 14xt + 15t^2 = \boxed{}$ **70.** $2x^2 - 5xr + 2r^2 = \boxed{}$

71. $8x^2 - 25xy + 3y^2 = \boxed{}$ **72.** $8y^2 - 25yt + 3t^2 = \boxed{}$ **73.** $6y^2r^2 + 15yr + 6 = \boxed{}$

74. $4r^2x^2 + 18rx + 14 = \boxed{}$ **75.** $12r^2t^2 + 18rt - 12 = \boxed{}$ **76.** $20t^2r^2 - 30tr - 20 = \boxed{}$

77. $8x^2 + 28xy + 12y^2 = \boxed{}$ **78.** $8x^2 + 28xy + 20y^2 = \boxed{}$

Factor Special Forms   Factor the given polynomial.

79. $x^2 - 36 = \boxed{}$ **80.** $x^2 - 4 = \boxed{}$ **81.** $100y^2 - 49 = \boxed{}$

82. $36y^2 - 49 =$ ⬚

83. $25 - r^2 =$ ⬚

84. $100 - 81r^2 =$ ⬚

85. $t^6 - 36 =$ ⬚

86. $t^{12} - 4 =$ ⬚

87. $x^{12} - 4y^{10} =$ ⬚

88. $x^{10} - 9y^{12} =$ ⬚

89. $144x^2 - 24x + 1 =$ ⬚

90. $81y^2 - 18y + 1 =$ ⬚

91. $4y^2 - 44yr + 121r^2 =$ ⬚

92. $9r^2 - 60rt + 100t^2 =$ ⬚

93. $x^3 - 729y^3 =$ ⬚

Hint: $-729 = -9^3$.

94. $x^3 - 1000y^3 =$ ⬚

Hint: $-1000 = -10^3$.

95. $81t^4 - 16 =$ ⬚

96. $16t^4 - 81 =$ ⬚

97. $4x^3 - 36x =$ ⬚

98. $10x^3 - 40x =$ ⬚

99. $4y^4r^4 - 4y^2r^2 =$ ⬚

100. $3y^4x^4 - 48y^2x^2 =$ ⬚

101. $24 - 6r^2 =$ ⬚

102. $9 - 9r^2 =$ ⬚

103. $72t^9 + 48t^8 + 8t^7 =$ ⬚

104. $18t^7 + 12t^6 + 2t^5 =$ ⬚

105. $t^8 + 216t^5 =$ ⬚

Hint: $216 = 6^3$.

106. $x^{11} + 8x^8 =$ ⬚

107. $x^2 + 81 =$ ⬚

108. $y^2 + 25 =$ ⬚

CHAPTER 8

Solving Quadratic Equations

8.1 Solving Quadratic Equations by Factoring

We have learned how to factor trinomials like $x^2 + 5x + 6$ into $(x + 2)(x + 3)$. This skill is needed to solve an equation like $x^2 + 5x + 6 = 0$, which is a quadratic equation. A **quadratic equation** is is an equation in the form $ax^2 + bx + c = 0$ with $a \neq 0$. We also consider equations such as $x^2 = x + 3$ and $5x^2 + 3 = (x + 1)^2 + (x + 1)(x - 3)$ to be quadratic equations, because we can expand any multiplication, add or subtract terms from both sides, and combine like terms to get the form $ax^2 + bx + c = 0$. The form $ax^2 + bx + c = 0$ is called the **standard form** of a quadratic equation.

Before we begin exploring the method of solving quadratic equations by factoring, we'll identify what types of equations are quadratic and which are not.

Checkpoint 8.1.2. Identify which of the items are quadratic equations.

a. The equation $2x^2 + 5x = 7$ (□ is   □ is not)   a quadratic equation.

b. The equation $5 - 2x = 3$ (□ is   □ is not)   a quadratic equation.

c. The equation $15 - x^3 = 3x^2 + 9x$ (□ is   □ is not)   a quadratic equation.

d. The equation $(x + 3)(x - 4) = 0$ (□ is   □ is not)   a quadratic equation.

e. The equation $x(x + 1)(x - 1) = 0$ (□ is   □ is not)   a quadratic equation.

f. The expression $x^2 - 5x + 6$ (□ is   □ is not)   a quadratic equation.

g. The equation $(2x - 3)(x + 5) = 12$ (□ is   □ is not)   a quadratic equation.

Explanation.

a. The equation $2x^2 + 5x = 7$ *is* a quadratic equation. To write it in standard form, simply subtract 7 from both sides.

b. The equation $5 - 2x = 3$ *is not* quadratic. It is a linear equation.

c. The equation $15 + x^3 = 3x^2 + 9x$ *is not* a quadratic equation because of the x^3 term.

d. The equation $(x + 3)(x - 4) = 0$ *is* a quadratic equation. If we expand the left-hand side of the equation, we would get something in standard form.

e. The equation $x(x + 1)(x - 1) = 0$ *is not* a quadratic equation. If we expanded the left-hand side of the equation, we would have an expression with an x^3 term, which automatically makes it not quadratic.

f. The expression $x^2 - 5x + 6$ *is not* a quadratic equation; it's not an *equation* at all. Instead, this is a quadratic *expression*.

g. The equation $(2x - 3)(x + 5) = 12$ *is* a quadratic equation. Multiplying out the left-hand side, and subtracting 12 form both sides, we would have a quadratic equation in standard form.

Now we'll look at an application that demonstrates the need and method for solving a quadratic equation by factoring.

Nita is in a physics class that launches a tennis ball from a rooftop 80 feet above the ground. They fire it directly upward at a speed of 64 feet per second and measure the time it takes for the ball to hit the ground below. We can model the height of the tennis ball, h, in feet, with the quadratic equation

$$h = -16t^2 + 64t + 80,$$

where t represents the time in seconds after the launch. Using the model we can predict when the ball will hit the ground.

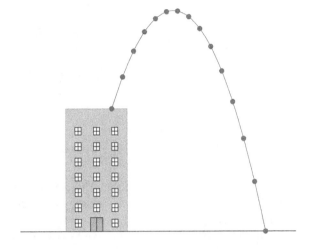

Figure 8.1.3: A Diagram of the Ball Thrown from the Roof

The ground has a height of 0, or $h = 0$. We will substitute 0 for h in the equation and we have

$$-16t^2 + 64t + 80 = 0$$

We need to solve this quadratic equation for t to find when the ball will hit the ground.

The key strategy for solving a *linear* equation is to separate the variable terms from the constant terms on either side of the equal sign. It turns out that this same method *will not work* for quadratic equations. Fortunately, we already have spent a good amount of time discussing a method that *will* work: factoring. If we can factor the polynomial on the left-hand side, we will be on the home stretch to solving the whole equation.

We will look for a common factor first, and see that we can factor out -16. Then we can finish factoring the trinomial:

$$-16t^2 + 64t + 80 = 0$$
$$-16(t^2 - 4t - 5) = 0$$
$$-16(t + 1)(t - 5) = 0$$

In order to finish solving the equation, we need to understand the following property. This property explains why it was *incredibly important* to *not* move the 80 in our example over to the other side of the equation before trying to factor.

Fact 8.1.4 Zero Product Property. *If the product of two or more numbers is equal to zero, then at least one of the numbers must be zero.*

One way to understand this property is to think about the equation $a \cdot b = 0$. Maybe $b = 0$, because that would certainly cause the equation to be true. But suppose that $b \neq 0$. Then it is safe to divide both sides by b, and the resulting equation says that $a = 0$. So no matter what, either $a = 0$ or $b = 0$.

To understand this property more, let's look at a few products:

$$4 \cdot 7 = 28 \qquad\qquad 4 \cdot 0 = 0 \qquad\qquad 4 \cdot 7 \cdot 3 = 84$$
$$0 \cdot 7 = 0 \qquad\qquad -4 \cdot 0 = 0 \qquad\qquad 4 \cdot 0 \cdot 3 = 0$$

When none of the factors are 0, the result is never 0. The only way to get a product of 0 is when one of the factors is 0. This property is unique to the number 0 and can be used no matter how many numbers are multiplied together.

Now we can see the value of factoring. We have three factors in our equation

$$-16(t + 1)(t - 5) = 0.$$

The first factor is the number -16. The second and third factors, $t+1$ and $t-5$, are expressions that represent numbers. Since the product of the three factors is equal to 0, one of the factors must be zero.

Since -16 is not 0, either $t + 1$ or $t - 5$ must be 0. This gives us two equations to solve:

$$\begin{array}{ccc} t + 1 = 0 & \text{or} & t - 5 = 0 \\ t + 1 - 1 = 0 - 1 & \text{or} & t - 5 + 5 = 0 + 5 \\ t = -1 & \text{or} & t = 5 \end{array}$$

We have found two solutions, -1 and 5. A quadratic expression will have at most two linear factors, not including any constants, so it can have up to two solutions.

Let's check each of our two solutions -1 and 5:

$$-16t^2 + 64t + 80 = 0 \qquad\qquad -16t^2 + 64t + 80 = 0$$
$$-16(-1)^2 + 64(-1) + 80 \overset{?}{=} 0 \qquad\qquad -16(5)^2 + 64(5) + 80 \overset{?}{=} 0$$
$$-16(1) - 64 + 80 \overset{?}{=} 0 \qquad\qquad -16(25) + 320 + 80 \overset{?}{=} 0$$
$$-16 - 64 + 80 \overset{?}{=} 0 \qquad\qquad -400 + 320 + 80 \overset{?}{=} 0$$
$$0 \overset{\checkmark}{=} 0 \qquad\qquad 0 \overset{\checkmark}{=} 0$$

We have verified our solutions. While there are two solutions to the equation, the solution -1 is not relevant to this physics model because it is a negative time which would tell us something about the ball's height *before* it was launched. The solution 5 does make sense. According to the model, the tennis ball will hit the ground 5 seconds after it is launched.

8.1.1 Further Examples

We'll now look at further examples of solving quadratic equations by factoring. The general process is outlined here:

Process 8.1.5 Solving Quadratic Equations by Factoring.

Simplify *Simplify the equation using distribution and by combining like terms.*

Isolate *Move all terms onto one side of the equation so that the other side has 0.*

Factor *Factor the quadratic expression.*

Apply the Zero Product Property *Apply the Zero Product Property.*

Solve *Solve the equation(s) that result after the zero product property was applied.*

Example 8.1.6 Solve $x^2 - 5x - 14 = 0$ by factoring.

Explanation.

$$x^2 - 5x - 14 = 0$$
$$(x - 7)(x + 2) = 0$$

$x - 7 = 0$	or	$x + 2 = 0$
$x - 7 + 7 = 0 + 7$	or	$x + 2 - 2 = 0 - 2$
$x = 7$	or	$x = -2$

The solutions are -2 and 7, so the solution set is written as $\{-2, 7\}$.

If the two factors of a polynomial happen to be the same, the equation will only have one solution. Let's look at an example of that.

Example 8.1.7 A Quadratic Equation with Only One Solution. Solve $x^2 - 10x + 25 = 0$ by factoring.

Explanation.

$$x^2 - 10x + 25 = 0$$
$$(x - 5)(x - 5) = 0$$
$$(x - 5)^2 = 0$$
$$x - 5 = 0$$
$$x - 5 + 5 = 0 + 5$$
$$x = 5$$

The solution is 5, so the solution set is written as $\{5\}$.

Example 8.1.8 Factor Out a Common Factor. Solve $5x^2 + 55x + 120 = 0$ by factoring.

Explanation. Note that the terms are all divisible by 5, so we can factor that out to start.

$$5x^2 + 55x + 120 = 0$$
$$5(x^2 + 11x + 24) = 0$$
$$5(x + 8)(x + 3) = 0$$

$x + 8 = 0$	or	$x + 3 = 0$
$x = -8$	or	$x = -3$

The solution set is $\{-8, -3\}$.

Example 8.1.9 Factoring Using the AC Method. Solve $3x^2 - 7x + 2 = 0$ by factoring.

Explanation. Recall that we multiply $3 \cdot 2 = 6$ and find a factor pair that multiplies to 6 and adds to

−7. The factors are −6 and −1. We use the two factors to replace the middle term with −6*x* and −*x*.

$$3x^2 - 7x + 2 = 0$$
$$3x^2 - 6x - x + 2 = 0$$
$$(3x^2 - 6x) + (-x + 2) = 0$$
$$3x(x - 2) - 1(x - 2) = 0$$
$$(3x - 1)(x - 2) = 0$$

$3x - 1 = 0$	or	$x - 2 = 0$
$3x = 1$	or	$x = 2$
$x = \dfrac{1}{3}$	or	$x = 2$

The solution set is $\left\{\frac{1}{3}, 2\right\}$.

So far the equations have been written in standard form, which is

$$ax^2 + bx + c = 0$$

If an equation is not given in standard form then we must rearrange it in order to use the Zero Product Property.

Example 8.1.10 Writing in Standard Form. Solve $x^2 - 10x = 24$ by factoring.

Explanation. There is nothing like the Zero Product Property for the number 24. We must have a 0 on one side of the equation to solve quadratic equations using factoring.

$$x^2 - 10x = 24$$
$$x^2 - 10x - 24 = 24 - 24$$
$$x^2 - 10x - 24 = 0$$
$$(x - 12)(x + 2) = 0$$

$x - 12 = 0$	or	$x + 2 = 0$
$x = 12$	or	$x = -2$

The solution set is $\{-2, 12\}$.

Example 8.1.11 Writing in Standard Form. Solve $(x + 4)(x - 3) = 18$ by factoring.

Explanation. Again, there is nothing like the Zero Product Property for a number like 18. We must expand the left side and subtract 18 from both sides.

$$(x + 4)(x - 3) = 18$$
$$x^2 + x - 12 = 18$$
$$x^2 + x - 12 - 18 = 18 - 18$$
$$x^2 + x - 30 = 0$$

$$(x + 6)(x - 5) = 0$$

$x + 6 = 0$	or	$x - 5 = 0$
$x = -6$	or	$x = 5$

The solution set is $\{-6, 5\}$.

Example 8.1.12 A Quadratic Equation with No Constant Term. Solve $2x^2 = 5x$ by factoring.

Explanation. It may be tempting to divide both sides of the equation by x. But x is a variable, and for all we know, maybe $x = 0$. So it is not safe to divide by x. As a general rule, never divide an equation by a variable in the solving process. Instead, we will put the equation in standard form.

$$2x^2 = 5x$$
$$2x^2 - 5x = 5x - 5x$$
$$2x^2 - 5x = 0$$

We can factor out x.

$$x(2x - 5) = 0$$

$x = 0$	or	$2x - 5 = 0$
$x = 0$	or	$2x = 5$
$x = 0$	or	$x = \dfrac{5}{2}$

The solution set is $\left\{0, \frac{5}{2}\right\}$. In general, if a quadratic equation does not have a constant term, then 0 will be one of the solutions.

Example 8.1.13 Factoring a Special Polynomial. Solve $x^2 = 9$ by factoring.

Explanation. We can put the equation in standard form and use factoring. In this case, we find a difference of squares.

$$x^2 = 9$$
$$x^2 - 9 = 0$$
$$(x + 3)(x - 3) = 0$$

$x + 3 = 0$	or	$x - 3 = 0$
$x = -3$	or	$x = 3$

The solution set is $\{-3, 3\}$.

Example 8.1.14 Solving an Equation with a Higher Degree. Solve $2x^3 - 10x^2 - 28x = 0$ by factoring.

Explanation. Although this equation is not quadratic, it does factor so we can solve it by factoring.

$$2x^3 - 10x^2 - 28x = 0$$
$$2x(x^2 - 5x - 14) = 0$$
$$2x(x - 7)(x + 2) = 0$$

$2x = 0$	or	$x - 7 = 0$	or	$x + 2 = 0$
$x = 0$	or	$x = 7$	or	$x = -2$

The solution set is $\{-2, 0, 7\}$.

8.1.2 Applications

Example 8.1.15 Kicking it on Mars. Some time in the recent past, Filip traveled to Mars for a vacation with his kids, Henrik and Karina, who wanted to kick a soccer ball around in the comparatively reduced gravity. Karina stood at point K and kicked the ball over her dad standing at point F to Henrik standing at point H. The height of the ball off the ground, h in feet, can be modeled by the equation $h = -0.01\left(x^2 - 70x - 1800\right)$, where x is how far to the right the ball is from Filip. Note that distances to the left of Filip will be negative.

    a. Find out how high the ball was above the ground when it passed over Filip's head.

    b. Find the distance from Karina to Henrik.

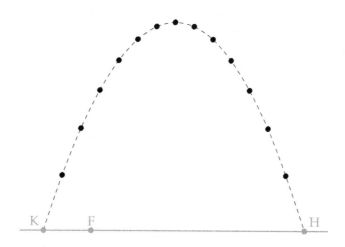

Figure 8.1.16: A Soccer Kick on Mars

Explanation.

    a. The ball was neither left nor right of Filip when it went over him, so $x = 0$. Plugging that value into our equation for x,

$$h = -0.01 \left(0^2 - 70(0) - 1800\right)$$
$$= -0.01(-1800)$$
$$= 18$$

It seems that the soccer ball was 18 feet above the ground when it flew over Filip.

b. The distance from Karina to Henrik is the same as the distance from point K to point H. These are the horizontal intercepts of the graph of the given formula: $h = -0.01 \left(x^2 - 70x - 1800\right)$. To find the horizontal intercepts, set $h = 0$ and solve for x.

$$0 = -0.01 \left(x^2 - 70x - 1800\right)$$

Note that we can divide by -0.01 on both sides of the equation to simplify.

$$0 = x^2 - 70x - 1800$$
$$0 = (x - 90)(x + 20)$$

So, either:

$$x - 90 = 0 \qquad\qquad \text{or} \qquad\qquad x + 20 = 0$$
$$x = 90 \qquad\qquad \text{or} \qquad\qquad x = -20$$

Since the x-values are how far right or left the points are from Filip, Karina is standing 20 feet left of Filip and Henrik is standing 90 feet right of Filip. Thus, the two kids are 110 feet apart.

It is worth noting that if this same kick, with same initial force at the same angle, took place on Earth, the ball would have traveled less than 30 feet from Karina before landing!

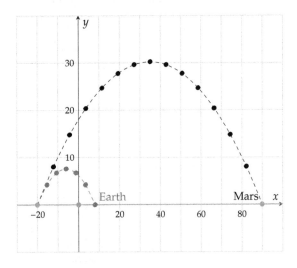

Figure 8.1.17: A Soccer Kick on Mars and the Same Kick on Earth

Example 8.1.18 An Area Application. Rajesh has a hot tub and he wants to build a deck around it. The hot tub is 7 ft by 5 ft and it is covered by a roof that is 99 ft². How wide can he make the deck so that it will be covered by the roof?

Explanation. We will define x to represent the width of the deck (in feet). Here is a diagram to help us understand the scenario.

Figure 8.1.19: Diagram for the Deck

The overall length is $7 + 2x$ feet, because Rajesh is adding x feet on each side. Similarly, the overall width is $5 + 2x$ feet.

The formula for the area of a rectangle is area = length $\cdot$ width. Since the total area of the roof is 99 ft^2, we can write and solve the equation:

$$(7 + 2x)(5 + 2x) = 99$$
$$4x^2 + 24x + 35 = 99$$
$$4x^2 + 24x + 35 - 99 = 99 - 99$$
$$4x^2 + 24x - 64 = 0$$
$$4(x^2 + 6x - 16) = 0$$
$$4(x + 8)(x - 2) = 0$$

$$\begin{array}{ccc} x + 8 = 0 & \text{or} & x - 2 = 0 \\ x = -8 & \text{or} & x = 2 \end{array}$$

Since a length cannot be negative, we take $x = 2$ as the only applicable solution. Rajesh should make the deck 2 ft wide on each side to fit under the roof.

Exercises

Warmup and Review   Factor the given polynomial.

1. $8r + 56 =$ [ ]

2. $5r + 10 =$ [ ]

3. $t^2 - 7t - 30 =$ [ ]

4. $t^2 - 3t - 70 =$ [ ]

5. $2t^2 - 13t + 21 =$ [ ]

6. $3x^2 - 11x + 8 =$ [ ]

7. $54x^2 + 9x + 54 = \boxed{}$ **8.** $24y^2 + 6y + 48 = \boxed{}$ **9.** $121y^4 - 144 = \boxed{}$

10. $64r^4 - 25 = \boxed{}$

Solve Quadratic Equations by Factoring Solve the equation.

11. $(x + 4)(x - 7) = 0$

12. $(x + 7)(x + 9) = 0$

13. $-74(x + 8)(11x - 7) = 0$

14. $-60(x + 10)(5x - 1) = 0$

15. $x^2 + 10x + 9 = 0$

16. $x^2 + 8x + 15 = 0$

17. $x^2 - 2x - 8 = 0$

18. $x^2 + 7x - 8 = 0$

19. $x^2 - 11x + 30 = 0$

20. $x^2 - 14x + 40 = 0$

21. $x^2 + 18x = -80$

22. $x^2 + 13x = -36$

23. $x^2 + 9x = 10$

24. $x^2 - 2x = 63$

25. $x^2 - 15x = -56$

26. $x^2 - 8x = -7$

27. $x^2 = 2x$

28. $x^2 = -x$

29. $7x^2 = -63x$

30. $8x^2 = -16x$

31. $9x^2 = 5x$

32. $10x^2 = -9x$

33. $x^2 - 4x + 4 = 0$

34. $x^2 - 6x + 9 = 0$

35. $x^2 = 8x - 16$

36. $x^2 = 12x - 36$

37. $49x^2 = -42x - 9$

38. $4x^2 = -28x - 49$

39. $5x^2 = -36x - 36$

40. $5x^2 = -52x - 20$

41. $x^2 - 100 = 0$

42. $x^2 - 81 = 0$

43. $9x^2 - 1 = 0$

44. $16x^2 - 81 = 0$

45. $9x^2 = 16$

46. $36x^2 = 1$

47. $x(x - 5) = 36$

48. $x(x + 1) = 30$

49. $x(5x + 62) = -120$

50. $x(5x + 54) = -40$

51. $(x - 7)(x - 1) = -8$

52. $(x + 2)(x + 4) = 3$

53. $(x + 2)(4x - 7) = 6 + 3x^2$ **54.** $(x - 2)(2x - 3) = x^2 - 6$ **55.** $x(x + 4) = 2x - 1$

56. $x(x + 12) = 3(2x - 3)$ **57.** $64x^2 + 48x + 9 = 0$ **58.** $81x^2 + 198x + 121 = 0$

59. $(x + 7)(x^2 + 16x + 60) = 0$.   **60.** $(x - 5)(x^2 + 3x + 2) = 0$ **61.** $x(x^2 - 4) = 0$

62. $x(x^2 - 16) = 0$ **63.** $x^3 - 10x^2 + 16x = 0$ **64.** $x^3 + 15x^2 + 54x = 0$

Quadratic Equation Application Problems

65. There is a rectangular lot in the garden, with 9 ft in length and 5 ft in width. You plan to expand the lot by an equal length around its four sides, and make the area of the expanded rectangle 117 ft². How long should you expand the original lot in four directions?

You should expand the original lot by [ ] in four directions.

66. There is a rectangular lot in the garden, with 9 ft in length and 3 ft in width. You plan to expand the lot by an equal length around its four sides, and make the area of the expanded rectangle 135 ft². How long should you expand the original lot in four directions?

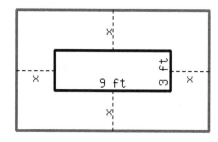

You should expand the original lot by [ ] in four directions.

67. Two numbers' sum is 16, and their product is 63. Find these two numbers.

These two numbers are [ ].

68. Two numbers' sum is 4, and their product is −60. Find these two numbers.

These two numbers are [ ].

69. A rectangle's base is 5 cm longer than its height. The rectangle's area is 50 cm². Find this rectangle's dimensions.

The rectangle's height is [ ].

The rectangle's base is [ ].

70. A rectangle's base is 8 cm longer than its height. The rectangle's area is 65 cm². Find this rectangle's dimensions.

The rectangle's height is [ ].

The rectangle's base is [ ].

71. A rectangle's base is 8 in shorter than three times its height. The rectangle's area is 3 in^2. Find this rectangle's dimensions.

The rectangle's height is [   ].

The rectangle's base is [   ].

72. A rectangle's base is 3 in shorter than twice its height. The rectangle's area is 9 in^2. Find this rectangle's dimensions.

The rectangle's height is [   ].

The rectangle's base is [   ].

Challenge

73. Give an example of a cubic equation that has three solutions: one solution is $x = 6$, the second solution is $x = -4$, and the third solution is $x = \frac{2}{3}$.

74. Solve for x in the equation $27x^{14} - 3x^{12} = 0$.

8.2 Square Root Properties

In this chapter, we will learn how to both simplify square roots and to do operations with square roots.

Definition 8.2.2 The Definition of the Square Root of a Number. If $y^2 = x$ for a positive number y, then y is called the **square root** of x, and we write $y = \sqrt{x}$, where the $\sqrt{}$ symbol is called the **radical** or the **root**. We call expressions with a root symbol **radical expressions**. The number inside the radical is called the **radicand**.

For example, since $4^2 = 16$, then $\sqrt{16} = 4$. Both $\sqrt{2}$ and $3\sqrt{2}$ are radical expressions. In both expressions, the number 2 is the radicand. You can review the square root basics in Section 1.3.

The word "radical" means something like "on the fringes" when used in politics, sports, and other places. It actually has that same meaning in math, when you consider a square with area A as in Figure 8.2.3.

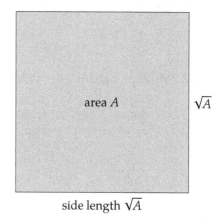

area A $\sqrt{A}$

side length $\sqrt{A}$

Figure 8.2.3: "Radical" means "off to the side."

8.2.1 Estimating Square Roots

When the radicand is a perfect square, its square root is a rational number. If the radicand is not a perfect square, the square root is irrational. We want to be able to estimate square roots without using a calculator.

To estimate $\sqrt{10}$, we can find the nearest perfect squares that are whole numbers on either side of 10. Recall that the perfect squares are $1, 4, 9, 16, 25, 36, 49, 64, \ldots$ The perfect square that is just below 10 is 9 and the perfect square just above 10 is 16. This tells us that $\sqrt{10}$ is between $\sqrt{9}$ and $\sqrt{16}$, or between 3 and 4. We can also say that $\sqrt{10}$ is much closer to 3 than 4 because 10 is closer to 9, so we think 3.1 or 3.2 would be a good estimate.

To check our estimate, let's find $\sqrt{10}$ with a calculator:

$$\sqrt{10} \approx 3.162$$

The actual value is just above 3 as we estimated, and between 3.1 and 3.2. Let's look at some more examples.

Checkpoint 8.2.4. Estimate $\sqrt{19}$ without a calculator.

Explanation. The radicand, 19, is between 16 and 25, so $\sqrt{19}$ is between $\sqrt{16}$ and $\sqrt{25}$, or between 4 and 5.

To be more precise, we notice that 19 is in the middle between 16 and 25 but closer to 16. We estimate $\sqrt{19}$

to be about 4.4.

We will check our estimate with a calculator:

$$\sqrt{19} \approx 4.358$$

Example 8.2.5 Estimate $\sqrt{3.2}$ without a calculator.

Explanation. The radicand 3.2 is between 1 and 4, so $\sqrt{3.2}$ is between $\sqrt{1}$ and $\sqrt{4}$, or between 1 and 2.

To be more precise, we notice that 3.2 is much closer to 4 than 1. We estimate $\sqrt{3.2}$ to be about 1.8.

We will check our estimate with a calculator:

$$\sqrt{3.2} \approx 1.788$$

8.2.2 Multiplication and Division Properties of Square Roots

Here is an example using perfect squares and the rules of exponents to show a relationship between the product of two square roots:

$$\sqrt{9 \cdot 16} = \sqrt{3^2 \cdot 4^2} = \sqrt{(3 \cdot 4)^2} = 3 \cdot 4 = 12$$

and

$$\sqrt{9} \cdot \sqrt{16} = \sqrt{3^2} \cdot \sqrt{4^2} = 3 \cdot 4 = 12$$

Whether we multiply the radicands first or take the square roots first, we get the same result. This tells us that in multiplication with radicals, we can combine factors into a single radical or separate them as needed.

Now let's look at division. When we learned how to find the square root of a fraction in Section 1.3, we saw that the numerators and denominators could be simplified separately. We multiply the numerators and denominators independently. Here is an example of two different ways to simplify a fraction in a square root:

$$\sqrt{\frac{25}{9}} = \sqrt{\left(\frac{5}{3}\right)^2} = \frac{5}{3}$$

and

$$\frac{\sqrt{25}}{\sqrt{9}} = \frac{\sqrt{5^2}}{\sqrt{3^2}} = \frac{5}{3}$$

Just like with multiplication, we can separate the numerators and denominators in a radical expression or combine them as needed. Note that we worked with expressions that were perfect squares, but these properties will work regardless of the number inside the radical. Let's summarize these properties.

Fact 8.2.6 Multiplication and Division Properties of Square Roots. *For any positive real numbers x and y we have the following properties:*

Multiplication Property of Square Roots $\sqrt{x \cdot y} = \sqrt{x} \cdot \sqrt{y}$

Division Property of Square Roots $\sqrt{\frac{x}{y}} = \frac{\sqrt{x}}{\sqrt{y}}$

8.2.3 Simplifying Square Roots

We can use Multiplication and Division Properties of Square Roots to simplify a radicand that is not a perfect square. Simplifying radicals is similar to simplifying fractions because we want the radicand to be as small as possible.

To understand why we can simplify radicals, let's use a calculator to compare $\sqrt{12}$ and $2\sqrt{3}$.

$$\sqrt{12} = 3.4641\ldots \qquad \text{and} \qquad 2\sqrt{3} = 3.4641\ldots$$

These are equivalent expressions so let's see how we can simplify $\sqrt{12}$ to $2\sqrt{3}$.

First, we will make a table of factor pairs for the number 12, as we did in Section 7.3.

$$1 \cdot 12$$
$$2 \cdot 6$$
$$3 \cdot 4$$

The factor pair with the largest perfect square is $3 \cdot 4$. We will use the property of multiplying radicals to separate the perfect square from the other factor. We write the perfect square first because it will end up in front of the radical.

$$\sqrt{12} = \sqrt{4} \cdot \sqrt{3}$$
$$= 2 \cdot \sqrt{3}$$
$$= 2\sqrt{3}$$

This process can be used to simplify any square root, or to determine that it is fully simplified. Let's look at a few more examples.

Example 8.2.7 Simplify $\sqrt{72}$.

Explanation.

Here is a table of factor pairs for the number 72.

$1 \cdot 72$	$4 \cdot 18$
$2 \cdot 36$	$6 \cdot 12$
$3 \cdot 24$	$8 \cdot 9$

The largest perfect square is 36 so we will rewrite 72 as $36 \cdot 2$.

$$\sqrt{72} = \sqrt{36 \cdot 2}$$
$$= \sqrt{36} \cdot \sqrt{2}$$
$$= 6\sqrt{2}$$

Notice that if we had chosen $4 \cdot 18$ we could simplify the radical partially but we would need to continue and find the perfect square of 9 in 18.

Checkpoint 8.2.8. Simplify $\sqrt{125}$.

Explanation. Here is a table of factor pairs for the number 125.

$$1 \cdot 125$$
$$5 \cdot 25$$

The largest perfect square is 25 so we will rewrite 125 as $25 \cdot 5$.

$$\sqrt{125} = \sqrt{25 \cdot 5}$$
$$= \sqrt{25} \cdot \sqrt{5}$$
$$= 5\sqrt{5}$$

Example 8.2.9 Simplify $\sqrt{30}$.

Explanation.

Here is a table of factor pairs for the number 30.

$1 \cdot 30 \qquad 3 \cdot 10$
$2 \cdot 15 \qquad 5 \cdot 6$

The number 30 does not have any factors that are perfect squares so it cannot be simplified further.

We can also use Division Property of Square Roots to simplify expressions.

Example 8.2.10

a. Simplify $\sqrt{\frac{9}{16}}$.

b. Simplify $\frac{\sqrt{50}}{\sqrt{2}}$.

Explanation.

a. For the first expression, we will use the Division Property of Square Roots:

$$\sqrt{\frac{9}{16}} = \frac{\sqrt{9}}{\sqrt{16}}$$
$$= \frac{3}{4}$$

b. For the second expression, we use the same property in reverse: $\frac{\sqrt{x}}{\sqrt{y}} = \sqrt{\frac{x}{y}}$:

$$\frac{\sqrt{50}}{\sqrt{2}} = \sqrt{\frac{50}{2}}$$
$$= \sqrt{25}$$
$$= 5$$

8.2.4 Multiplying Square Root Expressions

If we use the Multiplication Property of Square Roots and the Division Property of Square Roots in the reverse order as

$$\sqrt{x} \cdot \sqrt{y} = \sqrt{xy} \qquad \text{and} \qquad \frac{\sqrt{x}}{\sqrt{y}} = \sqrt{\frac{x}{y}},$$

we can use these properties to multiply and divide square root expressions. We want to simplify each radical first to keep the radicands as small as possible. Let's look at a few examples.

Example 8.2.11 Multiply $\sqrt{8} \cdot \sqrt{54}$.

Explanation. We will simplify each radical first, and then multiply them together. We do not want to multiply $8 \cdot 54$ because we will end up with a larger number that is harder to factor.

$$
\begin{aligned}
\sqrt{8} \cdot \sqrt{54} &= \sqrt{4 \cdot 2} \cdot \sqrt{9 \cdot 6} \\
&= 2\sqrt{2} \cdot 3\sqrt{6} \\
&= 2 \cdot 3\sqrt{2 \cdot 6} \\
&= 2 \cdot 3\sqrt{2 \cdot 2 \cdot 3} \\
&= 6 \cdot 2\sqrt{3} \\
&= 12\sqrt{3}
\end{aligned}
$$

We could have multiplied $2 \cdot 6$ inside the radical to get 12 and then factored 12 into $4 \cdot 3$. Whenever you find a pair of identical factors, this is a perfect square.

Checkpoint 8.2.12. Multiply $2\sqrt{7} \cdot 3\sqrt{21}$.

Explanation. First multiply the non-radical factors together and the radical factors together. Then look for further simplifications.

$$
\begin{aligned}
2\sqrt{7} \cdot 3\sqrt{21} &= 2 \cdot 3 \cdot \sqrt{7} \cdot \sqrt{21} \\
&= 6 \cdot \sqrt{7} \cdot \sqrt{7 \cdot 3} \\
&= 6\sqrt{7 \cdot 7 \cdot 3} \\
&= 6 \cdot 7 \cdot \sqrt{3} \\
&= 42\sqrt{3}
\end{aligned}
$$

Example 8.2.13 Multiply $\sqrt{\frac{6}{5}} \cdot \sqrt{\frac{3}{5}}$.

Explanation.

$$
\begin{aligned}
\sqrt{\frac{6}{5}} \cdot \sqrt{\frac{3}{5}} &= \sqrt{\frac{6}{5} \cdot \frac{3}{5}} \\
&= \sqrt{\frac{18}{25}} \\
&= \frac{\sqrt{18}}{\sqrt{25}} \\
&= \frac{\sqrt{9 \cdot 2}}{5} \\
&= \frac{3\sqrt{2}}{5}
\end{aligned}
$$

8.2.5 Adding and Subtracting Square Root Expressions

We learned the Multiplication Property of Square Roots previously and applied this to multiplication of square roots, but we cannot apply this property to the operations of addition or subtraction. Here are two examples to demonstrate this:

$$\sqrt{9+16} \overset{?}{=} \sqrt{9} + \sqrt{16}$$

$$\sqrt{25} \overset{?}{=} 3 + 4$$

$$5 \overset{\text{no}}{=} 7$$

$$\sqrt{169-25} \overset{?}{=} \sqrt{169} - \sqrt{25}$$

$$\sqrt{144} \overset{?}{=} 13 - 5$$

$$12 \overset{\text{no}}{=} 8$$

We do not get the same result if we separate the radicals, so we must complete any additions and subtractions inside the radical first.

To add and subtract radical expressions, we will need to recognize that we can only add and subtract like terms. In this case, we will call them **like radicals**. In fact, adding like radicals will work just like adding like terms

$$x + x = 2x$$

and

$$\sqrt{5} + \sqrt{5} = 2\sqrt{5}$$

We can verify that the second equation is true by replacing x with $\sqrt{5}$ in the second equation. Let's look at a few more examples.

Example 8.2.14 Simplify $\sqrt{2} + \sqrt{8}$.

Explanation.

$$\sqrt{2} + \sqrt{8} = \sqrt{2} + \sqrt{4 \cdot 2}$$
$$= \sqrt{2} + 2\sqrt{2}$$
$$= 3\sqrt{2}$$

To help understand $\sqrt{2} + 2\sqrt{2} = 3\sqrt{2}$, think of $x + 2x = 3x$ or "a taco plus two tacos is three tacos."

Checkpoint 8.2.15. Simplify $2\sqrt{3} - 3\sqrt{48}$.

Explanation. First we will simplify the radical term where 48 is the radicand, and we may see that we then have like radicals.

$$2\sqrt{3} - 3\sqrt{48} = 2\sqrt{3} - 3\sqrt{16 \cdot 3}$$
$$= 2\sqrt{3} - 3 \cdot 4\sqrt{3}$$
$$= 2\sqrt{3} - 12\sqrt{3}$$
$$= -10\sqrt{3}$$

Example 8.2.16 Simplify $\sqrt{2} + \sqrt{27}$.

Explanation.

$$\sqrt{2} + \sqrt{27} = \sqrt{2} + \sqrt{9 \cdot 3}$$
$$= \sqrt{2} + 3\sqrt{3}$$

We cannot simplify the expression further because $\sqrt{2}$ and $\sqrt{3}$ are not like radicals.

Example 8.2.17 Simplify $\sqrt{6} - \sqrt{18} \cdot \sqrt{12}$.

Explanation. In this example, we should multiply the latter two square roots first and then see if we have like radicals.

$$
\begin{aligned}
\sqrt{6} - \sqrt{18} \cdot \sqrt{12} &= \sqrt{6} - \sqrt{9 \cdot 2} \cdot \sqrt{4 \cdot 3} \\
&= \sqrt{6} - 3\sqrt{2} \cdot 2\sqrt{3} \\
&= \sqrt{6} - 3 \cdot 2 \cdot \sqrt{2} \cdot \sqrt{3} \\
&= \sqrt{6} - 6\sqrt{6} \\
&= -5\sqrt{6}
\end{aligned}
$$

8.2.6 Rationalizing the Denominator

When simplifying square root expressions, we have seen that we need to make the radicand as small as possible. Another rule is that we do not leave any irrational numbers, such as $\sqrt{3}$ or $2\sqrt{5}$, in the denominator of a fraction. In other words, we want the denominator to be rational. The process of dealing with such numbers in the denominator is called **rationalizing the denominator.**

Let's see how we can remove the square root symbol from the denominator in $\frac{1}{\sqrt{5}}$. If we multiply a radical by itself, the result is the radicand, by Definition 8.2.2. As an example:

$$\sqrt{5} \cdot \sqrt{5} = 5$$

To write $\frac{1}{\sqrt{5}}$ as an equivalent fraction, we must multiply both the numerator and denominator by the same number. If we multiply the numerator and denominator by $\sqrt{5}$, we have:

$$
\begin{aligned}
\frac{1}{\sqrt{5}} &= \frac{1}{\sqrt{5}} \cdot \frac{\sqrt{5}}{\sqrt{5}} \\
&= \frac{1 \cdot \sqrt{5}}{\sqrt{5} \cdot \sqrt{5}} \\
&= \frac{\sqrt{5}}{5}
\end{aligned}
$$

We can use a calculator to verify that $\frac{1}{\sqrt{5}} = \frac{\sqrt{5}}{5}$. They both are $0.4472\ldots$. Let's look at a few more examples.

Example 8.2.18 Rationalize the denominator in $\frac{6}{\sqrt{3}}$.

Explanation. We will rationalize this denominator by multiplying the numerator and denominator by $\sqrt{3}$:

$$\frac{6}{\sqrt{3}} = \frac{6}{\sqrt{3}} \cdot \frac{\sqrt{3}}{\sqrt{3}}$$

$$= \frac{6 \cdot \sqrt{3}}{\sqrt{3} \cdot \sqrt{3}}$$

$$= \frac{6\sqrt{3}}{3}$$

$$= 2\sqrt{3}$$

Note that we reduced any fractions that are outside the radical.

Checkpoint 8.2.19. Rationalize the denominator in $\frac{2}{\sqrt{10}}$.

Explanation. We will rationalize the denominator by multiplying the numerator and denominator by $\sqrt{10}$:

$$\frac{2}{\sqrt{10}} = \frac{2}{\sqrt{10}} \cdot \frac{\sqrt{10}}{\sqrt{10}}$$

$$= \frac{2 \cdot \sqrt{10}}{\sqrt{10} \cdot \sqrt{10}}$$

$$= \frac{2\sqrt{10}}{10}$$

$$= \frac{\sqrt{10}}{5}$$

Again note that the fraction was simplified in the last step.

Example 8.2.20 Rationalize the denominator in $\sqrt{\frac{2}{7}}$.

Explanation.

$$\sqrt{\frac{2}{7}} = \frac{\sqrt{2}}{\sqrt{7}}$$

$$= \frac{\sqrt{2}}{\sqrt{7}} \cdot \frac{\sqrt{7}}{\sqrt{7}}$$

$$= \frac{\sqrt{2} \cdot \sqrt{7}}{\sqrt{7} \cdot \sqrt{7}}$$

$$= \frac{\sqrt{14}}{7}$$

8.2.7 More Complicated Square Root Operations

In Section 6.4, we learned how to multiply polynomials like $2(x + 3)$ and $(x + 2)(x + 3)$. All the methods we learned apply when we multiply square root expressions. We will look at a few examples done with different methods.

Example 8.2.21 Multiply $\sqrt{5}\left(\sqrt{3} - \sqrt{2}\right)$.

Explanation. We will use the distributive property to do this problem:

$$\sqrt{5}\left(\sqrt{3} - \sqrt{2}\right) = \sqrt{5}\sqrt{3} - \sqrt{5}\sqrt{2}$$
$$= \sqrt{15} - \sqrt{10}$$

Example 8.2.22 Multiply $\left(\sqrt{6} + \sqrt{12}\right)\left(\sqrt{3} - \sqrt{2}\right)$.

Explanation. We will use the FOIL Method to do this problem:

$$\left(\sqrt{6} + \sqrt{12}\right)\left(\sqrt{3} - \sqrt{2}\right) = \sqrt{6}\sqrt{3} - \sqrt{6}\sqrt{2} + \sqrt{12}\sqrt{3} - \sqrt{12}\sqrt{2}$$
$$= \sqrt{18} - \sqrt{12} + \sqrt{36} - \sqrt{24}$$
$$= 3\sqrt{2} - 2\sqrt{3} + 6 - 2\sqrt{6}$$

When simplifying radicals it is useful to keep in mind that for any $x \geq 0$,

$$\sqrt{x} \cdot \sqrt{x} = x.$$

Example 8.2.23 Expand $\left(\sqrt{3} - \sqrt{2}\right)^2$.

Explanation. We will use the FOIL method to expand this expression:

$$\left(\sqrt{3} - \sqrt{2}\right)^2 = \left(\sqrt{3} - \sqrt{2}\right)\left(\sqrt{3} - \sqrt{2}\right)$$
$$= \left(\sqrt{3}\right)^2 - \sqrt{3}\sqrt{2} - \sqrt{2}\sqrt{3} + \left(\sqrt{2}\right)^2$$
$$= 3 - \sqrt{6} - \sqrt{6} + 2$$
$$= 5 - 2\sqrt{6}$$

Example 8.2.24 Multiply $\left(\sqrt{5} - \sqrt{7}\right)\left(\sqrt{5} + \sqrt{7}\right)$.

Explanation. We will again use the FOIL method to expand this expression, but will note that it is a special form $(a - b)(a + b)$ and will simplify to $a^2 - b^2$:

$$\left(\sqrt{5} - \sqrt{7}\right)\left(\sqrt{5} + \sqrt{7}\right) = \left(\sqrt{5}\right)^2 + \sqrt{5}\sqrt{7} - \sqrt{7}\sqrt{5} - \left(\sqrt{7}\right)^2$$
$$= 5 + \sqrt{35} - \sqrt{35} - 7$$
$$= -2$$

Exercises

Review and Warmup   Which of the following are square numbers? There may be more than one correct answer.

1.

   □ 16   □ 86   □ 40   □ 9   □ 132   □ 64

2.

   □ 53   □ 121   □ 25   □ 84   □ 93   □ 1

Evaluate the following.

3.   a. $\sqrt{49} = $ [____]

b. $\sqrt{36} = $ [____]

c. $\sqrt{25} = $ [____]

4.   a. $\sqrt{64} = $ [____]

b. $\sqrt{4} = $ [____]

c. $\sqrt{121} = $ [____]

5.   a. $\sqrt{\dfrac{81}{121}} = $ [____]

b. $\sqrt{-\dfrac{9}{100}} = $ [____]

6.   a. $\sqrt{\dfrac{121}{36}} = $ [____]

b. $\sqrt{-\dfrac{64}{25}} = $ [____]

7. Do not use a calculator.

a. $\sqrt{144} = $ [____]

b. $\sqrt{1.44} = $ [____]

c. $\sqrt{14400} = $ [____]

8. Do not use a calculator.

a. $\sqrt{4} = $ [____]

b. $\sqrt{0.04} = $ [____]

c. $\sqrt{400} = $ [____]

9. Do not use a calculator.

a. $\sqrt{9} = $ [____]

b. $\sqrt{900} = $ [____]

c. $\sqrt{90000} = $ [____]

10. Do not use a calculator.

a. $\sqrt{16} = $ [____]

b. $\sqrt{1600} = $ [____]

c. $\sqrt{160000} = $ [____]

11. Do not use a calculator.

a. $\sqrt{25} = $ [____]

b. $\sqrt{0.25} = $ [____]

c. $\sqrt{0.0025} = $ [____]

12. Do not use a calculator.

a. $\sqrt{49} = $ [____]

b. $\sqrt{0.49} = $ [____]

c. $\sqrt{0.0049} = $ [____]

13. Without using a calculator, estimate the value of $\sqrt{65}$:

    (□ 7.94 □ 8.94 □ 8.06 □ 7.06)

14. Without using a calculator, estimate the value of $\sqrt{78}$:

    (□ 9.83 □ 8.83 □ 9.17 □ 8.17)

Simplify Radical Expressions   Evaluate the following.

15. $\sqrt{\dfrac{100}{121}} = $ [_____].

16. $\sqrt{\dfrac{121}{144}} = $ [_____].

17. $-\sqrt{4} = $ [_____].

18. $-\sqrt{9} = $ [_____].

19. $\sqrt{-25} = $ [_____].

20. $\sqrt{-36} = $ [_____].

21. $\sqrt{-\dfrac{36}{49}} = $ [_____].

22. $\sqrt{-\dfrac{49}{144}} = $ [_____].

23. $-\sqrt{\dfrac{81}{100}} = $ [_____].

24. $-\sqrt{\dfrac{100}{121}} = $ [_____].

25. a. $\sqrt{169} - \sqrt{144} = $ [_____]

     b. $\sqrt{169 - 144} = $ [_____]

26. a. $\sqrt{25} - \sqrt{9} = $ [_____]

     b. $\sqrt{25 - 9} = $ [_____]

Simplify the radical expression or state that it is not a real number.

27. $\dfrac{\sqrt{8}}{\sqrt{2}} = $ [_____]

28. $\dfrac{\sqrt{54}}{\sqrt{6}} = $ [_____]

29. $\dfrac{\sqrt{4}}{\sqrt{36}} = $ [_____]

30. $\dfrac{\sqrt{2}}{\sqrt{32}} = $ [_____]

31. $\sqrt{343} = $ [_____]

32. $\sqrt{90} = $ [_____]

33. $\sqrt{360} = $ [_____]

34. $\sqrt{2156} = $ [_____]

35. $\sqrt{231} = $ [_____]

36. $\sqrt{70} = $ [_____]

Multiplying Square Root Expressions   Simplify the expression.

37. $4\sqrt{7} \cdot 7\sqrt{5} = $ [_____]

38. $5\sqrt{7} \cdot 4\sqrt{2} = $ [_____]

39. $6\sqrt{13} \cdot 2\sqrt{25} = $ [_____]

40. $6\sqrt{3} \cdot 7\sqrt{121} = $ [_____]

41. $4\sqrt{6} \cdot 3\sqrt{12} = $ [_____]

42. $5\sqrt{3} \cdot 5\sqrt{150} = $ [_____]

43. $\sqrt{2} \cdot 4\sqrt{50} = $ [_____]

44. $\sqrt{2} \cdot 6\sqrt{8} = $ [_____]

45. $\sqrt{\dfrac{7}{3}} \cdot \sqrt{\dfrac{5}{3}} = $ [_____]

46. $\sqrt{\dfrac{3}{4}} \cdot \sqrt{\dfrac{1}{4}} = \boxed{}$ **47.** $\sqrt{\dfrac{6}{19}} \cdot \sqrt{\dfrac{3}{19}} = \boxed{}$ **48.** $\sqrt{\dfrac{28}{11}} \cdot \sqrt{\dfrac{4}{11}} = \boxed{}$

Adding and Subtracting Square Root Expressions   Simplify the expression.

49. $16\sqrt{10} - 17\sqrt{10} = \boxed{}$ **50.** $17\sqrt{3} - 18\sqrt{3} = \boxed{}$

51. $19\sqrt{2} - 19\sqrt{2} + 14\sqrt{2} = \boxed{}$ **52.** $20\sqrt{23} - 15\sqrt{23} + 19\sqrt{23} = \boxed{}$

53. $\sqrt{8} + \sqrt{18} = \boxed{}$ **54.** $\sqrt{50} + \sqrt{18} = \boxed{}$

55. $\sqrt{48} - \sqrt{12} = \boxed{}$ **56.** $\sqrt{12} - \sqrt{75} = \boxed{}$

57. $\sqrt{180} + \sqrt{125} + \sqrt{27} + \sqrt{75} = \boxed{}$ **58.** $\sqrt{54} + \sqrt{24} + \sqrt{108} + \sqrt{12} = \boxed{}$

59. $\sqrt{294} - \sqrt{54} - \sqrt{72} - \sqrt{50} = \boxed{}$ **60.** $\sqrt{175} - \sqrt{63} - \sqrt{180} - \sqrt{20} = \boxed{}$

Rationalizing the Denominator   Evaluate the following.

61. $\dfrac{9}{\sqrt{25}} = \boxed{}$.   **62.** $\dfrac{1}{\sqrt{4}} = \boxed{}$.

Rationalize the denominator and simplify the expression.

63. $\dfrac{1}{\sqrt{2}} = \boxed{}$ **64.** $\dfrac{1}{\sqrt{3}} = \boxed{}$ **65.** $\dfrac{4}{\sqrt{5}} = \boxed{}$ **66.** $\dfrac{7}{\sqrt{5}} = \boxed{}$

67. $\dfrac{7}{8\sqrt{6}} = \boxed{}$ **68.** $\dfrac{10}{3\sqrt{7}} = \boxed{}$ **69.** $\dfrac{12}{\sqrt{30}} = \boxed{}$ **70.** $\dfrac{15}{\sqrt{35}} = \boxed{}$

71. $\dfrac{10}{\sqrt{2}} = \boxed{}$ **72.** $\dfrac{6}{\sqrt{2}} = \boxed{}$ **73.** $\dfrac{1}{\sqrt{18}} = \boxed{}$ **74.** $\dfrac{1}{\sqrt{63}} = \boxed{}$

75. $\dfrac{8}{\sqrt{180}} = \boxed{}$ **76.** $\dfrac{10}{\sqrt{252}} = \boxed{}$ **77.** $\sqrt{\dfrac{13}{64}} = \boxed{}$ **78.** $\sqrt{\dfrac{7}{81}} = \boxed{}$

79. $\sqrt{\dfrac{9}{2}} = \boxed{}$　　80. $\sqrt{\dfrac{81}{2}} = \boxed{}$　　81. $\sqrt{\dfrac{3}{7}} = \boxed{}$　　82. $\sqrt{\dfrac{5}{2}} = \boxed{}$

83. $\sqrt{\dfrac{54}{7}} = \boxed{}$　　84. $\sqrt{\dfrac{48}{11}} = \boxed{}$

More Complicated Square Root Operations　Expand and simplify the expression.

85. $\sqrt{7}\left(\sqrt{3} + \sqrt{13}\right) = \boxed{}$

86. $\sqrt{7}\left(\sqrt{19} + \sqrt{5}\right) = \boxed{}$

87. $\left(5 + \sqrt{13}\right)\left(6 + \sqrt{13}\right) = \boxed{}$

88. $\left(10 + \sqrt{13}\right)\left(9 + \sqrt{13}\right) = \boxed{}$

89. $\left(8 - \sqrt{5}\right)\left(7 - 3\sqrt{5}\right) = \boxed{}$

90. $\left(5 - \sqrt{5}\right)\left(6 - 4\sqrt{5}\right) = \boxed{}$

91. $\left(3 + \sqrt{7}\right)^2 = \boxed{}$

92. $\left(4 + \sqrt{5}\right)^2 = \boxed{}$

93. $\left(\sqrt{3} - 5\right)^2 = \boxed{}$

94. $\left(\sqrt{7} - 6\right)^2 = \boxed{}$

95. $\left(\sqrt{35} - \sqrt{5}\right)^2 = \boxed{}$

96. $\left(\sqrt{14} + \sqrt{7}\right)^2 = \boxed{}$

97. $\left(9 - 5\sqrt{7}\right)^2 = \boxed{}$

98. $\left(6 - 3\sqrt{5}\right)^2 = \boxed{}$

99. $\left(4 - \sqrt{5}\right)\left(4 + \sqrt{5}\right) = \boxed{}$

100. $\left(9 - \sqrt{6}\right)\left(9 + \sqrt{6}\right) = \boxed{}$

101. $\left(\sqrt{6} + \sqrt{7}\right)\left(\sqrt{6} - \sqrt{7}\right) = \boxed{}$

102. $\left(\sqrt{7} + \sqrt{5}\right)\left(\sqrt{7} - \sqrt{5}\right) = \boxed{}$

103. $\left(5\sqrt{5} + 6\sqrt{7}\right)\left(5\sqrt{5} - 6\sqrt{7}\right) = \boxed{}$

104. $\left(6\sqrt{6} + 4\sqrt{5}\right)\left(6\sqrt{6} - 4\sqrt{5}\right) = \boxed{}$

8.3 Solving Quadratic Equations by Using a Square Root

In Section 8.1, we learned how to solve quadratic equations by factoring. In this section, we will learn how to solve some specific types of quadratic equations using the **square root property**. We will also learn how to use the **Pythagorean Theorem** to find the length of one side of a right triangle when the other two lengths are known.

8.3.1 Solving Quadratic Equations Using the Square Root Property

When we learned how to solve linear equations, we used inverse operations to isolate the variable. For example, we use subtraction to remove an unwanted term that is added to one side of a linear equation. We can't quite do the same thing with squaring and using square roots, but we can do something very similar. Taking the square root is the inverse of squaring *if you happen to know the original number was positive*. In general, we have to remember that the original number may have been negative, and that usually leads to *two* solutions to a quadratic equation.

For example, if $x^2 = 9$, we can think of undoing the square with a square root, and $\sqrt{9} = 3$. However, there are *two* numbers that we can square to get 9: -3 and 3. So we need to include both solutions. This brings us to the Square Root Property.

Fact 8.3.2 The Square Root Property. *If k is positive, and $x^2 = k$ then $x = -\sqrt{k}$ or $x = \sqrt{k}$. The positive solution, $\sqrt{k}$, is called the **principal root** of k.*

Example 8.3.3 Solve for y in $y^2 - 49 = 0$.

Explanation. While we could factor and use the Zero Product Property, here we are demonstrating The Square Root Property instead. We need to isolate the squared quantity.

$$y^2 - 49 = 0$$
$$y^2 - 49 + 49 = 0 + 49$$
$$y^2 = 49$$

$$y = -\sqrt{49} \qquad \text{or} \qquad y = \sqrt{49}$$
$$y = -7 \qquad \text{or} \qquad y = 7$$

To check these solutions, we will replace y with -7 and with 7:

$$y^2 - 49 = 0 \qquad\qquad y^2 - 49 = 0$$
$$(-7)^2 - 49 \stackrel{?}{=} 0 \qquad\qquad (7)^2 - 49 \stackrel{?}{=} 0$$
$$49 - 49 \stackrel{\checkmark}{=} 0 \qquad\qquad 49 - 49 \stackrel{\checkmark}{=} 0$$

The solution set is $\{-7, 7\}$.

Remark 8.3.4. Every solution to a quadratic equation can be checked, as shown in Example 8.3.3. In general, the process of checking is omitted from this section.

Remark 8.3.5. Factoring will generally be a possible approach to solving a quadratic equation when the solution(s) are rational, but won't be a possible approach when the solution(s) are irrational.

For example, we could have solved the quadratic equation in Example 8.3.3 by factoring in this way:

$$y^2 - 49 = 0$$
$$(y + 7)(y - 7) = 0$$

$y + 7 = 0$	or	$y - 7 = 0$
$y = -7$	or	$y = 7$

However, as we'll see in Example 8.3.9, we *cannot* solve $2n^2 - 3 = 0$ by factoring but we *can* use the square root property.

Checkpoint 8.3.6. Solve for z in $4z^2 - 81 = 0$.

Explanation. Before we use the square root property we need to isolate the squared quantity.

$$4z^2 - 81 = 0$$
$$4z^2 = 81$$
$$z^2 = \frac{81}{4}$$

$z = -\sqrt{\dfrac{81}{4}}$	or	$z = \sqrt{\dfrac{81}{4}}$
$z = -\dfrac{9}{2}$	or	$z = \dfrac{9}{2}$

The solution set is $\left\{-\frac{9}{2}, \frac{9}{2}\right\}$.

We can also use the square root property to solve an equation that has a squared expression.

Example 8.3.7 Solve for p in $50 = 2(p - 1)^2$.

Explanation. It's important here to suppress any urge you may have to expand the squared binomial. We begin by isolating the squared expression.

$$50 = 2(p - 1)^2$$
$$\frac{50}{2} = \frac{2(p - 1)^2}{2}$$
$$25 = (p - 1)^2$$

Now that we have the squared expression isolated, we can use the square root property.

$(p - 1) = -\sqrt{25}$	or	$(p - 1) = \sqrt{25}$
$p - 1 = -5$	or	$p - 1 = 5$
$p = -4$	or	$p = 6$

The solution set is $\{-4, 6\}$.

This method of solving quadratic equations is not limited to equations that have rational solutions, or when the radicands are perfect squares. Here are a few examples where the solutions are irrational numbers.

Checkpoint 8.3.8. Solve for q in $(q+2)^2 - 12 = 0$.

Explanation. It's important here to suppress any urge you may have to expand the squared binomial.

$$(q+2)^2 - 12 = 0$$
$$(q+2)^2 = 12$$

$$(q+2) = -\sqrt{12} \quad \text{or} \quad (q+2) = \sqrt{12}$$
$$q+2 = -2\sqrt{3} \quad \text{or} \quad q+2 = 2\sqrt{3}$$
$$q = -2\sqrt{3} - 2 \quad \text{or} \quad q = 2\sqrt{3} - 2$$

The solution set is $\left\{ -2\sqrt{3} - 2, 2\sqrt{3} - 2 \right\}$.

To check the solution, we would replace q with each of $-2\sqrt{3} - 2$ and $2\sqrt{3} - 2$ in the original equation, as shown here:

$$\left(\left(-2\sqrt{3} - 2 \right) + 2 \right)^2 - 12 \overset{?}{=} 0 \qquad \left(\left(2\sqrt{3} - 2 \right) + 2 \right)^2 - 12 \overset{?}{=} 0$$
$$\left(-2\sqrt{3} \right)^2 - 12 \overset{?}{=} 0 \qquad\qquad \left(2\sqrt{3} \right)^2 - 12 \overset{?}{=} 0$$
$$(-2)^2 \left(\sqrt{3} \right)^2 - 12 \overset{?}{=} 0 \qquad\qquad (2)^2 \left(\sqrt{3} \right)^2 - 12 \overset{?}{=} 0$$
$$(4)(3) - 12 \overset{?}{=} 0 \qquad\qquad (4)(3) - 12 \overset{?}{=} 0$$
$$12 - 12 \overset{\checkmark}{=} 0 \qquad\qquad\qquad 12 - 12 \overset{\checkmark}{=} 0$$

Note that these simplifications relied on exponent rules and the multiplicative property of square roots.

Remember that if a square root is in the denominator then we need to rationalize it like we learned in Section 8.2. We will need to rationalize the denominator in the next example.

Example 8.3.9 Solve for n in $2n^2 - 3 = 0$.

Explanation.

$$2n^2 - 3 = 0$$
$$2n^2 = 3$$
$$n^2 = \frac{3}{2}$$

$$n = -\sqrt{\frac{3}{2}} \qquad\qquad \text{or} \qquad\qquad n = \sqrt{\frac{3}{2}}$$
$$n = -\frac{\sqrt{3}}{\sqrt{2}} \qquad\qquad \text{or} \qquad\qquad n = \frac{\sqrt{3}}{\sqrt{2}}$$
$$n = -\frac{\sqrt{3} \cdot \sqrt{2}}{\sqrt{2} \cdot \sqrt{2}} \qquad\qquad \text{or} \qquad\qquad n = \frac{\sqrt{3} \cdot \sqrt{2}}{\sqrt{2} \cdot \sqrt{2}}$$

$$n = -\frac{\sqrt{6}}{2} \qquad \text{or} \qquad n = \frac{\sqrt{6}}{2}$$

The solution set is $\left\{-\frac{\sqrt{6}}{2}, \frac{\sqrt{6}}{2}\right\}$.

When the radicand is a negative number, there is no real solution. Here is an example of an equation with no real solution.

Example 8.3.10 Solve for x in $x^2 + 49 = 0$.

Explanation.

$$x^2 + 49 = 0$$
$$x^2 = -49$$

Since $\sqrt{-49}$ is not a real number, we say the equation has no real solution.

8.3.2 The Pythagorean Theorem

Right triangles have an important property called the **Pythagorean Theorem**.

Theorem 8.3.11 The Pythagorean Theorem. *For any right triangle, the lengths of the three sides have the following relationship: $a^2 + b^2 = c^2$. The sides a and b are called **legs** and the longest side c is called the **hypotenuse**.*

Figure 8.3.12: In a right triangle, the length of its three sides satisfy the equation $a^2 + b^2 = c^2$

Keisha is designing a wooden frame in the shape of a right triangle, as shown in Figure 8.3.14. The legs of the triangle are 3 ft and 4 ft. How long should she make the diagonal side? Use the Pythagorean Theorem to find the length of the hypotenuse.

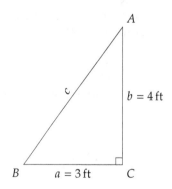

Figure 8.3.14

Example 8.3.13 According to Pythagorean Theorem, we have:

$$c^2 = a^2 + b^2$$
$$c^2 = 3^2 + 4^2$$
$$c^2 = 9 + 16$$
$$c^2 = 25$$

Now we have a quadratic equation that we need to solve. We need to find the number that has a square of 25. That is what the square root operation does.

$$c = \sqrt{25}$$
$$c = 5$$

The diagonal side Keisha will cut is 5 ft long.

Note that -5 is also a solution of $c^2 = 25$ because $(-5)^2 = 25$ but a length cannot be a negative number. We will need to include both solutions when they are relevant.

Example 8.3.15 A 16.5ft ladder is leaning against a wall. The distance from the base of the ladder to the wall is 4.5 feet. How high on the wall can the ladder reach?

The Pythagorean Theorem says:

$$a^2 + b^2 = c^2$$
$$4.5^2 + b^2 = 16.5^2$$
$$20.25 + b^2 = 272.25$$

Now we need to isolate b^2 in order to solve for b:

$$20.25 + b^2 - 20.25 = 272.25 - 20.25$$
$$b^2 = 252$$

To remove the square, we use the square root property. Because this is a geometric situation we only need to use the principal root:

$$b = \sqrt{252}$$

Now simplify this radical and then approximate it:

$$b = \sqrt{36 \cdot 7}$$
$$b = 6\sqrt{7}$$
$$b \approx 15.87$$

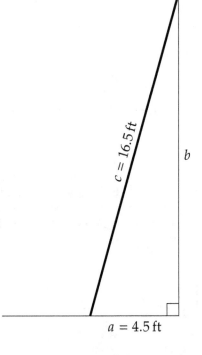

Figure 8.3.16: Leaning Ladder

The ladder can reach about 15.87 feet high on the wall.

Here are some more examples using the Pythagorean Theorem to find sides of triangles. Note that in many contexts, only the principal root will be relevant.

Example 8.3.17 Find the missing length in this right triangle.

Figure 8.3.18: A Right Triangle

Explanation. We will use the Pythagorean Theorem to solve for x:

$$5^2 + x^2 = 10^2$$
$$25 + x^2 = 100$$
$$x^2 = 75$$
$$x = \sqrt{75} \qquad \text{(no need to consider } -\sqrt{75} \text{ in this context)}$$
$$x = \sqrt{25 \cdot 3}$$
$$x = 5\sqrt{3}$$

The missing length is $x = 5\sqrt{3}$.

Example 8.3.19 Sergio is designing a 50-inch TV, which implies the diagonal of the TV's screen will be 50 inches long. He needs the screen's width to height ratio to be 4 : 3. Find the TV screen's width and height.

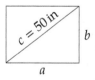

Figure 8.3.20: Pythagorean Theorem Problem

Explanation. Let's let x represent the height of the screen, in inches. Since the screen's width to height ratio will be 4 : 3, then the width is $\frac{4}{3}$ times as long as the height, or $\frac{4}{3}x$ inches. We will draw a diagram.

Figure 8.3.21: Pythagorean Theorem Problem

Now we can use the Pythagorean Theorem to write and solve an equation:

$$a^2 + b^2 = c^2$$

$$\left(\frac{4}{3}x\right)^2 + x^2 = 50^2$$

$$\frac{16}{9}x^2 + \frac{9}{9}x^2 = 2500$$

$$\frac{25}{9}x^2 = 2500$$

$$\frac{9}{25} \cdot \frac{25}{9}x^2 = \frac{9}{25} \cdot 2500$$

$$x^2 = 900$$

$$x = 30$$

Since the screen's height is 30 inches, its width is $\frac{4}{3}x = \frac{4}{3}(30) = 40$ inches.

Example 8.3.22 Luca wanted to make a bench.

Figure 8.3.23: Sketch of a Bench with Highlighted Back

He wanted the top of the bench back to be a perfect portion of a circle, in the shape of an arc, as in Figure 8.3.24. (Note that this won't be a half-circle, just a small portion of a circular edge.) He started with a rectangular board 3 inches wide and 48 inches long, and a piece of string, like a compass, to draw a circular arc on the board. How long should the string be so that it can be swung round to draw the arc?

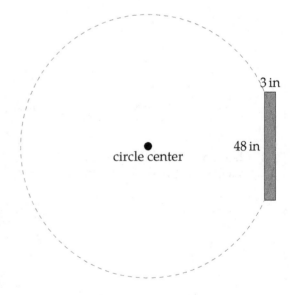

Figure 8.3.24: Bench Back Board

Explanation. Let's first define x to be the radius of the circle in question, in inches. The circle should go through the bottom corners of the board and just barely touch the top of the board. That means that the line from the middle of the bottom of the board to the center of the circle will be 3 inches shorter than the radius.

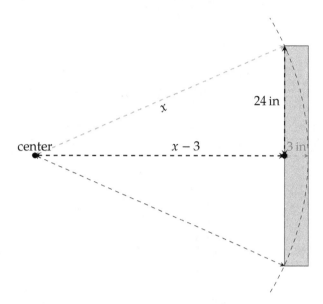

Figure 8.3.25: Bench Back Board Diagram

Now we can set up the Pythagorean Theorem based on the scenario. The equation $a^2 + b^2 = c^2$ turns into...

$$(x - 3)^2 + 24^2 = x^2$$
$$x^2 - 6x + 9 + 576 = x^2$$
$$-6x + 585 = 0$$

Note that at this point the equation is no longer quadratic! Solve the linear equation by isolating x

$$6x = 585$$
$$x = 97.5$$

So, the circle radius required is 97.5 inches. Luca found a friend to stand on the string end and drew a circular segment on the board to great effect.

Exercises

Solving Quadratic Equations with the Square Root Property Solve the equation.

1. $x^2 = 4$

2. $x^2 = 9$

3. $x^2 = \frac{1}{16}$

4. $x^2 = \frac{1}{25}$

5. $x^2 = 12$

6. $x^2 = 20$

7. $x^2 = 43$

8. $x^2 = 53$

9. $10x^2 = 160$

10. $2x^2 = 8$

11. $x^2 = \frac{144}{121}$

12. $x^2 = \frac{64}{25}$

13. $81x^2 = 16$

14. $121x^2 = 144$

15. $29x^2 - 37 = 0$

16. $5x^2 - 43 = 0$

17. $-5 - 7t^2 = -8$

18. $14 - 3t^2 = 7$

19. $5x^2 + 2 = 0$

20. $47x^2 + 5 = 0$

21. $(x - 5)^2 = 36$

22. $(x - 3)^2 = 4$

23. $(11x + 6)^2 = 100$

24. $(11x + 6)^2 = 16$

25. $26 - 2(r + 8)^2 = 8$

26. $8 - 5(t + 8)^2 = 3$

27. $(x - 4)^2 = 61$

28. $(x + 10)^2 = 2$

29. $(x + 6)^2 = 75$

30. $(x + 7)^2 = 63$

31. $7 = 154 - (y - 6)^2$

32. $7 = 154 - (y - 8)^2$

Pythagorean Theorem Applications

33.

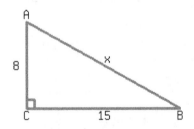

Find the value of x.

$x = $ [   ]

34.

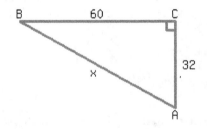

Find the value of x.

$x = $ [   ]

35.

Find the value of x.

$x =$ [ ]

36.

Find the value of x.

$x =$ [ ]

37.

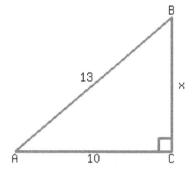

Find the value of x, accurate to at least two decimal places.

$x \approx$ [ ]

38.

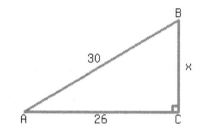

Find the value of x, accurate to at least two decimal places.

$x \approx$ [ ]

39.

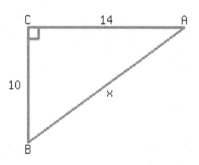

Find the exact value of x.

$x =$ [ ]

40.

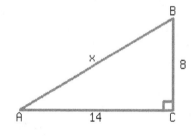

Find the exact value of x.

$x =$ [ ]

41.

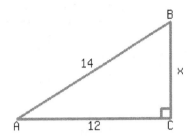

Find the exact value of x.

$x =$ _____

42.

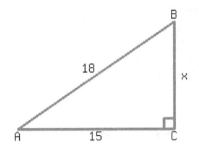

Find the exact value of x.

$x =$ _____

43. Rebecca is designing a rectangular garden. The garden's diagonal must be 18.5 feet, and the ratio between the garden's base and height must be 4 : 3. Find the length of the garden's base and height.

The garden's base is _____

feet and its height is _____ .

44. Jessica is designing a rectangular garden. The garden's diagonal must be 44.2 feet, and the ratio between the garden's base and height must be 15 : 8. Find the length of the garden's base and height.

The garden's base is _____

feet and its height is _____ .

45. Ashley is designing a rectangular garden. The garden's base must be 6 feet, and the ratio between the garden's hypotenuse and height must be 5 : 3. Find the length of the garden's hypotenuse and height.

The garden's hypotenuse is _____

feet and its height is _____ .

46. Neil is designing a rectangular garden. The garden's base must be 4.8 feet, and the ratio between the garden's hypotenuse and height must be 13 : 5. Find the length of the garden's hypotenuse and height.

The garden's hypotenuse is _____

feet and its height is _____ .

Challenge

47. Imagine that you are in Math Land, where roads are perfectly straight, and Mathlanders can walk along a perfectly straight line between any two points. One day, you bike 2 miles west, 4 miles north, and 8 miles east. Then, your bike gets a flat tire and you have to walk home. How far do you have to walk? You have to walk _____ miles home.

8.4 The Quadratic Formula

We have learned how to solve quadratic equations using factoring and the square root property. In this section, we will learn a third method, the quadratic formula. We will also learn when to use each method and to distinguish between linear and quadratic equations.

8.4.1 Solving Quadratic Equations with the Quadratic Formula

The standard form of a quadratic equation is

$$ax^2 + bx + c = 0$$

Let's look at two examples as a review of solving quadratic equations.

First, let's look at when $b = 0$ and the equation looks like $ax^2 + c = 0$. One way to solve this type of equation is with the square root property.

> **Example 8.4.2** Solve for x in $x^2 - 4 = 0$.
>
> **Explanation.**
>
> $$x^2 - 4 = 0$$
> $$x^2 = 4$$
>
> $$x = -2 \qquad \text{or} \qquad x = 2$$
>
> The solution set is $\{-2, 2\}$.

Second, if we can factor the left side of the equation in $ax^2 + bx + c = 0$, then we can solve the equation by factoring.

> **Example 8.4.3** Solve for x in $x^2 - 4x - 12 = 0$.
>
> **Explanation.**
>
> $$x^2 - 4x - 12 = 0$$
> $$(x - 6)(x + 2) = 0$$
>
> $$x - 6 = 0 \qquad \text{or} \qquad x + 2 = 0$$
> $$x = 6 \qquad \text{or} \qquad x = -2$$
>
> The solution set is $\{-2, 6\}$.

A third method for solving a quadratic equation is to use what is known as the quadratic formula.

Fact 8.4.4 The Quadratic Formula. *For any quadratic equation $ax^2 + bx + c = 0$, the solutions are given by*

$$x = \frac{-b \pm \sqrt{b^2 - 4ac}}{2a}$$

As we have seen from solving quadratic equations, there can be at most two solutions. Both of the solutions are included in the quadratic formula with the $\pm$ symbol. We could write the two solutions separately as

$$x = \frac{-b - \sqrt{b^2 - 4ac}}{2a} \qquad \text{or} \qquad x = \frac{-b + \sqrt{b^2 - 4ac}}{2a}$$

but it is more efficient to simplify them together.

This method for solving quadratic equations will work to solve *every* quadratic equation. It is most helpful when $b \neq 0$ and when $ax^2 + bx + c$ cannot be factored. In this section, we will only focus on how to use the formula.

In Section 8.1, we saw an example where a physics class launched a tennis ball off the roof of a building. In that example, the numbers were simplified so we could solve it by factoring. Now we will solve a similar example with more realistic numbers.

> **Example 8.4.5** Linh is in another physics class that launches a tennis ball from a rooftop that is 90.2 feet above the ground. They fire it directly upward at a speed of 14.4 feet per second and measure the time it takes for the ball to hit the ground below. We can model the height of the tennis ball, h, in feet, with the quadratic equation $h = -16x^2 + 14.4x + 90.2$, where x represents the time in seconds after the launch. According to the model, when should the ball hit the ground? Round the time to one decimal place.
>
> The ground has a height of 0 feet. Substituting 0 for h in the equation, we have this quadratic equation:
>
> $$0 = -16x^2 + 14.4x + 90.2$$
>
> We cannot solve this equation with factoring or the square root property, so we will use the quadratic formula. First we will identify that $a = -16$, $b = 14.4$ and $c = 90.2$, and substitute them into the formula:
>
> $$x = \frac{-b \pm \sqrt{b^2 - 4ac}}{2a}$$
>
> $$x = \frac{-(14.4) \pm \sqrt{(14.4)^2 - 4(-16)(90.2)}}{2(-16)}$$
>
> $$x = \frac{-14.4 \pm \sqrt{207.36 - (-5772.8)}}{-32}$$
>
> $$x = \frac{-14.4 \pm \sqrt{207.36 + 5772.8}}{-32}$$
>
> $$x = \frac{-14.4 \pm \sqrt{5980.16}}{-32}$$
>
> These are the exact solutions but because we have a context we want to approximate the solutions with decimals.
>
> $$x \approx -2.0 \text{ or } x \approx 2.9$$
>
> We don't use the negative solution because a negative time does not make sense in this context. The ball will hit the ground approximately 2.9 seconds after it is launched.

The quadratic formula can be used to solve any quadratic equation, but it requires that you don't make *any* slip-up with remembering the formula, that you correctly identify a, b, and c, and that you don't make any arithmetic mistakes when you calculate and simplify. We recommend that you always check whether you can factor or use the square root property before using the quadratic formula. Here is another example.

Example 8.4.6 Solve for x in $2x^2 - 9x + 5 = 0$.

Explanation. First, we check and see that we cannot factor the left side (because we can't find two numbers that multiply to 10 and add to -9) or use the square root property (because $b \neq 0$) so we must use the quadratic formula. Next we identify that $a = 2$, $b = -9$ and $c = 5$. We will substitute them into the quadratic formula:

$$x = \frac{-b \pm \sqrt{b^2 - 4ac}}{2a}$$

$$x = \frac{-(-9) \pm \sqrt{(-9)^2 - 4(2)(5)}}{2(2)}$$

$$x = \frac{-9 \pm \sqrt{81 - 40}}{4}$$

$$x = \frac{-9 \pm \sqrt{41}}{4}$$

This is fully simplified because we cannot simplify $\sqrt{41}$ or reduce the fraction. The solution set is $\left\{ \frac{-9-\sqrt{22}}{4}, \frac{-9+\sqrt{22}}{4} \right\}$. We do not have a context here so we leave the solutions in their exact form.

When a quadratic equation is not in standard form we must convert it before we can identify the values of a, b and c. We will show that in the next example.

Example 8.4.7 Solve for x in $x^2 = -10x - 3$.

Explanation. First, we convert the equation into standard form by adding $10x$ and 3 to each side of the equation:

$$x^2 + 10x + 3 = 0$$

Next, we check and see that we cannot factor the left side or use the square root property so we must use the quadratic formula. We identify that $a = 1$, $b = 10$ and $c = 3$. We will substitute them into the quadratic formula:

$$x = \frac{-b \pm \sqrt{b^2 - 4ac}}{2a}$$

$$x = \frac{-10 \pm \sqrt{(10)^2 - 4(1)(3)}}{2(1)}$$

$$x = \frac{-10 \pm \sqrt{100 - 12}}{2}$$

$$x = \frac{-10 \pm \sqrt{88}}{2}$$

Now we need to simplify the square root:

$$x = \frac{-10 \pm 2\sqrt{22}}{2}$$

Lastly we need to reduce the fractions, which can be done by separating terms:

$$x = \frac{-10}{2} \pm \frac{2\sqrt{22}}{2}$$
$$x = -5 \pm \sqrt{22}$$

The solution set is $\{-5 - \sqrt{22}, -5 + \sqrt{22}\}$.

Remark 8.4.8. The irrational solutions to quadratic equations can be checked, although doing so can sometimes involve a lot of simplification and is not shown throughout this section. As an example, to check the solution of $-5 + \sqrt{22}$ from Example 8.4.7, we would replace x with $-5 + \sqrt{22}$ and check that the two sides of the equation are equal. This check is shown here:

$$x^2 = -10x - 3$$
$$(-5 + \sqrt{22})^2 \stackrel{?}{=} -10(-5 + \sqrt{22}) - 3$$
$$(-5)^2 + 2(-5)(\sqrt{22}) + (\sqrt{22})^2 \stackrel{?}{=} -10(-5 + \sqrt{22}) - 3$$
$$25 - 10\sqrt{22} + 22 \stackrel{?}{=} 50 - 10\sqrt{22} - 3$$
$$47 - 10\sqrt{22} \stackrel{\checkmark}{=} 47 - 10\sqrt{22}$$

When the radicand from the quadratic formula (which is called the **discriminant**) is a negative number, the quadratic equation has no real solution. Example 8.4.9 shows what happens in this case.

Example 8.4.9 Solve for y in $y^2 - 4y + 8 = 0$.

Explanation. Identify that $a = 1$, $b = -4$ and $c = 8$. We will substitute them into the quadratic formula:

$$y = \frac{-b \pm \sqrt{b^2 - 4ac}}{2a}$$
$$= \frac{-(-4) \pm \sqrt{(-4)^2 - 4(1)(8)}}{2(1)}$$
$$= \frac{4 \pm \sqrt{16 - 32}}{2}$$
$$= \frac{4 \pm \sqrt{-16}}{2}$$

The square root of a negative number is not a real number, so we will simply state that this equation has no real solutions.

8.4.2 Recognizing Linear and Quadratic Equations

Now that we have solved both linear and quadratic equations, it is important to identify each type of equation. Recall that a **linear equation** has a degree of one and a **quadratic equation** has a degree of two. If there is any other operation on the variable such as division or a square root then it is not linear or quadratic. We can have other operations on numbers, but not the variables. Let's look at an example.

Checkpoint 8.4.10. Identify whether each equation is linear, quadratic or neither.

a. The equation $3 = 7y^2 - 8y$ is   (□ linear   □ quadratic   □ neither)   .

b. The equation $5x + 3 = 7x - 8$ is   (□ linear   □ quadratic   □ neither)   .

c. The equation $r^3 - 7 = 4$ is   (□ linear   □ quadratic   □ neither)   .

d. The equation $\sqrt{7}x + 4 = 10$ is   (□ linear   □ quadratic   □ neither)   .

Explanation.   We will check the degree of each equation:

a. This is a quadratic equation because there is a y^2 term.

b. This is a linear equation because the highest exponent is one.

c. This is neither linear nor quadratic because it has a degree of three.

d. This is a linear equation because it has a degree of one. The coefficient is irrational but the variable is not in the square root.

8.4.3 Solving Linear and Quadratic Equations

When an equation is **linear**, we move all variable terms to one side of the equation and all constant terms to the other side. Then we use division if needed to solve for the variable. This is outlined in List 3.1.4.

When an equation is **quadratic**, we have three different methods we can use. Here is an outline of the general process for determining which method to use.

Process 8.4.11 Solving Quadratic Equations.

1. *First, check whether there is a linear term, or whether there is only a squared expression and a constant. If there is only a squared expression and a constant, isolate the squared quantity and use the **square root method**.*

2. *If there is a linear term, put the equation in **standard form** with all of the terms on one side and zero on the other side.*

   a) *If the polynomial factors, solve the equation by **factoring**.*

   b) *If the polynomial does not factor, use the **quadratic formula**.*

Here are some examples:

Example 8.4.12 Solve for x in $x^2 = 7x^2 - 12$.

Explanation.   This is a quadratic equation because there are x^2 terms. There are no x terms so we will use the square root method. We start by combining the x^2 terms on the left.

$$x^2 = 7x^2 - 12$$
$$-6x^2 = -12$$
$$x^2 = 2$$

$$x = -\sqrt{2} \qquad \text{or} \qquad x = \sqrt{2}$$

The solution set is $\{-\sqrt{2}, \sqrt{2}\}$.

Example 8.4.13 Solve for p in $p^2 = -2p + 2$.

Explanation. This is a quadratic equation that also contains a linear term so we will put the equation in standard form. Then we see that the left side does not factor so we will use the quadratic formula.

$$p^2 + 2p - 2 = 0$$

$$p = \frac{-(2) \pm \sqrt{(2)^2 - 4(1)(-2)}}{2(1)}$$

$$p = \frac{-2 \pm \sqrt{4 + 8}}{2}$$

$$p = \frac{-2 \pm \sqrt{12}}{2}$$

$$p = \frac{-2 \pm 2\sqrt{3}}{2}$$

$$p = \frac{-2}{2} \pm \frac{2\sqrt{3}}{2}$$

$$p = -1 \pm \sqrt{3}$$

The solution set is $\{1 - \sqrt{3}, 1 + \sqrt{3}\}$.

Example 8.4.14 Solve for t in $7 - t = 9t + 11$.

Explanation. This is a linear equation so we will combine the linear terms on one side and the constant terms on the other side.

$$-10t = 4$$

$$t = \frac{4}{-10}$$

$$t = -\frac{2}{5}$$

The solution set is $\left\{-\frac{2}{5}\right\}$.

Example 8.4.15 Solve for z in $z^2 - 10 = 3z + 30$.

Explanation. This is a quadratic equation with a linear term so we will put it in standard form. Then we see that the left side factors so we solve by factoring.

$$z^2 - 10 = 3z + 30$$

$$z^2 - 3z - 40 = 0$$

$$(z + 5)(z - 8) = 0$$

$$z + 5 = 0 \qquad \text{or} \qquad z - 8 = 0$$
$$z = -5 \qquad \text{or} \qquad z = 8$$

The solution set is $\{-5, 8\}$.

Exercises

Review and Warmup

1. Evaluate $\dfrac{2a - 2B + 10}{-10a + 10B}$ for $a = 6$ and $B = -5$.

2. Evaluate $\dfrac{2c + 8b - 9}{-c - 5b}$ for $c = -6$ and $b = 10$.

3. Evaluate the expression $\dfrac{1}{3}(x + 3)^2 - 9$ when $x = -6$.

4. Evaluate the expression $\dfrac{1}{6}(x + 3)^2 - 7$ when $x = -9$.

5. Evaluate the expression $-16t^2 + 64t + 128$ when $t = 3$.

6. Evaluate the expression $-16t^2 + 64t + 128$ when $t = 5$.

7. Evaluate the expression x^2:

    a. When $x = 6$, $x^2 = $ ⬚

    b. When $x = -4$, $x^2 = $ ⬚

8. Evaluate the expression x^2:

    a. When $x = 3$, $x^2 = $ ⬚

    b. When $x = -7$, $x^2 = $ ⬚

9. Evaluate each algebraic expression for the given value(s):

$\dfrac{\sqrt{x}}{y} - \dfrac{y}{x}$, for $x = 16$ and $y = 3$: ⬚

10. Evaluate each algebraic expression for the given value(s):

$\dfrac{y}{4x} - \dfrac{\sqrt{x}}{2y}$, for $x = 9$ and $y = 8$: ⬚

Solve Quadratic Equations Using the Quadratic Formula Solve the equation.

11. $x^2 - 8x + 5 = 0$

12. $x^2 - 10x + 17 = 0$

13. $10x^2 + 17x - 20 = 0$

14. $15x^2 + 4x - 32 = 0$

15. $x^2 = -5x - 3$

16. $x^2 = x + 1$

17. $x^2 + 9x + 9 = 0$

18. $x^2 - 9x + 9 = 0$

19. $3x^2 + 5x + 1 = 0$

20. $2x^2 + 3x - 1 = 0$

21. $7x^2 - 2x - 1 = 0$

22. $4x^2 - 10x - 5 = 0$

23. $5x^2 - 10x + 10 = 0$

24. $5x^2 - 5x + 7 = 0$

Solve Quadratic Equations Using an Appropriate Method   Solve the equation.

25. $2x^2 - 18 = 0$

26. $3x^2 - 75 = 0$

27. $9t + 7 = t + 23$

28. $7a + 2 = a + 50$

29. $9x^2 - 16 = 0$

30. $49x^2 - 64 = 0$

31. $-2 - 7r^2 = -4$

32. $8 - 7t^2 = 3$

33. $x^2 + 70x = 0$

34. $x^2 - 90x = 0$

35. $9y + 10 = -9y + 10 - 3y$

36. $6t + 5 = -6t + 5 - 3t$

37. $x^2 - 2x = 8$

38. $x^2 + 8x = 9$

39. $(x + 1)^2 = 49$

40. $(x + 3)^2 = 9$

41. $x^2 = -3x - 1$

42. $x^2 = -3x + 2$

43. $4x^2 = 9x - 4$

44. $3x^2 = 5x - 1$

45. $-6 - 3(x - 10)^2 = -9$

46. $14 - 2(y + 10)^2 = 6$

47. $-10 = -8c - 10 - c$

48. $31 = -5A - 5 - A$

Quadratic Formula Applications

49. Two numbers' sum is 16, and their product is 60. Find these two numbers.

These two numbers are ☐ .

50. Two numbers' sum is 14, and their product is 48. Find these two numbers.

These two numbers are ☐ .

51. Two numbers' sum is 11.7, and their product is 28.22. Find these two numbers.

These two numbers are ☐ .
(Use a comma to separate your numbers.)

52. Two numbers' sum is 2.1, and their product is −95.92. Find these two numbers.

These two numbers are ☐ .
(Use a comma to separate your numbers.)

53. A rectangle's base is 8 cm longer than its height. The rectangle's area is 65 cm². Find this rectangle's dimensions.

The rectangle's height is ☐ .

The rectangle's base is ☐ .

54. A rectangle's base is 6 cm longer than its height. The rectangle's area is 72 cm². Find this rectangle's dimensions.

The rectangle's height is ☐ .

The rectangle's base is ☐ .

55. A rectangle's base is 3 in shorter than five times its height. The rectangle's area is 36 in². Find this rectangle's dimensions.

The rectangle's height is [   ].

The rectangle's base is [   ].

56. A rectangle's base is 7 in shorter than four times its height. The rectangle's area is 15 in². Find this rectangle's dimensions.

The rectangle's height is [   ].

The rectangle's base is [   ].

57. You will build a rectangular sheep pen next to a river. There is no need to build a fence along the river, so you only need to build three sides.

You have a total of 480 feet of fence to use, and the area of the pen must be 28600 square feet. Find the dimensions of the pen.

There should be two solutions: When the width is [   ] feet, the length is [   ] feet.

When the width is [   ] feet, the length is [   ] feet.

58. You will build a rectangular sheep pen next to a river. There is no need to build a fence along the river, so you only need to build three sides.

You have a total of 550 feet of fence to use, and the area of the pen must be 37700 square feet. Find the dimensions of the pen.

There should be two solutions: When the width is [   ] feet, the length is [   ] feet.

When the width is [   ] feet, the length is [   ] feet.

59. There is a rectangular lot in the garden, with 7 ft in length and 3 ft in width. You plan to expand the lot by an equal length around its four sides, and make the area of the expanded rectangle 77 ft². How long should you expand the original lot in four directions?

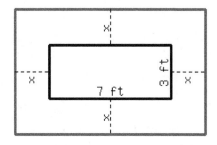

You should expand the original lot by [   ] in four directions.

60. There is a rectangular lot in the garden, with 10 ft in length and 6 ft in width. You plan to expand the lot by an equal length around its four sides, and make the area of the expanded rectangle 192 ft². How long should you expand the original lot in four directions?

You should expand the original lot by [   ] in four directions.

61. One car started at Town A, and traveled due north at 70 miles per hour. 3.5 hours later, another car started at the same spot and traveled due east at 65 miles per hour. Assume both cars don't stop, after how many hours since the second car starts would the distance between them be 469 miles? Round your answer to two decimal places if needed.

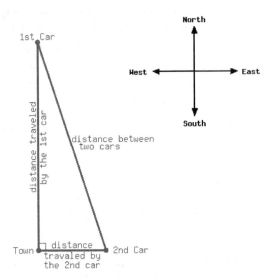

Approximately [   ] hours since the second car starts, the distance between those two cars would be 469 miles.

62. One car started at Town A, and traveled due north at 40 miles per hour. 1.5 hours later, another car started at the same spot and traveled due east at 55 miles per hour. Assume both cars don't stop, after how many hours since the second car starts would the distance between them be 229 miles? Round your answer to two decimal places if needed.

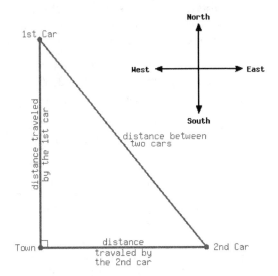

Approximately [   ] hours since the second car starts, the distance between those two cars would be 229 miles.

63. An object is launched upward at the height of 240 meters. It's height can be modeled by

$$h = -4.9t^2 + 50t + 240,$$

where h stands for the object's height in meters, and t stands for time passed in seconds since its launch. The object's height will be 280 meters twice before it hits the ground. Find how many seconds since the launch would the object's height be 280 meters. Round your answers to two decimal places if needed.

The object's height would be 280 meters the first time at [   ] seconds, and then the second time at [   ] seconds.

64. An object is launched upward at the height of 260 meters. It's height can be modeled by

$$h = -4.9t^2 + 90t + 260,$$

where h stands for the object's height in meters, and t stands for time passed in seconds since its launch. The object's height will be 270 meters twice before it hits the ground. Find how many seconds since the launch would the object's height be 270 meters. Round your answers to two decimal places if needed.

The object's height would be 270 meters the first time at [   ] seconds, and then the second time at [   ] seconds.

65. Currently, an artist can sell 230 paintings every year at the price of $60.00 per painting. Each time he raises the price per painting by $5.00, he sells 10 fewer paintings every year.

Assume he will raise the price per painting x times, then he will sell $230 - 10x$ paintings every year at the price of $60 + 5x$ dollars. His yearly income can be modeled by the equation:

$$i = (60 + 5x)(230 - 10x)$$

where i stands for his yearly income in dollars. If the artist wants to earn $14,700.00 per year from selling paintings, what new price should he set?

To earn $14,700.00 per year, the artist could sell his paintings at two different prices. The lower price is [   ] per painting, and the higher price is [   ] per painting.

66. Currently, an artist can sell 250 paintings every year at the price of $110.00 per painting. Each time he raises the price per painting by $10.00, he sells 10 fewer paintings every year.

Assume he will raise the price per painting x times, then he will sell $250 - 10x$ paintings every year at the price of $110 + 10x$ dollars. His yearly income can be modeled by the equation:

$$i = (110 + 10x)(250 - 10x)$$

where i stands for his yearly income in dollars. If the artist wants to earn $32,000.00 per year from selling paintings, what new price should he set?

To earn $32,000.00 per year, the artist could sell his paintings at two different prices. The lower price is [   ] per painting, and the higher price is [   ] per painting.

67. Solve for x in the equation $mx^2 + nx + p = 0$.

8.5 Complex Solutions to Quadratic Equations

8.5.1 Imaginary Numbers

Let's look at how to simplify a square root that has a negative radicand. Remember that $\sqrt{16} = 4$ because $4^2 = 16$. So what could $\sqrt{-16}$ be equal to? There is no real number that we can square to get -16, because when you square a real number, the result is either positive or 0. You might think about 4 and -4, but:

$$4^2 = 16 \text{ and } (-4)^2 = 16$$

so neither of those could be $\sqrt{-16}$. To handle this situation, mathematicians separate a factor of $\sqrt{-1}$ and represent it with the letter i, which stands for **imaginary unit**.

Definition 8.5.2 Imaginary Numbers. The **imaginary unit**, i, is defined by $i = \sqrt{-1}$. The imaginary unit[1] satisfies the equation $i^2 = -1$. A real number times i, such as $4i$, is called an **imaginary number**.

Now we can simplify square roots with negative radicands like $\sqrt{-16}$.

$$\sqrt{-16} = \sqrt{-1 \cdot 16}$$
$$= \sqrt{-1} \cdot \sqrt{16}$$
$$= i \cdot 4$$
$$= 4i$$

Imaginary numbers are widely used in electrical engineering, physics, computer science and other fields. Let's look some more examples.

Example 8.5.3 Simplify $\sqrt{-2}$.

Explanation.

$$\sqrt{-2} = \sqrt{-1 \cdot 2}$$
$$= \sqrt{-1} \cdot \sqrt{2}$$
$$= i\sqrt{2}$$

We write the i first because it's difficult to tell the difference between $\sqrt{2}i$ and $\sqrt{2i}$.

Example 8.5.4 Simplify $\sqrt{-72}$.

Explanation.

$$\sqrt{-72} = \sqrt{-1 \cdot 36 \cdot 2}$$
$$= \sqrt{-1} \cdot \sqrt{36} \cdot \sqrt{2}$$
$$= 6i\sqrt{2}$$

---

[1]en.wikipedia.org/wiki/Imaginary_number

8.5.2 Solving Quadratic Equations with Imaginary Solutions

Example 8.5.5 Solve for x in $x^2 + 49 = 0$, where x is an imaginary number.

Explanation. There is no x term so we will use the square root method.

$$x^2 + 49 = 0$$
$$x^2 = -49$$

$x = -\sqrt{-49}$	or	$x = \sqrt{-49}$
$x = -\sqrt{-1} \cdot \sqrt{49}$	or	$x = \sqrt{-1} \cdot \sqrt{49}$
$x = -7i$	or	$x = 7i$

The solution set is $\{-7i, 7i\}$.

Example 8.5.6 Solve for p in $p^2 + 75 = 0$, where p is an imaginary number.

Explanation. There is no p term so we will use the square root method.

$$p^2 + 75 = 0$$
$$p^2 = -75$$

$p = -\sqrt{-75}$	or	$p = \sqrt{-75}$
$p = -\sqrt{-1} \cdot \sqrt{25} \cdot \sqrt{3}$	or	$p = \sqrt{-1} \cdot \sqrt{25} \cdot \sqrt{3}$
$p = -5i\sqrt{3}$	or	$p = 5i\sqrt{3}$

The solution set is $\left\{-5i\sqrt{3}, 5i\sqrt{3}\right\}$.

8.5.3 Solving Quadratic Equations with Complex Solutions

A **complex number** is a combination of a real number and an imaginary number, like $3 + 2i$ or $-4 - 8i$.

Definition 8.5.7 Complex Number. A **complex number** is a number that can be expressed in the form $a + bi$, where a and b are real numbers and i is the imaginary unit. In this expression, a is the **real part** and b (not bi) is the **imaginary part** of the complex number[2].

Here are some examples of equations that have complex solutions.

Example 8.5.8 Solve for m in $(m - 1)^2 + 18 = 0$, where m is a complex number.

Explanation. This equation has a squared expression so we will use the square root method.

$$(m - 1)^2 + 18 = 0$$
$$(m - 1)^2 = -18$$

---
[2]en.wikipedia.org/wiki/Complex_number

$$m - 1 = -\sqrt{-18} \qquad \text{or} \qquad m - 1 = \sqrt{-18}$$
$$m - 1 = -\sqrt{-1} \cdot \sqrt{9} \cdot \sqrt{2} \qquad \text{or} \qquad m - 1 = \sqrt{-1} \cdot \sqrt{9} \cdot \sqrt{2}$$
$$m - 1 = -3i\sqrt{2} \qquad \text{or} \qquad m - 1 = 3i\sqrt{2}$$
$$m = 1 - 3i\sqrt{2} \qquad \text{or} \qquad m = 1 + 3i\sqrt{2}$$

The solution set is $\left\{1 - 3i\sqrt{2}, 1 + 3i\sqrt{2}\right\}$.

Example 8.5.9 Solve for y in $y^2 - 4y + 13 = 0$, where y is a complex number.

Explanation.   Note that there is a y term, but the left side does not factor. We will use the quadratic formula. We identify that $a = 1$, $b = -4$ and $c = 13$ and substitute them into the quadratic formula.

$$
\begin{aligned}
y &= \frac{-b \pm \sqrt{b^2 - 4ac}}{2a} \\
&= \frac{-(-4) \pm \sqrt{(-4)^2 - 4(1)(13)}}{2(1)} \\
&= \frac{4 \pm \sqrt{16 - 52}}{2} \\
&= \frac{4 \pm \sqrt{-36}}{2} \\
&= \frac{4 \pm \sqrt{-1} \cdot \sqrt{36}}{2} \\
&= \frac{4 \pm 6i}{2} \\
&= 2 \pm 3i
\end{aligned}
$$

The solution set is $\{2 - 3i, 2 + 3i\}$.

Note that in Example 8.5.9, the expressions $2 + 3i$ and $2 - 3i$ are fully simplified. In the same way that the terms 2 and $3x$ cannot be combined, the terms 2 and $3i$ can not be combined.

Remark 8.5.10. Each complex solution can be checked, just as every real solution can be checked. For example, to check the solution of $2 + 3i$ from Example 8.5.9, we would replace y with $2 + 3i$ and check that the two sides of the equation are equal. In doing so, we will need to use the fact that $i^2 = -1$. This check is shown here:

$$y^2 - 4y + 13 = 0$$
$$(2 + 3i)^2 - 4(2 + 3i) + 13 \overset{?}{=} 0$$
$$(2^2 + 2(3i) + 2(3i) + (3i)^2) - 4 \cdot 2 - 4 \cdot (3i) + 13 \overset{?}{=} 0$$
$$4 + 6i + 6i + 9i^2 - 8 - 12i + 13 \overset{?}{=} 0$$
$$4 + 9(-1) - 8 + 13 \overset{?}{=} 0$$
$$4 - 9 - 8 + 13 \overset{?}{=} 0$$
$$0 \overset{\checkmark}{=} 0$$

Exercises

Simplifying Square Roots with Negative Radicands Simplify the radical and write it into a complex number.

1. $\sqrt{-105} =$ [   ]

2. $\sqrt{-42} =$ [   ]

3. $\sqrt{-72} =$ [   ]

4. $\sqrt{-56} =$ [   ]

5. $\sqrt{-126} =$ [   ]

6. $\sqrt{-200} =$ [   ]

Quadratic Equations with Imaginary and Complex Solutions Solve the quadratic equation. Solutions could be complex numbers.

7. $y^2 = -16$

8. $r^2 = -100$

9. $-6r^2 + 2 = 296$

10. $-3t^2 + 2 = 50$

11. $2t^2 + 10 = 6$

12. $2t^2 - 2 = -8$

13. $-6x^2 + 4 = 124$

14. $-6x^2 - 7 = 293$

15. $9(y + 4)^2 + 1 = -323$

16. $6(y - 5)^2 - 7 = -61$

17. $r^2 + 2r + 5 = 0$

18. $r^2 + 6r + 10 = 0$

19. $t^2 + 4t + 9 = 0$

20. $t^2 + 4t + 11 = 0$

8.6 Strategies for Solving Quadratic Equations

In this section, we will review how to solve quadratic equations using three different methods: The square root method, factoring and the quadratic formula.

8.6.1 How to Choose a Method for Solving a Quadratic Equation

Process 8.6.2. *So far, we have learned three methods for solving quadratic equations in standard form, $ax^2 + bx + c = 0$:*

1. *When $b = 0$, as in $x^2 - 4 = 0$, we can use the square root property, Property 8.3.2.*

2. *If we can easily factor the polynomial, as in $x^2 - 4x - 12 = 0$, we will solve the equation by factoring and use the zero product property, Property 8.1.4.*

3. *If we cannot solve the equation by the first two methods, we must use the Quadratic Formula, Property 8.4.4. This formula works for any quadratic equation, but the first two methods are usually easier.*

Let's look at a few examples for how to choose which method to use.

Example 8.6.3 Solve for y in $y^2 - 49 = 0$.

Explanation. In this equation, $b = 0$, so it is easiest to use the square root method. We isolate the squared quantity and then use the square root property.

$$y^2 - 49 = 0$$
$$y^2 = 49$$

$y = -\sqrt{49}$	or	$y = \sqrt{49}$
$y = -7$	or	$y = 7$

The solution set is $\{-7, 7\}$.

Because 49 is a perfect square, we could also solve this equation by factoring.

$$y^2 - 49 = 0$$
$$(y + 7)(y - 7) = 0$$

$y + 7 = 0$	or	$y - 7 = 0$
$y = -7$	or	$y = 7$

We get the same solution set, $\{-7, 7\}$.

We can also use the square root method when a binomial is squared, like $(p - 1)^2$, as we will see in the next example.

Example 8.6.4 Solve for p in $-40 = 10 - 2(p - 1)^2$.

Explanation. We isolate the squared binomial and then use the square root property.

$$-40 = 10 - 2(p - 1)^2$$

$$-50 = -2(p-1)^2$$
$$25 = (p-1)^2$$

$p - 1 = -5$	or	$p - 1 = 5$
$p = -4$	or	$p = 6$

The solution set is $\{-4, 6\}$

Let's check the solution $p = -4$:

$$-40 = 10 - 2(p-1)^2$$
$$-40 \stackrel{?}{=} 10 - 2(-4-1)^2$$
$$-40 \stackrel{?}{=} 10 - 2(-5)^2$$
$$-40 \stackrel{?}{=} 10 - 2(25)$$
$$-40 \stackrel{\checkmark}{=} 10 - 50$$

The solution $p = -4$ is verified. Checking $p = 6$ is left as an exercise.

When we have a middle term in $ax^2 + bx + c = 0$, we cannot use the square root property. We look first to see if we can solve the equation by factoring. Here are some examples.

Example 8.6.5 Solve for x in $x^2 - 4x - 12 = 0$.

Explanation. The equation is already in standard form and we can factor the polynomial on the left side of the equation. We will factor it and then use the zero product property to solve the equation.

$$x^2 - 4x - 12 = 0$$
$$(x-6)(x+2) = 0$$

$x - 6 = 0$	or	$x + 2 = 0$
$x = 6$	or	$x = -2$

The solution set is $\{-2, 6\}$.

Example 8.6.6 Solve for t in $2t^2 - 30t + 28 = 0$.

Explanation. First we factor out the common factor of 2. Then we can see that the polynomial is factorable. We solve it using the zero product property.

$$2t^2 - 30t + 28 = 0$$
$$2(t^2 - 15t + 14) = 0$$
$$2(t-1)(t-14) = 0$$

$t - 1 = 0$	or	$t - 14 = 0$
$t = 1$	or	$t = 14$

The solution set is $\{1, 14\}$.

If the equation is not in standard form, we must rewrite it before we can solve it. Let's look at the next example.

Example 8.6.7 Solve for x in $(x + 4)(x - 3) = 18$.

Explanation.   We need to have one side equal to 0 in order to use the zero product property, so we will multiply the left side and subtract 18 from both sides.

$$(x + 4)(x - 3) = 18$$
$$x^2 + x - 12 = 18$$
$$x^2 + x - 12 - 18 = 18 - 18$$
$$x^2 + x - 30 = 0$$
$$(x + 6)(x - 5) = 0$$

$$x + 6 = 0 \qquad\qquad \text{or} \qquad\qquad x - 5 = 0$$
$$x = -6 \qquad\qquad \text{or} \qquad\qquad x = 5$$

The solution set is $\{-6, 5\}$.

When it's difficult or impossible to factor the trinomial in $ax^2 + bx + c = 0$, we have to resort to the Quadratic Formula:
$$x = \frac{-b \pm \sqrt{b^2 - 4ac}}{2a}$$

Here is an example.

Example 8.6.8 Solve for x in $x^2 - 10x + 3 = 0$.

Explanation.   We identify that $a = 1$, $b = -10$ and $c = 3$ and substitute them into the Quadratic Formula:

$$x = \frac{-b \pm \sqrt{b^2 - 4ac}}{2a}$$
$$x = \frac{-(-10) \pm \sqrt{(-10)^2 - 4(1)(3)}}{2(1)}$$
$$x = \frac{10 \pm \sqrt{100 - 12}}{2}$$
$$x = \frac{10 \pm \sqrt{88}}{2}$$
$$x = \frac{10 \pm 2\sqrt{22}}{2}$$
$$x = 5 \pm \sqrt{22}$$

The solution set is $\left\{5 - \sqrt{22}, 5 + \sqrt{22}\right\}$.

If a quadratic equation is not in standard form we need to rewrite it to identify the values of a, b and c. Let's look at an example.

Example 8.6.9 Solve for x in $-3x^2 - 1 = -8x$.

Explanation. First, we convert the equation into standard form:

$$-3x^2 - 1 = -8x$$
$$-3x^2 + 8x - 1 = 0$$

Now we can identify that $a = -3$, $b = 8$ and $c = -1$. We will substitute them into the Quadratic Formula:

$$x = \frac{-b \pm \sqrt{b^2 - 4ac}}{2a}$$
$$x = \frac{-8 \pm \sqrt{8^2 - 4(-3)(-1)}}{2(-3)}$$
$$x = \frac{-8 \pm \sqrt{64 - 12}}{-6}$$
$$x = \frac{-8 \pm \sqrt{52}}{-6}$$
$$x = \frac{-8 \pm 2\sqrt{13}}{-6}$$
$$x = \frac{4 \pm \sqrt{13}}{3}$$

The solution set is $\left\{ \frac{4-\sqrt{13}}{3}, \frac{4+\sqrt{13}}{3} \right\}$

Also recall that if the radicand is negative, there is no real solution to the equation.

This was a brief review of solving quadratic equations. If you would like the full explanation of solving using the square root method, you can go to Section 8.3. For solving by factoring, you can go to Section 8.1. If you want more on the quadratic formula, you can go to Section 8.4.

Exercises

Solving Quadratic Equations Using the Square Root Method Solve the equation.

1. $x^2 = 144$

2. $x^2 = 4$

3. $3x^2 = 48$

4. $4x^2 = 36$

5. $2x^2 + 23 = 0$

6. $43x^2 + 31 = 0$

7. $9 - 3(r + 3)^2 = -3$

8. $48 - 6(t - 3)^2 = -6$

Solving Quadratic Equations by Factoring Solve the equation.

9. $82(x + 8)(14x - 5) = 0$

10. $-82(x + 10)(7x - 9) = 0$

11. $x^2 - 11x + 10 = 0$

12. $x^2 - 13x + 36 = 0$

13. $x^2 - 15x = -56$

14. $x^2 - 7x = -6$

15. $x^2 + 14x + 49 = 0$

16. $x^2 + 16x + 64 = 0$

17. $x(5x + 59) = -90$

18. $x(5x + 51) = -10$

19. $(x + 3)(x - 5) = -7$

20. $(x + 6)(x + 3) = -2$

Solving Quadratic Equations Using the Quadratic Formula   Solve the equation.

21. $x^2 - 2x - 11 = 0$

22. $x^2 + 6x - 2 = 0$

23. $x^2 = -3x - 1$

24. $x^2 = 5x - 2$

25. $5x^2 - 8x + 4 = 0$

26. $4x^2 + 10x + 7 = 0$

Choosing Which Method to Use   Solve the equation.

27. $x^2 + 4x = 32$

28. $x^2 + 9x = 10$

29. $2x^2 = 32$

30. $3x^2 = 27$

31. $4x^2 - 8x + 5 = 0$

32. $3x^2 + 3x + 10 = 0$

33. $13x^2 + 29 = 0$

34. $67x^2 + 41 = 0$

35. $4x^2 = -29x - 7$

36. $5x^2 = -39x - 28$

37. $x^2 - 10x + 8 = 0$

38. $x^2 - 2x - 19 = 0$

39. $29 - 3(x + 10)^2 = 2$

40. $2 - 6(y + 9)^2 = -4$

41. $x^2 + 11x = -30$

42. $x^2 + 8x = -12$

8.7 Solving Quadratic Equations Chapter Review

8.7.1 Solving Quadratic Equations by Factoring

In Section 8.1 we covered the zero product property and learned an algorithm for solving quadratic equations by factoring.

> **Example 8.7.1 Solving Using Factoring.** Solve the quadratic equations using factoring.
>
> a. $x^2 - 2x - 15 = 0$ c. $6x^2 + x - 12 = 0$ e. $x^3 - 64x = 0$
>
> b. $4x^2 - 40x = -96$ d. $(x - 3)(x + 2) = 14$
>
> **Explanation.**
>
> a. Use factor pairs.
>
> $$x^2 - 2x - 15 = 0$$
> $$(x - 5)(x + 3) = 0$$
>
> $$x - 5 = 0 \qquad \text{or} \qquad x + 3 = 0$$
> $$x = 5 \qquad \text{or} \qquad x = -3$$
>
> So the solution set is $\{5, -3\}$.
>
> b. Start by putting the equation in standard form and factoring out the greatest common factor.
>
> $$4x^2 - 40x = -96$$
> $$4x^2 - 40x + 96 = 0$$
> $$4\left(x^2 - 10x + 24\right) = 0$$
> $$4(x - 6)(x - 4) = 0$$
>
> $$x - 6 = 0 \qquad \text{or} \qquad x - 4 = 0$$
> $$x = 6 \qquad \text{or} \qquad x = 4$$
>
> So the solution set is $\{4, 6\}$.
>
> c. Use the AC method.
>
> $$6x^2 + x - 12 = 0$$
>
> Note that $a \cdot c = -72$ and that $9 \cdot -8 = -72$ and $9 - 8 = 1$
>
> $$6x^2 + 9x - 8x - 12 = 0$$
> $$\left(6x^2 + 9x\right) + (-8x - 12) = 0$$
> $$3x\left(2x + 3\right) - 4\left(2x + 3\right) = 0$$
> $$(2x + 3)\left(3x - 4\right) = 0$$

$$2x + 3 = 0 \qquad \text{or} \qquad 3x - 4 = 0$$

$$x = -\frac{3}{2} \qquad \text{or} \qquad x = \frac{4}{3}$$

So the solution set is $\left\{-\frac{3}{2}, \frac{4}{3}\right\}$.

d. Start by putting the equation in standard form.

$$(x - 3)(x + 2) = 14$$
$$x^2 - x - 6 = 14$$
$$x^2 - x - 20 = 0$$
$$(x - 5)(x + 4) = 0$$

$$x - 5 = 0 \qquad \text{or} \qquad x + 4 = 0$$
$$x = 5 \qquad \text{or} \qquad x = -4$$

So the solution set is $\{5, -4\}$.

e. Even though this equation has a power higher than 2, we can still find all of its solutions by following the algorithm. Start by factoring out the greatest common factor.

$$x^3 - 64x = 0$$
$$x\left(x^2 - 64\right) = 0$$
$$x(x - 8)(x + 8) = 0$$

$$x = 0 \qquad \text{or} \qquad x - 8 = 0 \qquad \text{or} \qquad x + 8 = 0$$
$$x = 0 \qquad \text{or} \qquad x = 8 \qquad \text{or} \qquad x = -8$$

So the solution set is $\{0, 8, -8\}$.

8.7.2 Square Root Properties

In Section 8.2 we covered the definition of a square root, how to estimate and simplify square roots, multiplication and division properties of square roots, and rationalizing the denominator.

Example 8.7.2 Estimating Square Roots. Estimate the value of $\sqrt{28}$ without a calculator.

Explanation. To estimate $\sqrt{28}$, we can find the nearest perfect squares that are whole numbers on either side of 28. Recall that the perfect squares are $1, 4, 9, 16, 25, 36, 49, 64, \ldots$ The perfect square that is just below 28 is 25 and the perfect square just above 28 is 36. This tells us that $\sqrt{28}$ is between $\sqrt{25}$ and $\sqrt{36}$, or between 5 and 6. We can also say that $\sqrt{28}$ is closer to 5 than 6 because 28 is closer to 25, so we think 5.2 or 5.3 would be a good estimate.

On the calculator we can see that $\sqrt{28} \approx 5.29$, so our guess was very close to reality.

Example 8.7.3 Multiplication and Division Properties of Square Roots. Simplify the expressions using the multiplication and division properties of square roots.

a $\sqrt{18} \cdot \sqrt{2}$.

b $\frac{\sqrt{18}}{\sqrt{2}}$.

Explanation.

a $\sqrt{18} \cdot \sqrt{2} = \sqrt{18 \cdot 2}$

$= \sqrt{36}$

$= 6$

b $\frac{\sqrt{18}}{\sqrt{2}} = \sqrt{\frac{18}{2}}$

$= \sqrt{9}$

$= 3$

Example 8.7.4 Simplifying Square Roots. Simplify the expression $\sqrt{54}$.

Explanation. Recall that the perfect squares are $1, 4, 9, 16, 25, 36, 49, 64, \ldots$ To simplify the $\sqrt{54}$, we need to look at that list and find the largest perfect square the goes into 54 evenly. In this case, it is 9. We then break up 54 into two factors 9 and 6, and we have:

$$\sqrt{54} = \sqrt{9 \cdot 6}$$
$$= \sqrt{9} \cdot \sqrt{6}$$
$$= 3\sqrt{6}$$

Since 6 has no perfect square factors, we can stop.

Example 8.7.5 Multiplying Square Root Expressions. Simplify the expression $\sqrt{50} \cdot \sqrt{27}$.

Explanation. Note that 25 is a perfect-square factor of 50, and that 9 is a perfect-square factor of 27. Now we have:

$$\sqrt{50} \cdot \sqrt{27} = \sqrt{25 \cdot 2} \cdot \sqrt{9 \cdot 3}$$
$$= \sqrt{25} \cdot \sqrt{2} \cdot \sqrt{9} \cdot \sqrt{3}$$
$$= 5 \cdot \sqrt{2} \cdot 3 \cdot \sqrt{3}$$
$$= 15\sqrt{6}$$

Example 8.7.6 Adding and Subtracting Square Root Expressions. Simplify the expression $\sqrt{32} + \sqrt{50}$.

Explanation. Recall that radicals can only be added if the radicands match identically, so we cannot initially combine these two terms. However, if we simplify first, we may be able to add terms later. Note that 16 is a perfect-square factor of 32, and that 25 is a perfect-square factor of 50.

$$\sqrt{32} + \sqrt{50} = \sqrt{16 \cdot 2} + \sqrt{25 \cdot 2}$$
$$= \sqrt{16} \cdot \sqrt{2} + \sqrt{25} \cdot \sqrt{2}$$
$$= 4\sqrt{2} + 5\sqrt{2}$$
$$= 9\sqrt{2}$$

Example 8.7.7 Rationalizing the Denominator. Rationalize the denominator in the expression $\frac{2}{\sqrt{6}}$.

Explanation.

$$\frac{2}{\sqrt{6}} = \frac{2}{\sqrt{6}} \cdot \frac{\sqrt{6}}{\sqrt{6}}$$
$$= \frac{2 \cdot \sqrt{6}}{\sqrt{6} \cdot \sqrt{6}}$$
$$= \frac{2\sqrt{6}}{6}$$
$$= \frac{\sqrt{6}}{3}$$

Example 8.7.8 More Complicated Square Roots. Expand $\left(\sqrt{5} + \sqrt{3}\right)^2$.

Explanation. We will use the FOIL method to expand this expression:

$$\left(\sqrt{5} + \sqrt{3}\right)^2 = \left(\sqrt{5} + \sqrt{3}\right)\left(\sqrt{5} + \sqrt{3}\right)$$
$$= \left(\sqrt{5}\right)^2 + \sqrt{5}\sqrt{3} + \sqrt{3}\sqrt{5} + \left(\sqrt{3}\right)^2$$
$$= 5 + \sqrt{15} + \sqrt{15} + 3$$
$$= 8 + 2\sqrt{15}$$

8.7.3 Solving Quadratic Equations by Using a Square Root

In Section 8.3 we covered how to solve quadratic equations using the square root property and how to use the Pythagorean Theorem.

Example 8.7.9 Solving Quadratic Equations Using the Square Root Property. Solve for w in $3(2-w)^2 - 24 = 0$.

Explanation. It's important here to suppress any urge you may have to expand the squared binomial. We begin by isolating the squared expression.

$$3(2 - w)^2 - 24 = 0$$
$$3(2 - w)^2 = 24$$
$$(2 - w)^2 = 8$$

Now that we have the squared expression isolated, we can use the square root property.

$2 - w = -\sqrt{8}$	or	$2 - w = \sqrt{8}$
$2 - w = -\sqrt{4 \cdot 2}$	or	$2 - w = \sqrt{4 \cdot 2}$
$2 - w = -\sqrt{4} \cdot \sqrt{2}$	or	$2 - w = \sqrt{4} \cdot \sqrt{2}$

$$2 - w = -2\sqrt{2} \qquad \text{or} \qquad 2 - w = 2\sqrt{2}$$
$$-w = -2\sqrt{2} - 2 \qquad \text{or} \qquad -w = 2\sqrt{2} - 2$$
$$w = 2\sqrt{2} + 2 \qquad \text{or} \qquad w = -2\sqrt{2} + 2$$

The solution set is $\left\{ 2\sqrt{2} + 2, -2\sqrt{2} + 2 \right\}$.

Example 8.7.10 The Pythagorean Theorem. Faven was doing some wood working in her garage. She needed to cut a triangular piece of wood for her project that had a hypotenuse of 16 inches, and the sides of the triangle should be equal in length. How long should she make her sides?

Explanation.

Let's start by representing the length of the triangle, measured in inches, by the letter x. That would also make the other side x inches long.

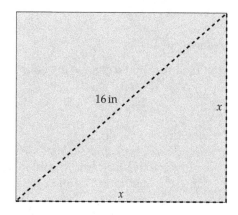

Figure 8.7.11: Piece of wood with labels for Faven

Faven should now set up the Pythagorean theorem regarding the picture. That would be

$$x^2 + x^2 = 16^2$$

Solving this equation, we have:

$$x^2 + x^2 = 16^2$$
$$x^2 + x^2 = 256$$
$$2x^2 = 256$$
$$x^2 = 128$$
$$\sqrt{x^2} = \sqrt{128}$$
$$x = \sqrt{64 \cdot 2}$$
$$x = \sqrt{64} \cdot \sqrt{2}$$
$$x = 8\sqrt{2}$$

$$x \approx 11.3$$

Faven should make the sides of her triangle about 11.3 inches long to force the hypotenuse to be 16 inches long.

8.7.4 The Quadratic Formula

In Section 8.4 we covered how to use the quadratic formula to solve any quadratic equation, as well as an algorithm to help solve linear and quadratic equations.

Example 8.7.12 Solving Quadratic Equations with the Quadratic Formula. Solve the equations using the quadratic formula.

a. $x^2 + 4x = 6$
b. $5x^2 - 2x + 1 = 0$

Explanation.

a. First we should change the equation into standard form.

$$x^2 + 4x = 6$$
$$x^2 + 4x - 6 = 0$$

Next, we check and see that we cannot factor the left side or use the square root property so we must use the quadratic formula. We identify that $a = 1$, $b = 4$, and $c = -6$. We will substitute them into the quadratic formula:

$$x = \frac{-b \pm \sqrt{b^2 - 4ac}}{2a}$$
$$x = \frac{-4 \pm \sqrt{(4)^2 - 4(1)(-6)}}{2(1)}$$
$$= \frac{-4 \pm \sqrt{16 + 24}}{2}$$
$$= \frac{-4 \pm \sqrt{40}}{2}$$
$$= \frac{-4 \pm \sqrt{4 \cdot 10}}{2}$$
$$= \frac{-4 \pm \sqrt{4} \cdot \sqrt{10}}{2}$$
$$= \frac{-4 \pm 2\sqrt{10}}{2}$$
$$= -\frac{4}{2} \pm \frac{2\sqrt{10}}{2}$$
$$= -2 \pm \sqrt{10}$$

So the solution set is $\left\{-2 + \sqrt{10}, -2 - \sqrt{10}\right\}$.

b. Since the equation $5x^2 - 2x + 1 = 0$ is already in standard form, we check and see that we cannot factor the left side or use the square root property so we must use the quadratic formula. We identify that $a = 5$, $b = -2$, and $c = 1$. We will substitute them into the quadratic formula:

$$x = \frac{-b \pm \sqrt{b^2 - 4ac}}{2a}$$
$$x = \frac{-(-2) \pm \sqrt{(-2)^2 - 4(5)(1)}}{2(5)}$$
$$= \frac{2 \pm \sqrt{4 - 20}}{10}$$
$$= \frac{2 \pm \sqrt{-16}}{10}$$

Since the solutions have square roots of negative numbers, we must conclude that there are no real solutions.

Example 8.7.13 Recognizing Linear and Quadratic Equations. Identify which equations are linear, which are quadratic, and which are neither.

a. $2(x - 3)^2 - 5x = 6$

b. $2(x - 3) - 5x = 6$

c. $2x - 6 = 7x^3$

d. $2x^2 - 6 = 7x^2$

e. $2\sqrt{x} - x - 6 = 0$

f. $2x - (x - 6) = 0$

Explanation.

a. $2(x - 3)^2 - 5x = 6$ is quadratic.

b. $2(x - 3) - 5x = 6$ is linear.

c. $2x - 6 = 7x^3$ is neither linear or quadratic.

d. $2x^2 - 6 = 7x^2$ is quadratic.

e. $2\sqrt{x} - x - 6 = 0$ is neither linear or quadratic.

f. $2x - (x - 6) = 0$ is linear.

Example 8.7.14 Solving Linear and Quadratic Equations. Use 8.4.11 to help solve the equations after deciding if they are linear or quadratic.

a. $4x^2 - 11x + 6 = 0$

b. $2(x - 6)^2 - 16 = 0$

c. $2(x - 6) - 16 = 0$

d. $3(x - 4)^2 - 15x = 0$

Explanation.

a. To solve the equation $4x^2 - 11x + 6 = 0$ we first note that it is quadratic. Since there is a linear term $(-11x)$, we must use either factoring or the quadratic formula, and we will try factoring first. Since the leading coefficient is 4, we will try the AC method. In this case, $ac = 24$: numbers that multiply to be 24 and add to be -11 are -8 and -3. So we split up th equation like this:

$$4x^2 - 11x + 6 = 0$$
$$4x^2 - 8x - 3x + 6 = 0$$

$$(4x^2 - 8x) + (-3x + 6) = 0$$
$$4x(x - 2) - 3(x - 2) = 0$$
$$(x - 2)(4x - 3) = 0$$

$x - 2 = 0$	or	$4x - 3 = 0$
$x = 2$	or	$x = \dfrac{3}{4}$

So, the solution set is $\left\{2, \frac{3}{4}\right\}$.

b. To solve the equation $2(x - 6)^2 - 16 = 0$ we first note that it is quadratic. Since there is no linear term, we should try using the square root method.

$$2(x - 6)^2 - 16 = 0$$
$$2(x - 6)^2 = 16$$
$$(x - 6)^2 = 8$$

$x - 6 = \sqrt{8}$	or	$x - 6 = -\sqrt{8}$
$x - 6 = \sqrt{4 \cdot 2}$	or	$x - 6 = -\sqrt{4 \cdot 2}$
$x - 6 = \sqrt{4} \cdot \sqrt{2}$	or	$x - 6 = -\sqrt{4} \cdot \sqrt{2}$
$x - 6 = 2\sqrt{2}$	or	$x - 6 = -2\sqrt{2}$
$x = 6 + 2\sqrt{2}$	or	$x = 6 - 2\sqrt{2}$

So, the solution set is $\left\{6 + 2\sqrt{2}, 6 - 2\sqrt{2}\right\}$.

c. To solve the equation $2(x - 6) - 16 = 0$ we first we first note that it is linear. Since it is linear, we just need to isolate the terms with the variable on one side and all the other terms on the other side of the equals sign.

$$2(x - 6) - 16 = 0$$
$$2x - 12 - 16 = 0$$
$$2x - 28 = 0$$
$$2x = 28$$
$$x = 14$$

So, the solution set is $\{14\}$.

d. To solve the equation $2(x - 6)^2 - 15x = 0$ we first note that it is quadratic. Since there is a linear term, we must use either factoring or the quadratic formula. Before we can decide which to use, we need to put the equation in standard form:

$$3(x - 4)^2 - 15x = 0$$
$$3(x - 4)(x - 4) - 15x = 0$$
$$3(x^2 - 8x + 16) - 15x = 0$$
$$3x^2 - 24x + 48 - 15x = 0$$

$$3x^2 - 39x + 48 = 0$$

Now we can see that the left hand side does not factor easily, so we will fall back on the quadratic formula. We identify that $a = 3$, $b = -39$, and $c = 48$.

$$x = \frac{-b \pm \sqrt{b^2 - 4ac}}{2a}$$

$$x = \frac{-(-39) \pm \sqrt{(-39)^2 - 4(3)(48)}}{2(3)}$$

$$= \frac{39 \pm \sqrt{1521 - 576}}{6}$$

$$= \frac{39 \pm \sqrt{945}}{6}$$

$$= \frac{39 \pm \sqrt{9 \cdot 105}}{6}$$

$$= \frac{39 \pm \sqrt{9} \cdot \sqrt{105}}{6}$$

$$= \frac{39 \pm 3\sqrt{105}}{6}$$

So the solution set is $\left\{ \frac{39 + 3\sqrt{105}}{6}, \frac{39 - 3\sqrt{105}}{6} \right\}$.

8.7.5 Complex Solutions to Quadratic Equations

In Section 8.5 we covered what both imaginary numbers and complex numbers are, as well as how to solve quadratic equations where the solutions are imaginary numbers or complex numbers.

Example 8.7.15 Imaginary Numbers. Simplify the expression $\sqrt{-12}$ using the imaginary number, i.

Explanation. Start by splitting the -1 from the 12 and by looking for the largest perfect-square factor of -12, which happens to be 4.

$$\sqrt{-12} = \sqrt{4 \cdot -1 \cdot 3}$$

$$= \sqrt{4} \cdot \sqrt{-1} \cdot \sqrt{3}$$

$$= 2i\sqrt{3}$$

Example 8.7.16 Solving Quadratic Equations with Imaginary Solutions. Solve for m in $2m^2 + 16 = 0$, where p is an imaginary number.

Explanation. There is no m term so we will use the square root method.

$$2m^2 + 16 = 0$$

$$2m^2 = -16$$

$$m^2 = -8$$

$$m = -\sqrt{-8} \qquad \text{or} \qquad m = \sqrt{-8}$$
$$m = -\sqrt{4} \cdot \sqrt{-1} \cdot \sqrt{2} \qquad \text{or} \qquad m = \sqrt{4} \cdot \sqrt{-1} \cdot \sqrt{2}$$
$$m = -2i\sqrt{2} \qquad \text{or} \qquad m = 2i\sqrt{2}$$

The solution set is $\left\{-2i\sqrt{2}, 2i\sqrt{2}\right\}$.

Example 8.7.17 Solving Quadratic Equations with Complex Solutions. Solve the equation $3(v-2)^2 + 54 = 0$, where v is a complex number.

Explanation.

$$3(v-2)^2 + 54 = 0$$
$$3(v-2)^2 = -54$$
$$(v-2)^2 = -18$$

$$v - 2 = -\sqrt{-18} \qquad \text{or} \qquad v - 2 = \sqrt{-18}$$
$$v - 2 = -\sqrt{9 \cdot -1 \cdot 2} \qquad \text{or} \qquad v - 2 = \sqrt{9 \cdot -1 \cdot 2}$$
$$v - 2 = -\sqrt{9} \cdot \sqrt{-1} \cdot \sqrt{2} \qquad \text{or} \qquad v - 2 = \sqrt{9} \cdot \sqrt{-1} \cdot \sqrt{2}$$
$$v - 2 = -3i\sqrt{2} \qquad \text{or} \qquad v - 2 = 3i\sqrt{2}$$
$$v = 2 - 3i\sqrt{2} \qquad \text{or} \qquad v = 2 + 3i\sqrt{2}$$

So, the solution set is $\left\{2 + 3i\sqrt{2}, 2 - 3i\sqrt{2}\right\}$.

8.7.6 Strategies for Solving Quadratic Equations

In Section 8.6 we reviews all three methods for solving quadratic equations that we know. For the full explanation of solving using the factoring, visit Section 8.1, solving using the square root method, visit Section 8.3, and for more on the quadratic formula, visit Section 8.4.

Example 8.7.18 How to Choose a Method for Solving a Quadratic Equation. Solve the quadratic equations using an effective method.

a. $(x-4)^2 - 2 = 0$ b. $(x-4)^2 - 2x = 0$ c. $(x-4)^2 + 2x = 0$

Explanation. All three of the equations here are very similar, so we will need to examine them closely to choose the best method for solving them.

a. To solve the equation $(x-4)^2 - 2 = 0$, first note that there is no linear term: there is only a square and a constant. This leads us to consider the square root method. Before doing that, isolate the square:

$$(x-4)^2 - 2 = 0$$

$$(x - 4)^2 = 2$$

Now we can apply the square root method to the equation.

$$x - 4 = \sqrt{2} \qquad \text{or} \qquad x - 4 = -\sqrt{2}$$
$$x = 4 + \sqrt{2} \qquad \text{or} \qquad x = 4 - \sqrt{2}$$

So the solution set is $\left\{ 4 + \sqrt{2}, 4 - \sqrt{2} \right\}$

b. To solve the equation $(x - 4)^2 - 2x = 0$, first note that there is a linear term $(-2x)$, so we must use either factoring or the quadratic formula. To use either, we must first put the equation in standard form.

$$(x - 4)^2 - 2x = 0$$
$$(x - 4)(x - 4) - 2x = 0$$
$$x^2 - 8x + 16 - 2x = 0$$
$$x^2 - 10x + 16 = 0$$

Now that the equation is in standard form, we can decide whether to use factoring or the quadratic formula. While the quadratic formula *always* works, it can take more time than factoring if factoring is possible. In this case, factoring entails answering the question "are there two integers that multiply to be 16 and add to be -10?" The answer is "yes": -8 and -2 are such numbers.

$$x^2 - 10x + 16 = 0$$
$$(x - 8)(x - 2) = 0$$

$$x - 8 = 0 \qquad \text{or} \qquad x - 2 = 0$$
$$x = 8 \qquad \text{or} \qquad x = 2$$

So the solution set is $\{2, 8\}$.

c. To solve the equation $(x - 4)^2 + 2x = 0$, first note that there is a linear term $(+2x)$, so we must use either factoring or the quadratic formula. To use either, we must first put the equation in standard form.

$$(x - 4)^2 + 2x = 0$$
$$(x - 4)(x - 4) + 2x = 0$$
$$x^2 - 8x + 16 + 2x = 0$$
$$x^2 - 6x + 16 = 0$$

Now that the equation is in standard form, we can decide whether to use factoring or the quadratic formula. In this case, factoring entails answering the question "are there two integers that multiply to be 16 and add to be -6?" The answer is "no," so we must use the quadratic formula. First, identify that $a = 1$, $b = -6$, and $c = 16$.

$$x = \frac{-b \pm \sqrt{b^2 - 4ac}}{2a}$$

$$x = \frac{-(-6) \pm \sqrt{(-6)^2 - 4(1)(16)}}{2(1)}$$

$$= \frac{6 \pm \sqrt{36 - 48}}{2}$$

$$= \frac{6 \pm \sqrt{-12}}{2}$$

At this point, we notice that the solutions are complex. Continue to simplify until they are completely reduced.

$$x = \frac{6 \pm \sqrt{4 \cdot -1 \cdot 3}}{2}$$

$$= \frac{6 \pm \sqrt{4} \cdot \sqrt{-1} \cdot \sqrt{3}}{2}$$

$$= \frac{6 \pm 2i\sqrt{3}}{2}$$

$$= \frac{6}{2} \pm \frac{2i\sqrt{3}}{2}$$

$$= 3 \pm i\sqrt{3}$$

So the solution set is $\left\{3 + i\sqrt{3}, 3 - i\sqrt{3}\right\}$.

Exercises

Solving Quadratic Equations by Factoring Solve the equation.

1. $x^2 - 6x - 27 = 0$

2. $x^2 - 5x - 50 = 0$

3. $x^2 + 11x = -18$

4. $x^2 + 18x = -80$

5. $x^2 = 9x$

6. $x^2 = 7x$

7. $x^2 - 8x + 16 = 0$

8. $x^2 - 10x + 25 = 0$

9. $4x^2 = -41x - 10$

10. $4x^2 = -25x - 36$

11. $x(x + 20) = 5(2x - 5)$

12. $x(x + 36) = 9(2x - 9)$

13. A rectangle's base is 7 in shorter than four times its height. The rectangle's area is 65 in². Find this rectangle's dimensions.

The rectangle's height is [   ].

The rectangle's base is [   ].

14. A rectangle's base is 7 in shorter than four times its height. The rectangle's area is 2 in². Find this rectangle's dimensions.

The rectangle's height is [   ].

The rectangle's base is [   ].

Square Root Properties

15. Without using a calculator, estimate the value of $\sqrt{22}$:

($\square$ 5.69 $\square$ 4.69 $\square$ 4.31 $\square$ 5.31)

16. Without using a calculator, estimate the value of $\sqrt{38}$:

($\square$ 6.16 $\square$ 6.84 $\square$ 5.84 $\square$ 5.16)

Evaluate the following.

17. $\sqrt{\dfrac{25}{81}} = \boxed{}$.   **18.** $\sqrt{\dfrac{36}{49}} = \boxed{}$.   **19.** $\sqrt{-64} = \boxed{}$.   **20.** $\sqrt{-100} = \boxed{}$.

Simplify the radical expression or state that it is not a real number.

21. $\dfrac{\sqrt{6}}{\sqrt{216}} = \boxed{}$ **22.** $\dfrac{\sqrt{4}}{\sqrt{144}} = \boxed{}$ **23.** $\sqrt{8} = \boxed{}$ **24.** $\sqrt{147} = \boxed{}$

Simplify the expression.

25. $3\sqrt{5} \cdot 3\sqrt{40} = \boxed{}$ **26.** $3\sqrt{15} \cdot 2\sqrt{30} = \boxed{}$ **27.** $\sqrt{\dfrac{28}{13}} \cdot \sqrt{\dfrac{4}{13}} = \boxed{}$

28. $\sqrt{\dfrac{15}{19}} \cdot \sqrt{\dfrac{5}{19}} = \boxed{}$ **29.** $\sqrt{20} + \sqrt{45} = \boxed{}$ **30.** $\sqrt{175} + \sqrt{112} = \boxed{}$

31. $\sqrt{176} - \sqrt{44} = \boxed{}$ **32.** $\sqrt{8} - \sqrt{32} = \boxed{}$

Rationalize the denominator and simplify the expression.

33. $\dfrac{1}{\sqrt{3}} = \boxed{}$ **34.** $\dfrac{1}{\sqrt{3}} = \boxed{}$ **35.** $\dfrac{2}{5\sqrt{5}} = \boxed{}$

36. $\dfrac{7}{3\sqrt{6}} = \boxed{}$ **37.** $\dfrac{1}{\sqrt{75}} = \boxed{}$ **38.** $\dfrac{1}{\sqrt{175}} = \boxed{}$

Expand and simplify the expression.

39. $\left(7 - \sqrt{7}\right)\left(10 - 3\sqrt{7}\right) = \boxed{}$ **40.** $\left(5 - \sqrt{7}\right)\left(8 - 5\sqrt{7}\right) = \boxed{}$

41. $\left(\sqrt{5} + \sqrt{13}\right)\left(\sqrt{5} - \sqrt{13}\right) = \boxed{}$ **42.** $\left(\sqrt{5} + \sqrt{7}\right)\left(\sqrt{5} - \sqrt{7}\right) = \boxed{}$

Solving Quadratic Equations by Using a Square Root   Solve the equation.

43. $x^2 = 20$ **44.** $x^2 = 99$ **45.** $144x^2 = 49$ **46.** $16x^2 = 9$

47. $(x + 5)^2 = 121$ **48.** $(x + 7)^2 = 49$ **49.** $2 - 2(t - 5)^2 = -6$ **50.** $54 - 5(x - 5)^2 = 9$

51.

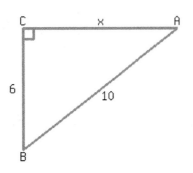

Find the value of x.

$x = $ _____

52.

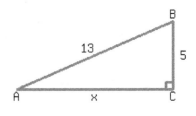

Find the value of x.

$x = $ _____

53. Parnell is designing a rectangular garden. The garden's diagonal must be 2 feet, and the ratio between the garden's base and height must be 4 : 3. Find the length of the garden's base and height.

The garden's base is _____

feet and its height is _____ .

54. Gregory is designing a rectangular garden. The garden's diagonal must be 42.9 feet, and the ratio between the garden's base and height must be 12 : 5. Find the length of the garden's base and height.

The garden's base is _____

feet and its height is _____ .

The Quadratic Formula   Solve the equation.

55. $14x^2 - 11x - 9 = 0$ **56.** $18x^2 - 37x - 20 = 0$ **57.** $x^2 = -3x - 1$

58. $x^2 = 5x - 3$ **59.** $2x^2 + 4x + 5 = 0$ **60.** $2x^2 - x + 1 = 0$

61. $10t + 7 = t + 88$ **62.** $9b + 10 = b + 66$ **63.** $-1 - 3y^2 = -3$

64. $-8 - 7r^2 = -10$ **65.** $x^2 + 39x = 0$ **66.** $x^2 + 59x = 0$

67. $x^2 + 8x = 9$

68. $x^2 - 3x = 54$

69. $(x - 7)^2 = 16$

70. $(x - 5)^2 = 121$

71. $x^2 = -9x - 19$

72. $x^2 = -7x - 8$

73. An object is launched upward at the height of 320 meters. It's height can be modeled by

$$h = -4.9t^2 + 90t + 320,$$

where h stands for the object's height in meters, and t stands for time passed in seconds since its launch. The object's height will be 350 meters twice before it hits the ground. Find how many seconds since the launch would the object's height be 350 meters. Round your answers to two decimal places if needed.

The object's height would be 350 meters the first time at [_____] seconds, and then the second time at [_____] seconds.

74. An object is launched upward at the height of 340 meters. It's height can be modeled by

$$h = -4.9t^2 + 70t + 340,$$

where h stands for the object's height in meters, and t stands for time passed in seconds since its launch. The object's height will be 350 meters twice before it hits the ground. Find how many seconds since the launch would the object's height be 350 meters. Round your answers to two decimal places if needed.

The object's height would be 350 meters the first time at [_____] seconds, and then the second time at [_____] seconds.

Complex Solutions to Quadratic Equations Simplify the radical and write it into a complex number.

75. $\sqrt{-48} =$ [_____]

76. $\sqrt{-98} =$ [_____]

Solve the quadratic equation. Solutions could be complex numbers.

77. $-2x^2 - 4 = 8$

78. $4x^2 + 7 = -5$

79. $-9(y - 4)^2 + 8 = 584$

80. $7(y + 7)^2 + 8 = -167$

Strategies for Solving Quadratic Equations Solve the equation.

81. $2x^2 + 29 = 0$

82. $41x^2 + 37 = 0$

83. $5x^2 = -31x - 44$

84. $5x^2 = -27x - 36$

85. $x^2 + 4x + 1 = 0$

86. $x^2 + 10x + 7 = 0$

87. $28 - 6(x + 6)^2 = 4$

88. $34 - 4(y + 6)^2 = -2$

89. $x^2 + 7x = -12$

90. $x^2 + 14x = -45$

CHAPTER 9

Graphs of Quadratic Functions

9.1 Introduction to Functions

In mathematics, we use functions to model real-life data. In this section, we will learn the definition of a function and related concepts.

9.1.1 Introduction to Functions

When working with two variables, we are interested in the relationship between those two variables. For example, consider the two variables of hare population and lynx population in a Canadian forest. If we know the value of one variable, are we able to determine the value of the second variable? If we know that one variable is increasing over time, do we know if the other is increasing or decreasing?

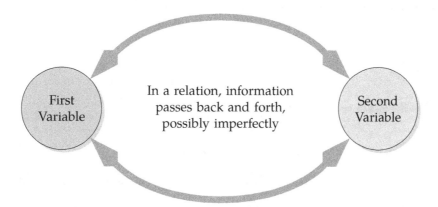

Figure 9.1.2: In a relation, some knowledge of one variable implies some knowledge about the other

Definition 9.1.3 Relation. A **relation** is a very general situation between two variables, where having a little bit of information about one variable could tell you something about the other variable. For example, if you know the hare population is high this year, you can say the lynx population is probably increasing. So "hare population" and "lynx population" make a relation. If one of the variables is identified as the "first" variable, the relation's **domain** is the set of all values that variable can take. Likewise, the relation's **range** is the set of all values that the second variable can take.

We are not so much concerned with relations in this book. But we are interested in a special type of relation called a function. Informally, a **function** is a device that takes input values for one variable one by one, thinks about them, and gives respective output values one by one for the other variable.

Example 9.1.4 Mariana has 5 chickens: Hazel, Yvonne, Georgia, Isabella, and Emma. For the relation "Chicken to Egg Color," the first variable (the input) is a chicken's name and the second variable (the output) is the color of that chicken's eggs. The relation's domain is the set of all of Mariana's chicken's names, and its range is the set of colors of her chicken's eggs. Figure 9.1.5 shows two inputs and their corresponding outputs.

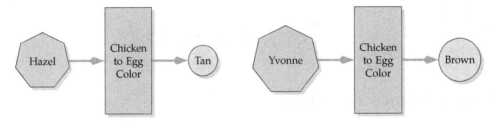

Figure 9.1.5: Two Pairs of Inputs and Outputs of the Relation "Chicken to Egg Color"

It would not be convenient to make diagrams like the ones in Figure 9.1.5 for all five chickens. There are too many inputs. Instead, Figure 9.1.6 represents the function graphically in a more concise way. The function's input variable is "chicken name," and its output variable is "egg color." Note that we are using the word "variable," because the chicken names and egg colors vary depending on which individual chicken you choose.

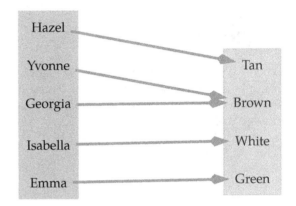

Figure 9.1.6: Diagram for the function "Chicken to Egg Color"

We can also use a set of ordered pairs to represent this function:

{(Hazel, Tan), (Yvonne, Brown), (Georgia, Brown), (Isabella, White), (Emma, Green)}

where you read the ordered pair left to right, with the first value as an input and the second value as its output.

Definition 9.1.7 Function. In mathematics, a function is a relation between a set of inputs and a set of outputs with the property that each input is related to exactly one output.

In Figure 9.1.6, we can see each chicken's name (input) is related to exactly one output, so the relation "Chicken to Egg Color" qualifies as a function. Note that it is irrelevant that multiple inputs might related to the same output, like in (Yvonne, brown) and (Georgia, brown). The point is that whichever chicken you are thinking about, you know exactly which color egg it lays.

Example 9.1.8 Next, we will look at the "inverse" relation named "Egg Color to Chicken." Here we consider the color of an egg to be the input, and want the output to be the name of a chicken. If the egg is green, we know it is Emma's. But what if the egg is brown?

In Figure 9.1.9, we can see the color brown (an input) is related to *two* outputs, Yvonne and Georgia. This disqualifies the relation "Egg Color to Chicken" from being a function. (It is still a *relation*, because in general knowing the egg color does give you *some* information about which chicken it may have come from.)

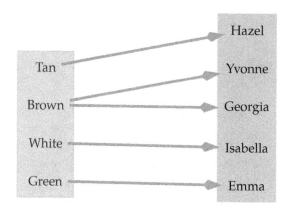

Figure 9.1.9: The relation "Egg Color to Chicken"

9.1.2 Functions as Predictors

Functions are useful because they describe our ability to accurately predict a result. If we can predict something precisely every time, then there is a function involved.

Example 9.1.10 If you go to the store and buy 5 two-dollar cans of soup, then you should predict that your total will be $10. No matter if you buy the soup in the morning, afternoon, or evening. If it doesn't total $10, then the cash register isn't *functioning*.

Example 9.1.11 A vending machine is like a function. You push a button and the item you desired pops out. In this case, the inputs are all of the buttons that you can press, and the outputs are the kinds of candy bars, chip bags, etc. that can come out. The mechanics and electronics that connect the buttons with the items represent the function.

Going in a little further, if the button A1 delivers a bag of *M&M's*, then you would be surprised if you pressed A1 and got anything other than *M&M's*. In this case, the machine wouldn't be functioning and you would get upset at your prediction ability being taken away.

What if buttons A1 and B3 both delivered *M&M's*? Would that violate the definition a function?

No, the vending machine is still a function even if two buttons generate the same output. Remember that to be a function, each *input* must generate a single output; that output doesn't have to be unique for each input. So as long as each button generates the same item each time you press it, there can be two buttons that deliver the same item.

Example 9.1.12 Some relations are not functions, because they can't be used to make 100% accurate predictions. For example, if you have a student's first name and want to determine their student ID number, you probably won't be able to look it up. For a common first name like "Michael," there will be many student ID number possibilities. Multiple outputs for a single input make "student ID number" not be a function of "first name."

On the other hand, if we exchange the variables and think of the ID number as the first variable, then there is only one student that ID number applies to, and there is only one official first name for that person. So "first name" *is* a function of "student ID number."

9.1.3 Algebraic Functions

Many functions have specific algebraic formulas to turn an input number into an output number. We explore some examples here.

> **Example 9.1.13** Aylen is an electrician who is hired to install a new circuit. She charges $111 to come to your house and then, in addition, $89 per hour to do the work. If x represents the total number of hours that the job takes and y represents the total cost of her work, in dollars, then the equation $y = 89x + 111$ relates the variables.
>
> We know that at the end of the job, if Aylen worked x hours, you are going to have *one* bill that totals $89x + 111$. This must mean that the bill is a function of the number of hours of labor. For every possible number of hours that she could work, you would only get one bill for that cost.
>
> Context aside, the expression $89x + 111$ represents a function of x, because it can be used to turn an input number x into a specific output. In fact, every algebraic expression in one variable represents a function for similar reasons.

> **Example 9.1.14** The equation $x^2 + y^2 = 25$ represents a relation that is not a function, where we view x as the "first variable" as usual. Remember that to *not* be a function, all we have to do is find one input that has two outputs. Let's pick a nice easy input number to test: $x = 4$. Substituting this value gives us
>
> $$x^2 + y^2 = 25$$
> $$(4)^2 + y^2 = 25$$
> $$16 + y^2 = 25$$
> $$y^2 = 9$$
>
> At this point, we see that y could be 3 *or* y could be -3. There are two y-values for the single x-value, so that must mean that $x^2 + y^2 = 25$ cannot represent a function.

Checkpoint 9.1.15. Identify which of the following represent functions and which do not.

a. The formula for the area of a circle is $A = \pi r^2$. With this equation, A (□ is □ is not) a function of r.

b. A quadratic equation can be written as $y = ax^2 + bx + c$. With this equation, y (□ is □ is not) a function of x.

c. With the equation $y^2 = x$, the variable y (□ is □ is not) a function of x.

Explanation.

a. Since each circle with a given radius has only one area, this must be a function. Another way to look at it is that for any one r, the formula tells you exactly what A must be.

b. If you plug in any one x-value into $y = ax^2 + bx + c$, you will know exactly what the value of y is. So y is a function of x.

c. For example, if $x = 9$, then y could be 3 or -3. Since there are two y-values for the single x-value, that must mean that the equation $y^2 = x$ cannot represent y as a function of x.

9.1.4 Function Notation

We know that the equation $y = 5x + 3$ represents y as a function of x, because for each x-value (input), there is only one y-value (output). If we want to determine the value of the output when the input is 2, we'd replace x with 2 and find the value of y:

$$y = 5(2) + 3$$
$$= 10 + 3$$
$$= 13$$

Our end result is that $y = 13$. Well, y *is* 13, but only in the situation when x is 2. In general, for other inputs, y is not going to be 13. So the equation $y = 13$ is lacking in the sense that it is not communicating everything we might want to say. It does not communicate the value of x that we used. **Function notation** will allow us to communicate *both* the input *and* the output at the same time. It will also allow us to give each function a name, which is helpful when we have multiple functions.

Functions can have names just like variables. The most common function name is f, since "f" stands for "function." A letter like f doesn't stand for a single number though. Instead, it represents an input-output relation like we've been discussing in this section.

We will write equations like $y = f(x)$, and what we mean is:

- "y equals f *of* x"
- the function's name is f
- the input variable is x
- the parentheses following the f surround the input; they do *not* indicate multiplication
- the output variable is y

Remark 9.1.16. Parentheses have a lot of uses in mathematics. Their use with functions is very specific, and it's important to note that f is *not* being multiplied by anything when we write $f(x)$. With function notation, the parentheses specifically are just meant to indicate the input by surrounding the input.

Example 9.1.17 The expression $f(x)$ is read as "f of x," and the expression $f(2)$ is read as "f of 2." Be sure to practice saying this correctly while reading.

The expression $f(2)$ means that 2 is being treated as an input, and the function f is turning it into an output. And then $f(2)$ represents that actual output number.

Remark 9.1.18. The most common letters used to represent functions are f, g, and h. The most common variables we use are x, y, and z. But we can use any function name and any input and output variable. When dealing with functions in context, it often makes sense to use meaningful function names and variables. For example, if we are modeling temperature of a cup of coffee as a function of time with a function C, we could use $T = C(t)$, where T is the temperature (in degrees Fahrenheit) after t minutes.

9.1.5 Evaluating Functions

When we determine a function's value for a specific input, this is known as *evaluating a function*. To do so, we replace the input with the numerical value given and determine the associated output.

When using function notation, instead of writing $5x + 3$ or $y = 5x + 3$, we often write something like $f(x) = 5x + 3$. We are saying that the rule for function f is to use the expression $5x + 3$. To find $f(2)$, wherever you see x in the formula $f(x) = 5x + 3$, substitute in 2:

$$f(x) = 5x + 3$$
$$f(2) = 5(2) + 3$$
$$= 10 + 3$$
$$= 13$$

Our end result is that $f(2) = 13$, which tells us that f turns 2 into 13. In other words, when the input is 2, the output will be 13.

Checkpoint 9.1.19 Functions with Algebraic Formulas. Find the given function values for a function g where $g(x) = 5x^2 - 3x + 4$.

a. $g(3) = \boxed{}$

b. $g(0) = \boxed{}$

c. $g(-2) = \boxed{}$

Explanation.

a. We will substitute $x = 3$ into $g(x)$:

$$g(3) = 5(3)^2 - 3(3) + 4$$
$$= 5(9) - 9 + 4$$
$$= 40$$

b. We will substitute $x = 0$ into $g(x)$:

$$g(0) = 5(0)^2 - 3(0) + 4$$
$$= 5(0) + 0 + 4$$
$$= 4$$

c. We will substitute $x = -2$ into $g(x)$. Especially when inputting negative numbers, be *certain* to put parentheses around the input values in the first step:

$$g(-2) = 5(-2)^2 - 3(-2) + 4$$
$$= 5(4) + 6 + 4$$
$$= 30$$

A function may be described by explicitly listing many inputs and their corresponding outputs in a table.

Example 9.1.20 Functions given in Table Form. Temperature readings for Portland, OR, on a given day are recorded in Table 9.1.21 . Let $f(x)$ be the temperature in degrees Fahrenheit x hours after midnight.

x, hours after midnight	0	1	2	3	4	5	6	7	8	9	10
$f(x)$, temperature in °F	45	44	42	42	43	44	45	48	49	50	53

Table 9.1.21: Recorded Temperatures in Portland, OR, on a certain day

a. What was the temperature at midnight?

b. Find $f(9)$. Explain what this function value represents in the context of the problem.

Explanation.

a. To determine the temperature at midnight, we look in the table where $x = 0$ and see that the output is 45. Using function notation, we would write:

$$f(0) = 45.$$

Thus at midnight the temperature was 45 °F.

b. To determine the value of $f(9)$, we look in the table where $x = 9$ and read the output:

$$f(9) = 50.$$

In context, this means that at 9AM the temperature was 50 °F.

A function may be described using a graph, where the horizontal axis corresponds to possible input values (the domain), and the vertical axis corresponds to possible output values (the range).

Example 9.1.22 Functions in Graphical Form. A colony of bees settled in Zahid's backyard in 2012. Let $B(x)$ be the number of bees in the colony (in thousands) x months after April 1, 2012. A graph of $y = B(x)$ is shown in Figure 9.1.23.

a. Determine the number of bees in the colony on July 1, 2012.

b. Find $B(0)$. Explain what this function value represents in the context of the problem.

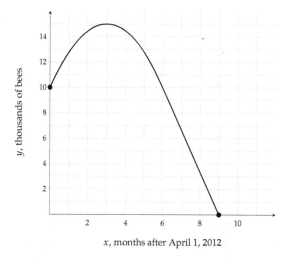

Figure 9.1.23: Bee Population

Explanation.

a. Since July 1 is 3 months after April 1, that means we need to use $x = 3$ as the input. On the horizontal axis, find 3, as in Figure 9.1.24. Looking straight up or down, we find the point $(3, 15)$ on the curve. That means that $B(3) = 15$. A value of $y = 15$ when $x = 3$ tells us that there were 15,000 bees 3 months after April 1, 2012 (on July 1, 2012).

b. To find $B(0)$, we recognize that this will be the output of the function when $x = 0$. The point $(0, 10)$

on the graph of $y = B(x)$ tells us that $B(0) = 10$. In the context of this problem, this means on April 1, 2012 there were 10,000 bees in the colony.

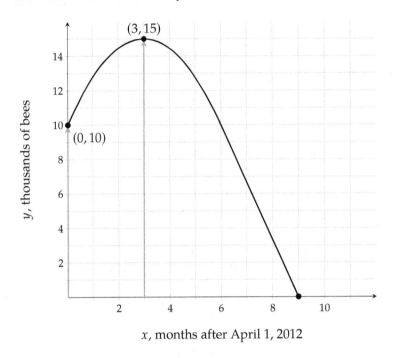

Figure 9.1.24: Bee Population

9.1.6 Solving Equations That Have Function Notation

Evaluating a function and *solving* an equation that has function notation in it are two separate things. Students understandably mix up these two tasks, because they are two sides of the same coin. Evaluating a function means that you know an input (typically, an x-value) and then you calculate an output (typically, a y-value). Solving an equation that has function notation in it is the opposite process. You know an output (typically, a y-value) and then you determine all the inputs that could have led to that output (typically, x-values).

Checkpoint 9.1.25 Functions with Algebraic Formulas. In Checkpoint 9.1.19, we found the function value when given the input, which we refer to as *evaluating* a function. To **solve** an equation that has function notation in it, we need to solve for the value of the variable that makes an equation true, just like in any other equation.

Solve the equations below for a function h where $h(x) = -4x + 7$. Check each answer and state the solution set.

a. $h(x) = -1$

The solution set is [ ].

b. $h(x) = 0$

The solution set is [ ].

c. $h(x) = 11$

The solution set is [ ].

Explanation.

a. To solve for x, we will substitute $h(x)$ with its formula, $-4x + 7$:

$$h(x) = -1$$
$$-4x + 7 = -1$$
$$-4x = -8$$
$$x = 2$$

The solution is 2, and the solution set is $\{2\}$.

b. To solve for x, we will substitute $h(x)$ with its formula:

$$h(x) = 0$$
$$-4x + 7 = 0$$
$$-4x = -7$$
$$x = \frac{7}{4}$$

The solution is $\frac{7}{4}$, and the solution set is $\left\{\frac{7}{4}\right\}$.

c. To solve for x, we will substitute $h(x)$ with its formula:

$$h(x) = 11$$
$$-4x + 7 = 11$$
$$-4x = 4$$
$$x = -1$$

The solution is -1, and the solution set is $\{-1\}$.

Example 9.1.26 Functions given in Table Form. In Example 9.1.20, we evaluated a function given in table form, using Table 9.1.21. Let's use that function to solve equations.

a. Solve $f(x) = 48$. Explain what this solution set represents in context.

b. When was the temperature 44 °F?

Explanation.

a. To solve $f(x) = 48$, we need to find the value of x that makes the equation true. Looking at the table, we look at the outputs and see that the output 48 occurs when $x = 7$. In abstract terms, the solution set is $\{7\}$. In context, this means that the temperature was 48 °F at 7AM.

b. To determine when the temperature was 44 °F, we look in the table to see where the output was 44 °F. This occurs when $x = 1$ and again when $x = 5$. In abstract terms, the solution set is $\{1, 5\}$. In context, this means that the temperature was 44 °F at 1AM and again at 5AM.

Example 9.1.27 Functions given in Graphical Form. In Example 9.1.22, we evaluated a function given in graphical form, using Figure 9.1.23. Now let's use that function to solve some equations.

a. Solve $B(x) = 0$. Explain what the solution set represents in the context of the problem.

    b. When did the population reach 13,000 bees?

Explanation.

    a. To solve $B(x) = 0$, we need to consider that number 0 as an output value, so it belongs on the vertical axis in Figure 9.1.28. Moving straight right or left, we find the point $(9, 0)$ is on the graph, and that tells us that $B(x) = 0$ when $x = 9$. Abstractly, the solution set is $\{9\}$. In the context of this problem, this means there were 0 bees in the colony 9 months after April 1, 2012 (on January 1, 2013).

    b. To determine when the number of bees reached 13,000, we need to recognize that 13 is an output value and locate it on the vertical axis. Moving straight right or left, we find (approximately) that the points $(1, 13)$ and $(5, 13)$ are on the graph. This means $x \approx 1$ and $x \approx 5$ are solutions. Abstractly, the solution set is approximately $\{1, 5\}$. In context, there will be 13,000 bees in the colony approximately 1 month after April 1, 2012 (on May 1, 2012) and again 5 months after April 1, 2012 (on September 1, 2012).

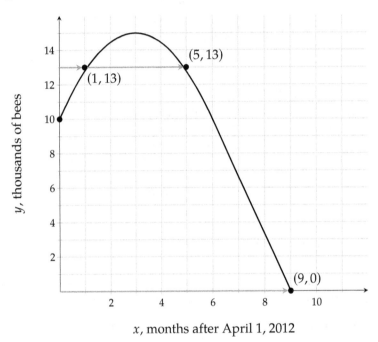

Figure 9.1.28: Bee Population

9.1.7 Domain and Range

Earlier we defined the domain and range of a relation. We repeat those definitions more formally here, specifically for functions.

Definition 9.1.29 Domain and Range. Given a function f, the **domain** of f is the collection of all valid input values for f. The **range** of f is the collection of all possible output values of f.

When working with functions, a common necessary task is to determine the function's domain and range. Also, the ability to identify domain and range is strong evidence that a person really understands the *concepts* of domain and range.

Example 9.1.30 Functions Defined by Ordered Pairs. The function f is defined by the ordered pairs

$$\{(1,2),(3,-2),(5,2),(7,-4),(9,6)\}.$$

Determine the domain and range of f.

Explanation. The ordered pairs tell us that $f(1) = 2$, $f(3) = -2$, etc. So the valid input values are 1, 3, 5, 7, and 9. This means the domain is the set $\{1,3,5,7,9\}$.

Similarly, the ordered pairs tell us that 2, −2, −4, and 6 are possible output values. Notice that the output 2 happened twice, but it only needs to be listed in this collection once. The range of f is $\{2,-2,-4,6\}$.

Example 9.1.31 Functions in Table Form. For each function defined using a table, state the domain and range.

   a. The function g is defined by:

x	−2	−1	0	1	2
y	5	5	5	5	5

   b. The function h is defined by:

x	0	1	2	3	4
y	8	6	4	2	0

Explanation.

   a. The table tells us that $g(-2) = 5$, $g(-1) = 5$, etc. So the valid input values are −2, −1, 0, 1, and 2. This means the domain of g is the set $\{-2,-1,0,1,2\}$.

     The only output evident from this table is 5, so the range of g is the set $\{5\}$.

   b. The table tells us that $h(0) = 8$, $h(1) = 6$, etc. So the valid input values are 0, 1, 2, 3, and 4. This means the domain of h is the set $\{0,1,2,3,4\}$.

     Similarly, the table shows us that the possible outputs are 8, 6, 4, 2, and 0. So the range of h is the set $\{8,6,4,2,0\}$.

Example 9.1.32 Functions in Graphical Form. Functions are graphed in Figure 9.1.33 through Figure 9.1.35. For each one, find its domain and range. In previous examples the domain and range were finite sets and we used set notation braces to communicate them. In these examples, the domain and range will be intervals, and so we use interval notation.

Figure 9.1.33: Function k

Figure 9.1.34: Function ℓ

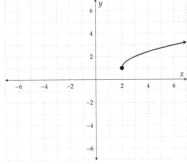

Figure 9.1.35: Function m

Explanation.

To find the domain of k, we look left and right across the x-axis. No matter where we look on the x-axis, we can look straight up or down and find a point on the graph. So no matter what input we imagine, there is always an corresponding output. Therefore the domain is all real numbers, which we write as $(-\infty, \infty)$.

Similarly, to find the range of k, we look up and down over the entire y-axis. No matter where we stop on the y-axis, we will be able to move straight left or right and find a point on the graph. So no matter what output we imagine, there is always an input that leads to that output. Therefore the range is also $(-\infty, \infty)$.

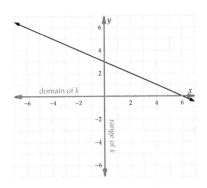

Figure 9.1.36: Function k

To find the domain of ℓ, we look left and right across the entire x-axis. No matter where we stop on the x-axis, we will be able to move straight up or down and find a point on the graph. (Although if we are far out left or right, we will have to look *very* far down to find that point.) So no matter what input we imagine, there is always an output to this function. Therefore the domain is $(-\infty, \infty)$.

To find the range of ℓ, we look up and down over the entire y-axis. For y-values larger than 4, if we look straight left or right, there is no point on the curve. Only with y-values 4 and under can we find an input that leads to such an output. So the range is all real numbers less than or equal to 4. In interval notation, that is $(-\infty, 4]$.

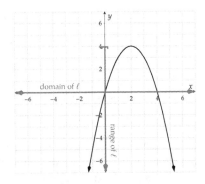

Figure 9.1.37: Function ℓ

To find the domain of *m*, we look left and right across the entire *x*-axis. For an *x*-value less than 2, there is no point on the curve directly above or below that *x*-value. So the graph is not telling you what *m(x)* would be. Only for *x*-values 2 and greater do we get an output value. So the domain is $[2, \infty)$.

To find the range of *m*, we look up and down over the entire *y*-axis. For *y*-values less than 1, if we look straight left or right, there is no point on the curve. Only with *y*-values 1 and greater can we find an input that leads to such an output. So the range is all real numbers greater than or equal to 1. In interval notation, that is $[1, \infty)$.

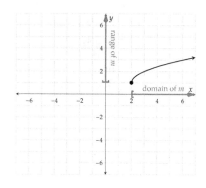

Figure 9.1.38: Function *m*

9.1.8 Determining if a Relation is a Function

We have seen functions that are defined using a verbal description, using an equation, using a list of ordered pairs, using a table, and using a graph. With all of these things, you might have a relation that is *not* actually a function. (We have seen a few examples of these as well.) Determining whether or not a given relation actually defines a function is another task that demonstrates actual understanding of the concept of a function.

Example 9.1.39 Verbally Described Relations. In the introduction, we discussed the relation between the variables of hare population and lynx population. These variables are related—for one thing, if the hare population is high, you will have information about the lynx population: it will probably be increasing because there is plenty of food.

But is this relation a function? Suppose that one year you know the hare population is 100,000 and the lynx population is 2000. Isn't it possible that some other year, the hare population is 100,000 again, but the lynx population is something different like 3000? Knowing one value of the first variable does not guarantee exactly one value of the second variable. So this relation is not a function.

Checkpoint 9.1.40 Sets of Ordered Pairs. The relations below are given in the form of ordered pairs. Determine if each is a function.

a. $\{(2,5), (3,6), (4,3), (4,-5), (8,0)\}$

(□ yes □ no)

b. $\{(1,2), (3,-2), (5,2), (7,-4), (9,6)\}$

(□ yes □ no)

Explanation.

a. The first relation *does not* represent a function because the input of 4 is associated with more than one output (3 and −5).

b. The second relation *does* represent a function as each input is associated with exactly one output. Note that it does not matter that both the inputs 1 and 5 lead to the same output. All that matters is that the input 1 has only one output. And the input 5 has only one output. And the same for each of the other inputs.

To determine if a relation that is represented graphically is a function, we need to visually determine if each input corresponds to exactly one output. How could that *not* happen? Somewhere on the horizontal axis, there would be a number, and looking straight up or straight down from there, you would find at least two points on the graph.

This thinking gives rise to the **vertical line test**.

Fact 9.1.41 Vertical Line Test. *If any vertical line passes through the graph of a relation more than once, then y is not a function of x in the graph.*

Example 9.1.42 Consider the relations given by Figure 9.1.43 and Figure 9.1.44.

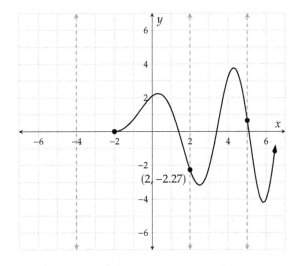

Figure 9.1.43: Fails the Vertical Line Test

Figure 9.1.44: Passes the Vertical Line Test

In Figure 9.1.43, we can see that there are vertical lines that pass through the graph more than once. This means that this graph does *not* represent a function. The issue is that supposing it *were* a function f, then what would $f(3)$ be? The graph suggests it could be three different things: ≈ 1.3, ≈ -1.6, and ≈ -3.7. With no clear single output for the input 3, we don't have a function.

However, in Figure 9.1.44, we see that all vertical lines pass through the graph one time (or not at all). Therefore, this graph does represent a function. If we name the function g, the domain of g is only numbers greater than or equal to 2. And for such numbers, the graphs shows us exactly one output for each input. For example, the graph shows us that $g(2) \approx -2.27$.

For relations in table form, we can determine if it makes a function by again checking to see if multiple outputs are ever associated with a single input.

Example 9.1.45 Functions in Table Form. For each relation shown in Table 9.1.46 through Table 9.1.48, determine if y is a function of x.

This relation represents a function as each input corresponds to exactly one output. For instance, the input -2 only corresponds to the output 5 and no other output. (Note that it does not matter that multiple inputs correspond to the same output 5.)

x	-2	-1	0	1	2
y	5	5	5	5	5

Table 9.1.46: A function?

This relation represents a function as each input corresponds to exactly one output. For instance, the input 0 only corresponds to the output 8 and no other output.

x	0	1	2	3	4
y	8	6	4	2	0

Table 9.1.47

This relation does not represent a function as some of the inputs correspond to more than one output. In particular, the input of 2 corresponds to both 7 and 9 for outputs.

x	−2	−2	0	2	2
y	1	3	5	7	9

Table 9.1.48

Exercises

Review and Warmup

1.

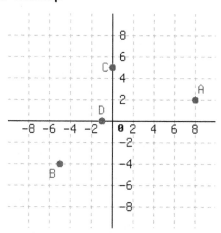

Write each point's position as an ordered pair, like $(1, 2)$.

$A =$ _____ $B =$ _____
$C =$ _____ $D =$ _____

2.

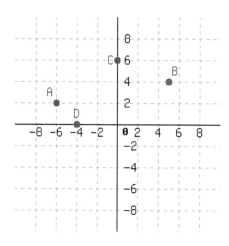

Write each point's position as an ordered pair, like $(1, 2)$.

$A =$ _____ $B =$ _____
$C =$ _____ $D =$ _____

3. Sketch the points $(8, 2)$, $(5, 5)$, $(−3, 0)$, and $(2, −6)$ on a Cartesian plane.

4. Sketch the points $(1, −4)$, $(−3, 5)$, $(0, 4)$, and $(−2, −6)$ on a Cartesian plane.

Determining Whether a Relation Is a Function of Not

5. Do these sets of ordered pairs make functions of x? What are their domains and ranges?

   a. $\left\{(9,8),(-10,4)\right\}$

   This set of ordered pairs   (□ describes   □ does not describe)   a function of x. This set of ordered pairs has domain [＿＿＿＿] and range [＿＿＿＿] .

   b. $\left\{(1,4),(0,6),(0,10)\right\}$

   This set of ordered pairs   (□ describes   □ does not describe)   a function of x. This set of ordered pairs has domain [＿＿＿＿] and range [＿＿＿＿] .

   c. $\left\{(3,0),(7,1),(-5,9),(-9,10)\right\}$

   This set of ordered pairs   (□ describes   □ does not describe)   a function of x. This set of ordered pairs has domain [＿＿＿＿] and range [＿＿＿＿] .

   d. $\left\{(5,8),(-3,8),(3,3),(9,9),(8,5)\right\}$

   This set of ordered pairs   (□ describes   □ does not describe)   a function of x. This set of ordered pairs has domain [＿＿＿＿] and range [＿＿＿＿] .

6. Do these sets of ordered pairs make functions of x? What are their domains and ranges?

   a. $\left\{(-3,1),(-10,6)\right\}$

   This set of ordered pairs   (□ describes   □ does not describe)   a function of x. This set of ordered pairs has domain [＿＿＿＿] and range [＿＿＿＿] .

   b. $\left\{(3,8),(-9,9),(2,5)\right\}$

   This set of ordered pairs   (□ describes   □ does not describe)   a function of x. This set of ordered pairs has domain [＿＿＿＿] and range [＿＿＿＿] .

   c. $\left\{(-2,7),(4,3),(-7,5),(-6,9)\right\}$

   This set of ordered pairs   (□ describes   □ does not describe)   a function of x. This set of ordered pairs has domain [＿＿＿＿] and range [＿＿＿＿] .

   d. $\left\{(-3,2),(-5,2),(-5,9),(10,4),(5,8)\right\}$

   This set of ordered pairs   (□ describes   □ does not describe)   a function of x. This set of ordered pairs has domain [＿＿＿＿] and range [＿＿＿＿] .

7. Does the following set of ordered pairs make for a function of x?

$$\{(-8,8),(5,9),(-3,8),(7,10),(3,4)\}$$

This set of ordered pairs   (□ describes   □ does not describe)   a function of x. This set of ordered pairs has domain [＿＿＿＿＿＿＿] and range [＿＿＿＿＿＿＿].

8. Does the following set of ordered pairs make for a function of x?

$$\{(8,3),(-6,5),(-1,6),(4,6),(1,10)\}$$

This set of ordered pairs   (□ describes   □ does not describe)   a function of x. This set of ordered pairs has domain [＿＿＿＿＿＿＿] and range [＿＿＿＿＿＿＿].

9. Decide whether each graph shows a relationship where y is a function of x.

The first graph   (□ does   □ does not)   give a function of x. The second graph   (□ does   □ does not)   give a function of x.

10. Decide whether each graph shows a relationship where y is a function of x.

The first graph   (□ does   □ does not)   give a function of x. The second graph   (□ does   □ does not)   give a function of x.

11. The following graphs show two relationships. Decide whether each graph shows a relationship where y is a function of x.

The first graph   (□ does   □ does not)   give a function of x. The second graph   (□ does   □ does not)   give a function of x.

12. The following graphs show two relationships. Decide whether each graph shows a relationship where y is a function of x.

The first graph   (□ does   □ does not)   give a function of x. The second graph   (□ does   □ does not)   give a function of x.

13. Which of the following graphs show y as a function of x?

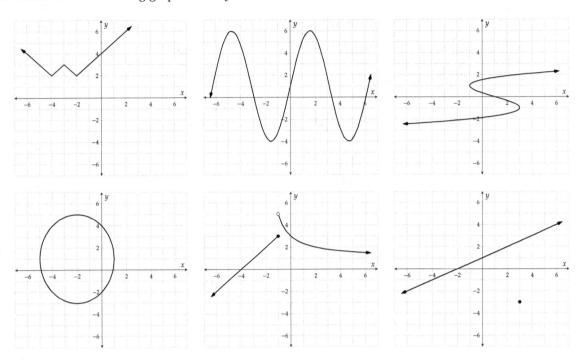

14. Which of the following graphs show y as a function of x?

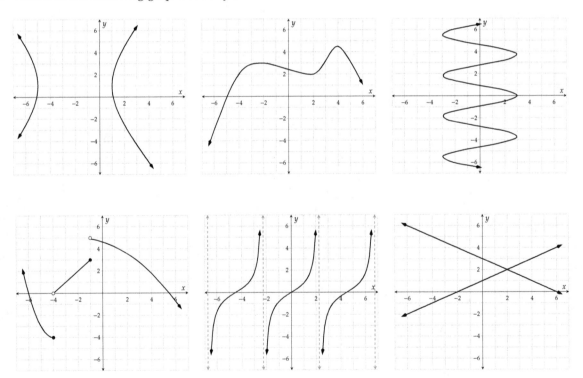

Evaluating Functions Algebraically   Evaluate the function at the given values.

15. $G(x) = x - 7$

    a. $G(4) =$ [   ]

    b. $G(-1) =$ [   ]

    c. $G(0) =$ [   ]

16. $h(x) = x - 5$

    a. $h(2) =$ [   ]

    b. $h(-4) =$ [   ]

    c. $h(0) =$ [   ]

17. $h(x) = 10x$

    a. $h(4) =$ [   ]

    b. $h(-5) =$ [   ]

    c. $h(0) =$ [   ]

18. $F(x) = 7x$

    a. $F(1) =$ [   ]

    b. $F(-1) =$ [   ]

    c. $F(0) =$ [   ]

19. $G(x) = -4x + 6$

    a. $G(3) =$ [   ]

    b. $G(-3) =$ [   ]

    c. $G(0) =$ [   ]

20. $G(x) = -2x + 1$

    a. $G(2) =$ [   ]

    b. $G(-3) =$ [   ]

    c. $G(0) =$ [   ]

21. $H(x) = -x + 6$

    a. $H(3) =$ [   ]

    b. $H(-5) =$ [   ]

    c. $H(0) =$ [   ]

22. $K(x) = -x + 3$

    a. $K(5) =$ [   ]

    b. $K(-2) =$ [   ]

    c. $K(0) =$ [   ]

23. $f(t) = t^2 - 5$

    a. $f(3) =$ [   ]

    b. $f(-3) =$ [   ]

    c. $f(0) =$ [   ]

24. $f(r) = r^2 + 4$

    a. $f(1) =$ [   ]

    b. $f(-3) =$ [   ]

    c. $f(0) =$ [   ]

25. $g(x) = -x^2 - 9$

    a. $g(5) =$ [   ]

    b. $g(-3) =$ [   ]

    c. $g(0) =$ [   ]

26. $h(t) = -t^2 + 3$

    a. $h(3) =$ [   ]

    b. $h(-2) =$ [   ]

    c. $h(0) =$ [   ]

27. $F(r) = 9$

    a. $F(2) =$ [   ]

    b. $F(9) =$ [   ]

    c. $F(0) =$ [   ]

28. $G(x) = -3$

    a. $G(1) =$ [   ]

    b. $G(-3) =$ [   ]

    c. $G(0) =$ [   ]

29. $G(x) = \dfrac{7x}{5x + 6}$

    a. $G(3) =$ [   ].

    b. $G(-7) =$ [   ].

30. $H(x) = \dfrac{5x}{-7x + 1}$

    a. $H(3) =$ [   ].

    b. $H(-2) =$ [   ].

31. $K(x) = -\dfrac{16}{x - 5}$.

    a. $K(7) =$ [   ].

    b. $K(5) =$ [   ].

32. $f(x) = -\dfrac{60}{x + 10}$.

    a. $f(-20) =$ [   ].

    b. $f(-10) =$ [   ].

33. $f(x) = x^2 - x - 1$

    a. $f(4) =$ [   ].

    b. $f(-4) =$ [   ].

34. $g(x) = x^2 - 5x + 4$

    a. $g(2) =$ [   ].

    b. $g(-4) =$ [   ].

35. $h(x) = -2x^2 - 3x - 5$

    a. $h(1) =$ [   ].

    b. $h(-5) =$ [   ].

36. $F(x) = -3x^2 + 2x + 5$

    a. $F(1) =$ [   ].

    b. $F(-2) =$ [   ].

37. $G(x) = \sqrt{x}$.

    a. $G(4) =$ [   ].

    b. $G\left(\frac{1}{49}\right) =$ [   ].

    c. $G(-8) =$ [   ].

38. $G(x) = \sqrt{x}$.

    a. $G(64) =$ [   ].

    b. $G\left(\frac{36}{25}\right) =$ [   ].

    c. $G(-8) =$ [   ].

39. $H(x) = \sqrt[3]{x}$

    a. $H(-64) =$ [   ].

    b. $H\left(\frac{125}{8}\right) =$ [   ].

40. $K(x) = \sqrt[3]{x}$

    a. $K(-1) =$ [   ].

    b. $K\left(\frac{8}{125}\right) =$ [   ].

Solving Equations with Function Notation

41. Solve for x, where $f(x) = -10x - 10$.

    a. If $f(x) = -60$, then $x =$ [   ].

    b. If $f(x) = -24$, then $x =$ [   ].

42. Solve for x, where $f(x) = 8x + 8$.

    a. If $f(x) = 40$, then $x =$ [   ].

    b. If $f(x) = 26$, then $x =$ [   ].

43. Solve for x, where $g(x) = x^2 + 9$.

    a. If $g(x) = 18$, then $x =$ [   ].

    b. If $g(x) = 8$, then $x =$ [   ].

44. Solve for x, where $h(x) = x^2 + 2$.

    a. If $h(x) = 3$, then $x =$ [   ].

    b. If $h(x) = -2$, then $x =$ [   ].

45. Solve for x, where $F(x) = x^2 - 6x + 2$.

    If $F(x) = -3$, then $x =$ [   ].

46. Solve for x, where $G(x) = x^2 + 15x + 47$.

    If $G(x) = -9$, then $x =$ [   ].

Functions Represented with Graphs

47. A function is graphed.

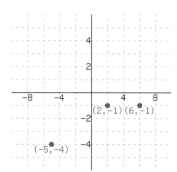

This function has domain ☐
and range ☐.

48. A function is graphed.

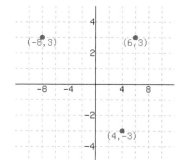

This function has domain ☐
and range ☐.

49. A function is graphed.

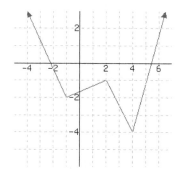

This function has domain ☐
and range ☐.

50. A function is graphed.

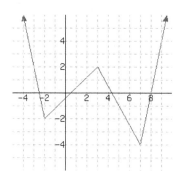

This function has domain ☐
and range ☐.

51. Function f is graphed.

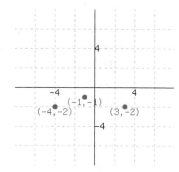

a. $f(-1) =$ ☐

b. Solve $f(x) = -2$.

52. Function f is graphed.

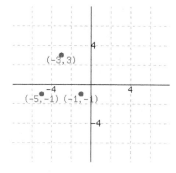

a. $f(-3) =$ ☐

b. Solve $f(x) = -1$.

53. Function f is graphed.

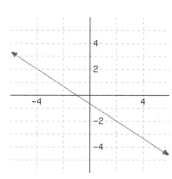

a. $f(-1) =$ ☐

b. Solve $f(x) = -2$.

54. Function f is graphed.

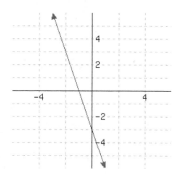

a. $f(0) =$ ☐

b. Solve $f(x) = 3$.

55. Function f is graphed.

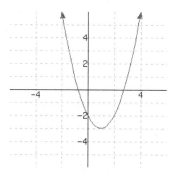

a. $f(3) =$ ☐

b. Solve $f(x) = -2$.

56. Function f is graphed.

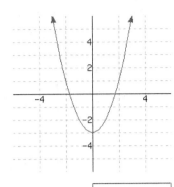

  a. $f(-1) =$ [   ]

  b. Solve $f(x) = 1$.

57. If $H(1) = 0$, then the point [   ] is on the graph of H.

If $(1, 6)$ is on the graph of H, then $H(1) =$ [   ].

58. If $K(8) = 12$, then the point [   ] is on the graph of K.

If $(8, 13)$ is on the graph of K, then $K(8) =$ [   ].

59. If $f(y) = t$, then the point [   ] is on the graph of f.

The answer is not a specific numerical point, but one with variables for coordinates.

60. If $f(x) = r$, then the point [   ] is on the graph of f.

The answer is not a specific numerical point, but one with variables for coordinates.

61. If (r, x) is on the graph of g, then $g(r) =$ [   ].

62. If (y, r) is on the graph of h, then $h(y) =$ [   ].

Function Notation in Context

63. Suppose that M is the function that computes how many miles are in x feet. Find the algebraic rule for M. (If you do not know how many feet are in one mile, you can look it up on Google.)

$M(x) =$ [   ]

Evaluate $M(19000)$ and interpret the result:

There are about [   ] miles in [   ] feet.

64. Suppose that K is the function that computes how many kilograms are in x pounds. Find the algebraic rule for K. (If you do not know how many pounds are in one kilogram, you can look it up on Google.)

$K(x) =$ [   ]

Evaluate $K(214)$ and interpret the result.

Something that weighs [   ] pounds would weigh about [   ] kilograms.

65. Stephanie started saving in a piggy bank on her birthday. The function $f(x) = 3x + 2$ models the amount of money, in dollars, in Stephanie's piggy bank. The independent variable represents the number of days passed since her birthday.

Interpret the meaning of $f(4) = 14$.

⊙ A. Four days after Stephanie started her piggy bank, there were $14 in it.

⊙ B. Fourteen days after Stephanie started her piggy bank, there were $4 in it.

⊙ C. The piggy bank started with $14 in it, and Stephanie saves $4 each day.

⊙ D. The piggy bank started with $4 in it, and Stephanie saves $14 each day.

66. Sherial started saving in a piggy bank on her birthday. The function $f(x) = 5x + 3$ models the amount of money, in dollars, in Sherial's piggy bank. The independent variable represents the number of days passed since her birthday.

Interpret the meaning of $f(1) = 8$.

⊙ A. The piggy bank started with $1 in it, and Sherial saves $8 each day.

⊙ B. The piggy bank started with $8 in it, and Sherial saves $1 each day.

⊙ C. One days after Sherial started her piggy bank, there were $8 in it.

⊙ D. Eight days after Sherial started her piggy bank, there were $1 in it.

67. An arcade sells multi-day passes. The function $g(x) = \frac{1}{4}x$ models the number of days a pass will work, where x is the amount of money paid, in dollars.

Interpret the meaning of $g(12) = 3$.

⊙ A. If a pass costs $12, it will work for 3 days.

⊙ B. Each pass costs $12, and it works for 3 days.

⊙ C. If a pass costs $3, it will work for 12 days.

⊙ D. Each pass costs $3, and it works for 12 days.

68. An arcade sells multi-day passes. The function $g(x) = \frac{1}{2}x$ models the number of days a pass will work, where x is the amount of money paid, in dollars.

Interpret the meaning of $g(4) = 2$.

⊙ A. Each pass costs $2, and it works for 4 days.

⊙ B. Each pass costs $4, and it works for 2 days.

⊙ C. If a pass costs $4, it will work for 2 days.

⊙ D. If a pass costs $2, it will work for 4 days.

69. Bobbi will spend $105 to purchase some bowls and some plates. Each bowl costs $2, and each plate costs $3. The function $p(b) = -\frac{2}{3}b + 35$ models the number of plates Bobbi will purchase, where b represents the number of bowls Bobbi will purchase.

   Interpret the meaning of $p(39) = 9$.

   ⊙ A. $9 will be used to purchase bowls, and $39 will be used to purchase plates.

   ⊙ B. If 9 bowls are purchased, then 39 plates will be purchased.

   ⊙ C. If 39 bowls are purchased, then 9 plates will be purchased.

   ⊙ D. $39 will be used to purchase bowls, and $9 will be used to purchase plates.

70. Lisa will spend $135 to purchase some bowls and some plates. Each bowl costs $2, and each plate costs $3. The function $p(b) = -\frac{2}{3}b + 45$ models the number of plates Lisa will purchase, where b represents the number of bowls Lisa will purchase.

   Interpret the meaning of $p(30) = 25$.

   ⊙ A. $25 will be used to purchase bowls, and $30 will be used to purchase plates.

   ⊙ B. If 25 bowls are purchased, then 30 plates will be purchased.

   ⊙ C. If 30 bowls are purchased, then 25 plates will be purchased.

   ⊙ D. $30 will be used to purchase bowls, and $25 will be used to purchase plates.

71. Barbara will spend $125 to purchase some bowls and some plates. Each plate costs $3, and each bowl costs $5. The function $q(x) = -\frac{3}{5}x + 25$ models the number of bowls Barbara will purchase, where x represents the number of plates to be purchased.

   Interpret the meaning of $q(5) = 22$.

   ⊙ A. 5 plates and 22 bowls can be purchased.

   ⊙ B. 22 plates and 5 bowls can be purchased.

   ⊙ C. $22 will be used to purchase bowls, and $5 will be used to purchase plates.

   ⊙ D. $5 will be used to purchase bowls, and $22 will be used to purchase plates.

72. Sharell will spend $360 to purchase some bowls and some plates. Each plate costs $9, and each bowl costs $8. The function $q(x) = -\frac{9}{8}x + 45$ models the number of bowls Sharell will purchase, where x represents the number of plates to be purchased.

   Interpret the meaning of $q(24) = 18$.

   ⊙ A. $18 will be used to purchase bowls, and $24 will be used to purchase plates.

   ⊙ B. 18 plates and 24 bowls can be purchased.

   ⊙ C. 24 plates and 18 bowls can be purchased.

   ⊙ D. $24 will be used to purchase bowls, and $18 will be used to purchase plates.

The following figure has the graph $y = d(t)$, which models a particle's distance from the starting line in feet, where t stands for time in seconds since timing started.

73.

74.

a. $d(8) =$ [   ]

b. Interpret the meaning of $d(8)$:

⊙ A. The particle was 3.33333 feet away from the starting line 8 seconds since timing started.

⊙ B. In the first 3.33333 seconds, the particle moved a total of 8 feet.

⊙ C. The particle was 8 feet away from the starting line 3.33333 seconds since timing started.

⊙ D. In the first 8 seconds, the particle moved a total of 3.33333 feet.

c. Solve $d(t) = 5$ for t. $t =$ [   ]

d. Interpret the meaning of part c's solution(s):

⊙ A. The article was 5 feet from the starting line 7 seconds since timing started.

⊙ B. The article was 5 feet from the starting line 1 seconds since timing started, and again 7 seconds since timing started.

⊙ C. The article was 5 feet from the starting line 1 seconds since timing started, or 7 seconds since timing started.

⊙ D. The article was 5 feet from the starting line 1 seconds since timing started.

a. $d(4) =$ [   ]

b. Interpret the meaning of $d(4)$:

⊙ A. The particle was 9 feet away from the starting line 4 seconds since timing started.

⊙ B. In the first 4 seconds, the particle moved a total of 9 feet.

⊙ C. The particle was 4 feet away from the starting line 9 seconds since timing started.

⊙ D. In the first 9 seconds, the particle moved a total of 4 feet.

c. Solve $d(t) = 6$ for t. $t =$ [   ]

d. Interpret the meaning of part c's solution(s):

⊙ A. The article was 6 feet from the starting line 2 seconds since timing started, or 8 seconds since timing started.

⊙ B. The article was 6 feet from the starting line 8 seconds since timing started.

⊙ C. The article was 6 feet from the starting line 2 seconds since timing started, and again 8 seconds since timing started.

⊙ D. The article was 6 feet from the starting line 2 seconds since timing started.

The function C models the the number of customers in a store t hours since the store opened.

75.

t	0	1	2	3	4	5	6	7
$C(t)$	0	48	78	92	100	78	38	0

a. $C(6) = $ [　　　　　]

b. Interpret the meaning of $C(6)$:

⊙ A. There were 38 customers in the store 6 hours after the store opened.

⊙ B. In 6 hours since the store opened, the store had an average of 38 customers per hour.

⊙ C. In 6 hours since the store opened, there were a total of 38 customers.

⊙ D. There were 6 customers in the store 38 hours after the store opened.

c. Solve $C(t) = 78$ for t. $t = $ [　　　　　]

d. Interpret the meaning of Part c's solution(s):

⊙ A. There were 78 customers in the store 2 hours after the store opened.

⊙ B. There were 78 customers in the store either 5 hours after the store opened, or 2 hours after the store opened.

⊙ C. There were 78 customers in the store 5 hours after the store opened.

⊙ D. There were 78 customers in the store 5 hours after the store opened, and again 2 hours after the store opened.

76.

t	0	1	2	3	4	5	6	7
$C(t)$	0	41	81	97	97	81	44	0

a. $C(7) = $ [　　　　　]

b. Interpret the meaning of $C(7)$:

⊙ A. There were 0 customers in the store 7 hours after the store opened.

⊙ B. There were 7 customers in the store 0 hours after the store opened.

⊙ C. In 7 hours since the store opened, there were a total of 0 customers.

⊙ D. In 7 hours since the store opened, the store had an average of 0 customers per hour.

c. Solve $C(t) = 97$ for t. $t = $ [　　　　　]

d. Interpret the meaning of Part c's solution(s):

⊙ A. There were 97 customers in the store 3 hours after the store opened.

⊙ B. There were 97 customers in the store either 3 hours after the store opened, or 4 hours after the store opened.

⊙ C. There were 97 customers in the store 3 hours after the store opened, and again 4 hours after the store opened.

⊙ D. There were 97 customers in the store 4 hours after the store opened.

9.2 Properties of Quadratic Functions

9.2.1 Introduction

In this section we will learn about quadratic functions and how to identify their key features on a graph. We will identify their direction, vertex, axis of symmetry and intercepts. We will also see how to graph a parabola by finding the vertex and making a table of function values. We will look at applications that involve the vertex of a quadratic function.

Definition 9.2.2. A **quadratic function** has the form $f(x) = ax^2 + bx + c$ where a, b, and c are real numbers, and $a \neq 0$. The graph of a quadratic function has the shape of a **parabola**.

Notice that a quadratic function has a squared term that linear functions do not have. If $a = 0$, the function is linear. To understand the shape and features of a quadratic function, let's look at an example.

9.2.2 Properties of Quadratic Functions

Hannah fired a toy rocket from the ground, which flew into the air at a speed of 64 feet per second. The path of the rocket can be modeled by the function f where $f(t) = -16t^2 + 64t$. To see the shape of the function we will make a table of values and plot the points. For the table we we will choose some values for t and then evaluate the function at each t-value:

t	$f(t) = -16t^2 + 64t$	Point
0	$f(0) = -16(0)^2 + 64(0) = 0$	$(0, 0)$
1	$f(1) = -16(1)^2 + 64(1) = 48$	$(1, 48)$
2	$f(2) = -16(2)^2 + 64(2) = 64$	$(2, 64)$
3	$f(3) = -16(3)^2 + 64(3) = 48$	$(3, 48)$
4	$f(4) = -16(4)^2 + 64(4) = 0$	$(4, 0)$

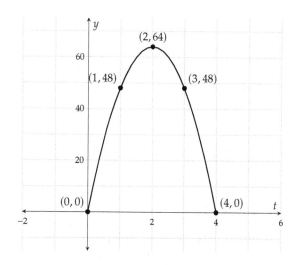

Table 9.2.3: Function values and points for $f(t) = -16t^2 + 64t$

Figure 9.2.4: Graph of $f(t) = -16t^2 + 64t$

Now that we have Table 9.2.3 and Figure 9.2.4, we can see the features of this parabola. Notice the symmetry in the shape of the graph and the y-values in the table. Consecutive y-values do not increase by a constant amount in the way that linear functions do.

The first feature that we will talk about is the *direction* that a parabola opens. All parabolas open either upward or downward. This parabola in the rocket example opens downward because a is negative. Here are some more quadratic functions graphed so we can see which way they open.

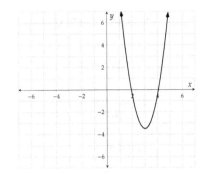

Figure 9.2.5: The graph of $y = x^2 - 2x + 2$ opens upward

Figure 9.2.6: The graph of $y = -\frac{1}{4}x^2 - \frac{1}{2}x - \frac{1}{4}$ opens downward

Figure 9.2.7: The graph of $y = 3x^2 - 18x + 23.5$ opens upward

Fact 9.2.8. *We only need to look at the sign of the leading coefficient to determine which way the graph opens. If the leading coefficient is positive, the parabola opens upward. If the leading coefficient is negative, the parabola opens downward.*

Checkpoint 9.2.9. Determine whether the graph of each quadratic function opens upward or downward.

a. The graph of the quadratic function $y = 3x^2 - 4x - 7$ opens   (☐ upward   ☐ downward)   .

b. The graph of the quadratic function $y = -5x^2 + x$ opens   (☐ upward   ☐ downward)   .

c. The graph of the quadratic function $y = 2 + 3x - x^2$ opens   (☐ upward   ☐ downward)   .

d. The graph of the quadratic function $y = \frac{1}{3}x^2 - \frac{2}{5}x + \frac{1}{4}$ opens   (☐ upward   ☐ downward)   .

Explanation.

a. The graph of the quadratic function $y = 3x^2 - 4x - 7$ opens upward as the leading coefficient is the positive number 3.

b. The graph of the quadratic function $y = -5x^2 + x$ opens downward as the leading coefficient is the negative number -5.

c. The graph of the quadratic function $y = 2 + 3x - x^2$ opens downward as the leading coefficient is -1. (Note that the leading coefficient is the coefficient on x^2.)

d. The graph of the quadratic function $y = \frac{1}{3}x^2 - \frac{2}{5}x + \frac{1}{4}$ opens upward as the leading coefficient is the positive number $\frac{1}{3}$.

The **vertex** is the highest or lowest point on the graph. In Figure 9.2.4, the vertex is $(2, 64)$. This tells us that Hannah's rocket reached its maximum height of 64 feet after 2 seconds. If the parabola opens downward, as in the rocket example, then the y-value of the vertex is the **maximum** y-value. If the parabola opens upward then the y-value of the vertex is the **minimum** y-value.

The **axis of symmetry** is a vertical line that passes through the vertex, dividing it in half. The vertex is the only point that does not have a symmetric point. We write the axis of symmetry as an equation of a vertical line so it always starts with "$x =$." In Figure 9.2.4, the equation for the axis of symmetry is $x = 2$.

The **vertical intercept** is the point where the parabola crosses the vertical axis. The vertical intercept is the y-intercept if the axes are labeled x and y. In Figure 9.2.4, the point $(0, 0)$ is the starting point of the rocket. The y-value of 0 means the rocket started on the ground.

The **horizontal intercept(s)** are the points where the parabola crosses the horizontal axis. They are the x-intercepts if the axes are labeled x and y. The point $(0, 0)$ on the path of the rocket is also a horizontal intercept. The *t*-value of 0 indicates the time when the rocket was launched from the ground. There is another horizontal intercept at the point $(4, 0)$, which means the rocket hit the ground after 4 seconds.

It is possible for a quadratic function to have 0, 1 or 2 horizontal intercepts. The figures below show an example of each.

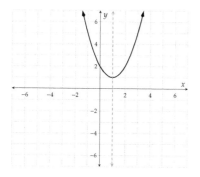

Figure 9.2.10: The graph of $y = x^2 - 2x + 2$ has no horizontal intercepts

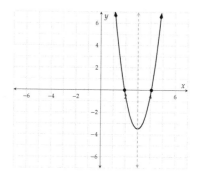

Figure 9.2.11: The graph of $y = -\frac{1}{4}x^2 - \frac{1}{2}x - \frac{1}{4}$ has one horizontal intercept

Figure 9.2.12: The graph of $y = 3x^2 - 18x + 23.5$ has two horizontal intercepts

Here is a summary of the properties of quadratic functions:

Direction A parabola opens upward if a is positive and opens downward of a is negative.

Vertex The vertex of a parabola is the maximum or minimum point on the graph.

Axis of Symmetry The axis of symmetry is the vertical line that passes through the vertex.

Vertical Intercept The vertical intercept is the point where the function intersects the vertical axis. There is exactly one vertical intercept.

Horizontal Intercept(s) The horizontal intercept(s) are the points where a function intersects the horizontal axis. The graph of a parabola can have 0, 1, or 2 horizontal intercepts.

List 9.2.13: Summary of Properties of Quadratic Functions

Example 9.2.14 Identify the key features of the quadratic function $y = x^2 - 2x - 8$ shown in Figure 9.2.15.

Explanation.

First, we see that this parabola opens upward because the leading coefficient is positive.

Then we locate the vertex which is the point $(1, -9)$. The axis of symmetry is the vertical line $x = 1$.

The vertical intercept or y-intercept is the point $(0, -8)$.

The horizontal intercepts are the points $(-2, 0)$ and $(4, 0)$.

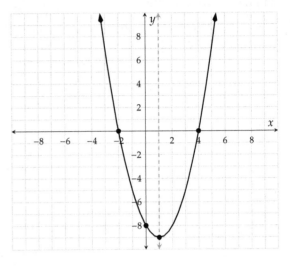

Figure 9.2.15: Graph of $y = x^2 - 2x - 8$

Checkpoint 9.2.16. Use the graph to answer the following questions.

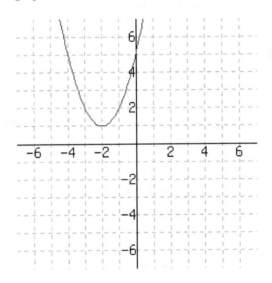

a. What are the coordinates of the vertex?

b. What is the equation of the axis of symmetry?

c. What are the coordinates of the x-intercept(s)?

d. What are the coordinates of the y-intercept?

Explanation.

a. The vertex is at $(-2, 1)$.

b. The equation of the axis of symmetry is $x = -2$.

c. There are no x-intercepts. (Answer None.)

d. The y-intercept is at $(0, 5)$.

9.2.3 Finding the Vertex and Axis of Symmetry Algebraically

The coordinates of the vertex are not easy to identify on a graph if they are not integers. Another way to find it is by using a formula.

Fact 9.2.17. *If we denote (h, k) as the coordinates of the vertex of a quadratic function $f(x) = ax^2 + bx + c$, then $h = -\frac{b}{2a}$.*

To understand why, we can look at the quadratic formula. The vertex is on the axis of symmetry, so it will always occur halfway between the two x-intercepts (if there are any). The quadratic formula shows that the x-intercepts happen at $-\frac{b}{2a}$ minus some number and at $-\frac{b}{2a}$ plus that same number:

$$x = \frac{-b \pm \sqrt{b^2 - 4ac}}{2a}$$

Example 9.2.18 Determine the vertex and axis of symmetry of the quadratic function $f(x) = x^2 - 4x - 12$.

We will find the x-value of the vertex using the formula $h = -\frac{b}{2a}$, for $a = 1$ and $b = -4$.

$$h = -\frac{b}{2a}$$
$$h = -\frac{(-4)}{2(1)}$$
$$h = 2$$

Now we know the x-value of the vertex is 2, so we may evaluate $f(2)$ to determine the y-value of the vertex:

$$k = f(2) = (2)^2 - 4(2) - 12$$
$$k = = 4 - 8 - 12$$
$$k = = -16$$

The vertex is the point $(2, -16)$ and the axis of symmetry is the line $x = 2$.

Example 9.2.19 Determine the vertex and axis of symmetry of the quadratic function $y = -3x^2 - 3x + 7$.

Explanation. Using the formula $h = -\frac{b}{2a}$ with $a = -3$ and $b = -3$, we have :

$$h = -\frac{b}{2a}$$
$$h = -\frac{(-3)}{2(-3)}$$
$$h = -\frac{1}{2}$$

Now that we've determined $h = -\frac{1}{2}$, we can substitute it for x to find the y-value of the vertex:

$$y = -3x^2 - 3x + 7$$

$$y = -3\left(-\frac{1}{2}\right)^2 - 3\left(-\frac{1}{2}\right) + 7$$

$$y = -3\left(\frac{1}{4}\right) + \frac{3}{2} + 7$$

$$y = -\frac{3}{4} + \frac{3}{2} + 7$$

$$y = -\frac{3}{4} + \frac{6}{4} + \frac{28}{4}$$

$$y = \frac{31}{4}$$

The vertex is the point $\left(-\frac{1}{2}, \frac{31}{4}\right)$ and the axis of symmetry is the line $x = -\frac{1}{2}$.

9.2.4 Graphing Quadratic Functions by Making a Table

When we learned how to graph lines, we could choose any x-values. For quadratic functions, though, we want to find the vertex and choose our x-values around it. Then we can use the property of symmetry to help us. Let's look at an example.

Example 9.2.20 Determine the vertex and axis of symmetry for the quadratic function $y = -x^2 - 2x + 3$. Then make a table of values and sketch the graph of the function.

Explanation. To determine the vertex of $y = -x^2 - 2x + 3$, we want to find the x-value of the vertex first. We use $h = -\frac{b}{2a}$ with $a = -1$ and $b = -2$:

$$h = -\frac{(-2)}{2(-1)}$$

$$h = \frac{2}{-2}$$

$$= -1$$

To find the y-coordinate of the vertex, we substitute $x = -1$ into the equation for our parabola.

$$y = -x^2 - 2x + 3$$

$$y = -(-1)^2 - 2(-1) + 3$$

$$= -1 + 2 + 3$$

$$= 4$$

Now we know that our axis of symmetry is the line $x = -1$ and the vertex is the point $(-1, 4)$. We will set up our table with two values on each side of $x = -1$. We choose $x = -3, -2, -1, 0$, and 1 as shown in Table 9.2.21.

Next, we'll determine the y-coordinates by replacing x with each value and we have the complete table as shown in Table 9.2.22. Notice that each pair of y-values on either side of the vertex match. This helps us to check that our vertex and y-values are correct.

x	$y = -x^2 - 2x + 3$	Point
-3		
-2		
-1		
0		
1		

x	$y = -x^2 - 2x + 3$	Point
-3	$y = -(-3)^2 - 2(-3) + 3 = 0$	$(-3, 0)$
-2	$y = -(-2)^2 - 2(-2) + 3 = 3$	$(-2, 3)$
-1	$y = -(-1)^2 - 2(-1) + 3 = 4$	$(-1, 4)$
0	$y = -(0)^2 - 2(0) + 3 = 3$	$(0, 3)$
1	$y = -(1)^2 - 2(1) + 3 = 0$	$(1, 0)$

Table 9.2.21: Setting up the table for $y = -x^2 - 2x + 3$

Table 9.2.22: Function values and points for $y = -x^2 - 2x + 3$

Now that we have our table, we will plot the points and draw in the axis of symmetry as shown in Figure 9.2.23. We complete the graph by drawing a smooth curve through the points and drawing an arrow on each end as shown in Figure 9.2.24

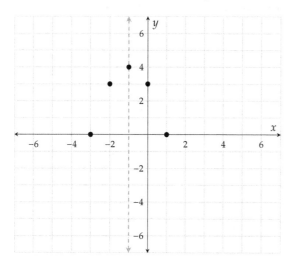

Figure 9.2.23: Plot of the points and axis of symmetry

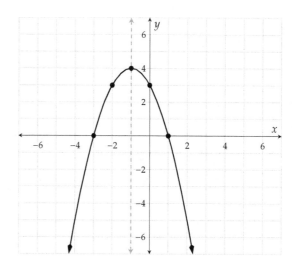

Figure 9.2.24: Graph of $y = -x^2 - 2x + 3$

The method we used works best when the x-value of the vertex is an integer. We can still make a graph if that is not the case as we will demonstrate in the next example.

Example 9.2.25 Determine the vertex and axis of symmetry for the quadratic function $y = 2x^2 - 3x - 4$. Use this to create a table of values and sketch the graph of this function.

Explanation. To determine the vertex of $y = 2x^2 - 3x - 4$, we'll find $h = -\frac{b}{2a}$ with $a = 2$ and $b = -3$:

$$h = -\frac{(-3)}{2(2)}$$
$$h = \frac{3}{4}$$

Next, we'll determine the y-coordinate by replacing x with $\frac{3}{4}$ in $y = 2x^2 - 3x - 4$:

$$y = 2\left(\frac{3}{4}\right)^2 - 3\left(\frac{3}{4}\right) - 4$$

$$y = 2\left(\frac{9}{16}\right) - \frac{9}{4} - 4$$

$$y = \frac{9}{8} - \frac{18}{8} - \frac{32}{8}$$

$$y = -\frac{41}{8}$$

Thus the vertex occurs at $\left(\frac{3}{4}, -\frac{41}{8}\right)$, or at $(0.75, -5.125)$. The axis of symmetry is then the line $x = \frac{3}{4}$, or $x = 0.75$.

Now that we know the x-value of the vertex, we will create a table. We will choose x-values on both sides of $x = 0.75$, but we will choose integers because it will be easier to find the function values.

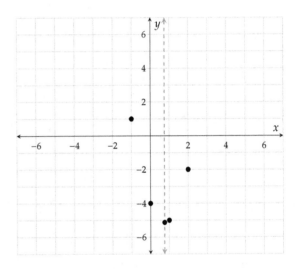

x	$y = 2x^2 - 3x - 4$	Point
-1	1	$(-1, 1)$
0	-4	$(0, -4)$
0.75	-5.125	$(0.75, -5.125)$
1	-5	$(1, -5)$
2	-2	$(2, -2)$

Table 9.2.26: Function values and points for $y = 2x^2 - 3x - 4$

Figure 9.2.27: Plot of initial points

The points graphed in Figure 9.2.27 don't have the symmetry we'd expect from a parabola. This is because the vertex occurs at an x-value that is not an integer, and all of the chosen values in the table are integers. We can use the axis of symmetry to determine more points on the graph (as shown in Figure 9.2.28), which will give it the symmetry we expect. From there, we can complete the sketch of this graph.

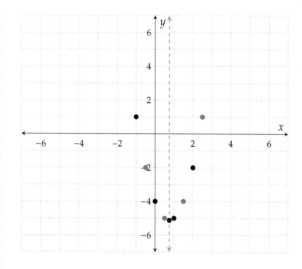

Figure 9.2.28: Plot of symmetric points

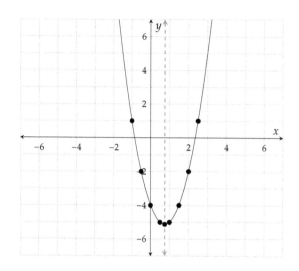

Figure 9.2.29: Graph of $y = 2x^2 - 3x - 4$

9.2.5 The Domain and Range of Quadratic Functions

In Example 9.1.32, we found the domain and range of different types of functions using their graphs. Now that we have graphed some quadratic functions, let's practice identifying the domain and range.

Example 9.2.30 We graphed the quadratic function $y = -x^2 - 2x + 3$ in Figure 9.2.24. The domain is the set of all possible inputs to the function. The function is a continuous curve and when we look horizontally, one arrow points to the left and the other arrow points to the right. This means all x-values can be used in the function. The domain is $\{x \mid x \text{ is a real number}\}$ which is equivalent to $(-\infty, \infty)$.

The range is the set of all outputs we can get from the function. For the range of this function we look vertically up and down the graph. This parabola opens downward, so both arrows point downward and the highest point along the graph is the vertex at $(-1, 4)$. The range is $\{y \mid y \leq 4\}$ which is equivalent to $(-\infty, 4]$.

Example 9.2.31 Use the graph of $y = 2x^2 - 3x - 4$ in Figure 9.2.29 and its vertex at $(0.75, -5.125)$ to identify the domain and range in set-builder and interval notation.

Explanation. For the domain, we look horizontally and see the graph is a continuous curve and one arrow points to the left and the other arrow points to the right. The domain is $\{x \mid x \text{ is a real number}\}$ which is equivalent to $(-\infty, \infty)$.

For the range we look vertically up and down the graph, which opens upward. Both arrows point upward and the lowest point on the graph is the vertex at $(0.75, -5.125)$. The range is $\{y \mid y \geq -5.125\}$ which is equivalent to $[-5.125, \infty)$.

Since all parabolas have the same shape, they all have the same domain of $\{x \mid x \text{ is a real number}\}$ which is equivalent to $(-\infty, \infty)$. The range depends on which way the parabola opens and the y-coordinate of the vertex. When we look at application problems, however, the domain and range will depend on the values that make sense in the given context. For example, times and lengths do not usually have negative values. We will revisit this after looking at some applications.

9.2.6 Applications of Quadratic Functions Involving the Vertex.

We looked at the height of Hannah's toy rocket with respect to time at the beginning of this section and saw that it reached a maximum height of 64 feet after 2 seconds. Let's look at some more applications that involve finding the **minimum** or **maximum** value of a quadratic function.

Example 9.2.32 Imagine that Jae got a new air rifle to shoot targets. The first thing they did with it was to sight the scope at a certain distance so the pellets consistently hit where the cross hairs are pointed. In Olympic 10-meter air rifle shooting[a], the bulls-eye is a 0.5 mm diameter dot, about the size of the head of a pin, so accuracy is key.

Jae would like to set up the air rifle scope to be accurate at a level distance of 35 yards (from the muzzle, which is the tip of the barrel), but they also need to know how much to correct for gravity at different distances. Since the projectile will be affected by gravity, knowing the distance that the target will be set up is essential to be accurate. After zeroing the scope reticule (cross-hairs) at 35 yards so that they can consistently hit the bulls-eye with the reticule directly over it, they set up targets at various distances to test the gun. Jae then shoots at the targets with the cross-hairs directly on the bulls-eye and measures the distance that the pellet hit above or below the bulls-eye when shot at those distances.

Distance to Target in Yards	5	10	20	30	35	40	50
Above/Below Bulls-eye	↓	↑	↑	↑	⊙	↓	↓
Distance Above/Below in Inches	0.1	0.6	1.1	0.6	0	0.8	3.2

Table 9.2.33: Shooting Distance vs Pellet Rise/Fall

Make a graph of the height above the bulls-eye that Jae shoots at the distances listed in the table and find the vertex. What does the vertex mean in this context?

Explanation.

(Note that values measured below the bulls-eye should be graphed as negative y-values. Keep in mind that the units on the axes are different: along the x-axis, the units are yards, whereas on the y-axis, the units are inches.)

Since the input values seem to be going up by 5s or 10s, we will scale the x-axis by 10s. The y-axis needs to be scaled by 1s.

From the graph we can see that the point $(20, 1.1)$ is our best guess for the real life vertex. This means the highest above the cross-hairs Jae hit was 1.1 inches when the target was 20 yards away.

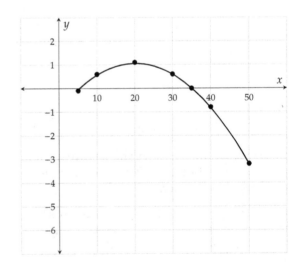

Figure 9.2.34: Graph of Target Data

[a]en.wikipedia.org/wiki/ISSF_10_meter_air_rifle

Example 9.2.35 We looked at the quadratic function $R = (13 + 0.25x)(1500 - 50x)$ in Example 6.4.2 of Section 6.4, where R was the revenue (in dollars) for x 25-cent price increases. This function had each jar of Avery's jam priced at 13 dollars, and simplified to

$$R = -12.5x^2 - 275x + 19500.$$

Find the vertex of this quadratic function and explain what it means in the context of this model.

Explanation. Note that if we tried to use $R = (13 + 0.25x)(1500 - 50x)$, we would not be able to immediately identify the values of a and b needed to determine the vertex. Using the expanded form of $R = -12.5x^2 - 275x + 19500$, we see that $a = -12.5$ and $b = -275$, so the vertex occurs at:

$$h = -\frac{b}{2a}$$
$$h = -\frac{-275}{2(-12.5)}$$
$$h = -11$$

We will now find the value of R for $x = -11$:

$$R = -12.5(-11)^2 - 275(-11) + 19500$$
$$R = 21012.5$$

Thus the vertex occurs at $(-11, 21012.5)$.

Literally interpreting this, we can state that -11 of the 25-cent price increases result in a maximum revenue of $21,012.50.

We can calculate "-11 of the 25-cent price increases" to be a decrease of $2.75. The price was set at $13 per jar, so the maximum revenue of $21,012.50 would occur when Avery sets the price at $10.25 per jar.

Example 9.2.36 Kali has 500 feet of fencing and she needs to build a rectangular pen for her goats. What are the dimensions of the rectangle that would give her goats the largest area?

Explanation. We will use ℓ for the length of the pen and w for the width, in feet. We know that the perimeter must be 500 feet so that gives us

$$2\ell + 2w = 500$$

First we will solve for the length:

$$2\ell + 2w = 500$$
$$2\ell = 500 - 2w$$
$$\ell = 250 - w$$

Now we can build a function for the rectangle's area, using the formula for area:

$$A(w) = \ell \cdot w$$
$$A(w) = (250 - w) \cdot w$$
$$A(w) = 250w - w^2$$
$$A(w) = -w^2 + 250w$$

The area is a quadratic function so we can identify $a = -1$ and $b = 250$ and find the vertex:

$$w = -\frac{(250)}{2(-1)}$$
$$w = \frac{250}{2}$$
$$w = 125$$

Since the width of the rectangle is 125 feet, we can find the length using our expression:

$$\ell = 250 - w$$
$$\ell = 250 - 125$$
$$\ell = 125$$

To find the maximum area we can either substitute the width into the area function or multiply the length by the width:

$$A = \ell \cdot w$$
$$A = 125 \cdot 125$$
$$A = 15{,}625$$

The maximum area that Kali can get is 15,625 square feet if she builds her pen to be a square with a length and width of 125 feet.

Returning to the domain and range, we will look at the path of Hannah's toy rocket in Graph 9.2.4. Looking horizontally, the t-values make sense from 0 seconds, when the rocket is fired, until 4 seconds, when it comes back to the ground. This give us a domain of $\{t \mid 0 \leq t \leq 4\}$ or $[0, 4]$. For the range, the height of the rocket goes from 0 feet on the ground and reaches a maximum height of 64 feet. The range is $\{f(t) \mid 0 \leq f(t) \leq 64\}$ or $[0, 64]$.

In the air-rifle application in Example 9.2.32, the x-values are connected from 5 to 50 yards. If we assume that Jae will never be competing in target shoots beyond 50 yards, the domain will be $[5, 50]$. The y-values go from -3.2 to 1.1 inches so the range is $[-3.2, 1.1]$.

In order to find the domain and range for many applications we need to know how to find the vertical and horizontal intercepts. We will look at that in the next section.

Exercises

Review and Warmup   Make a table for the equation.

1. The first row is an example.

x	$y = -x + 3$	Points
−3	6	(−3, 6)
−2	_____	_____
−1	_____	_____
0	_____	_____
1	_____	_____
2	_____	_____

2. The first row is an example.

x	$y = -x + 4$	Points
−3	7	(−3, 7)
−2	_____	_____
−1	_____	_____
0	_____	_____
1	_____	_____
2	_____	_____

3. The first row is an example.

x	$y = \frac{5}{8}x - 1$	Points
−24	−16	(−24, −16)
−16	_____	_____
−8	_____	_____
0	_____	_____
8	_____	_____
16	_____	_____

4. The first row is an example.

x	$y = \frac{5}{4}x + 7$	Points
−12	−8	(−12, −8)
−8	_____	_____
−4	_____	_____
0	_____	_____
4	_____	_____
8	_____	_____

5. Evaluate the expression $\frac{1}{5}(x + 3)^2 - 2$ when $x = -8$.

6. Evaluate the expression $\frac{1}{2}(x + 4)^2 - 7$ when $x = -6$.

7. Evaluate the expression $-16t^2 + 64t + 128$ when $t = 3$.

8. Evaluate the expression $-16t^2 + 64t + 128$ when $t = -5$.

Algebraically Determining the Vertex and Axis of Symmetry of Quadratic Functions   Find the axis of symmetry and vertex of the quadratic function.

9. $y = -4x^2 - 16x + 5$

Axis of symmetry: ☐

Vertex: ☐

10. $y = -3x^2 + 12x - 2$

Axis of symmetry: ☐

Vertex: ☐

11. $y = 3 - 40x - 4x^2$

Axis of symmetry: ☐

Vertex: ☐

12. $y = -3 - 6x - 3x^2$

Axis of symmetry: ⬚

Vertex: ⬚

13. $y = -3 - x^2 - 2x$

Axis of symmetry: ⬚

Vertex: ⬚

14. $y = 4 - x^2 - 4x$

Axis of symmetry: ⬚

Vertex: ⬚

15. $y = 3x^2 + 6x$

Axis of symmetry: ⬚

Vertex: ⬚

16. $y = 4x^2 - 24x$

Axis of symmetry: ⬚

Vertex: ⬚

17. $y = 4 - 5x^2$

Axis of symmetry: ⬚

Vertex: ⬚

18. $y = 2 - 4x^2$

Axis of symmetry: ⬚

Vertex: ⬚

19. $y = -3x^2 + 9x - 4$

Axis of symmetry: ⬚

Vertex: ⬚

20. $y = -2x^2 - 6x + 2$

Axis of symmetry: ⬚

Vertex: ⬚

21. $y = -2x^2 + 2x + 5$

Axis of symmetry: ⬚

Vertex: ⬚

22. $y = 2x^2 - 10x - 2$

Axis of symmetry: ⬚

Vertex: ⬚

23. $y = 3x^2$

Axis of symmetry: ⬚

Vertex: ⬚

24. $y = 0.4x^2$

Axis of symmetry: ⬚

Vertex: ⬚

25. $y = 0.5x^2 + 1$

Axis of symmetry: ⬚

Vertex: ⬚

26. $y = 5x^2 - 5$

Axis of symmetry: ⬚

Vertex: ⬚

27. $y = -0.5(x - 1)^2 + 3$

Axis of symmetry: ⬚

Vertex: ⬚

28. $y = -0.4(x + 4)^2 - 2$

Axis of symmetry: ⬚

Vertex: ⬚

Graphing Quadratic Functions Using the Vertex and a Table

29. For the given quadratic function, find the vertex. Then create a table of ordered pairs centered around the vertex and make a graph of the function.

$$f(x) = x^2 + 2$$

30. For the given quadratic function, find the vertex. Then create a table of ordered pairs centered around the vertex and make a graph of the function.

$$f(x) = x^2 + 1$$

31. For the given quadratic function, find the vertex. Then create a table of ordered pairs centered around the vertex and make a graph of the function.

    $f(x) = x^2 - 5$

32. For the given quadratic function, find the vertex. Then create a table of ordered pairs centered around the vertex and make a graph of the function.

    $f(x) = x^2 - 3$

33. For the given quadratic function, find the vertex. Then create a table of ordered pairs centered around the vertex and make a graph of the function.

    $f(x) = (x - 2)^2$

34. For the given quadratic function, find the vertex. Then create a table of ordered pairs centered around the vertex and make a graph of the function.

    $f(x) = (x - 4)^2$

35. For the given quadratic function, find the vertex. Then create a table of ordered pairs centered around the vertex and make a graph of the function.

    $f(x) = (x + 3)^2$

36. For the given quadratic function, find the vertex. Then create a table of ordered pairs centered around the vertex and make a graph of the function.

    $f(x) = (x + 2)^2$

Graphing Quadratic Functions Using the Vertex and a Table

37. For $f(x) = 4x^2 - 8x + 5$, determine the vertex, create a table of ordered pairs, and then graph the function.

38. For $f(x) = 2x^2 + 4x + 7$, determine the vertex, create a table of ordered pairs, and then graph the function.

39. For $f(x) = -x^2 + 4x + 2$, determine the vertex, create a table of ordered pairs, and then graph the function.

40. For $f(x) = -x^2 + 2x - 5$, determine the vertex, create a table of ordered pairs, and then graph the function.

41. For $f(x) = x^2 - 5x + 3$, determine the vertex, create a table of ordered pairs, and then graph the function.

42. For $f(x) = x^2 + 7x - 1$, determine the vertex, create a table of ordered pairs, and then graph the function.

43. For $f(x) = -2x^2 - 5x + 6$, determine the vertex, create a table of ordered pairs, and then graph the function.

44. For $f(x) = 2x^2 - 9x$, determine the vertex, create a table of ordered pairs, and then graph the function.

Domain and Range   A function is graphed.

45.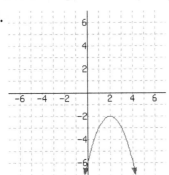

This function has domain ⬚ and range ⬚.

46.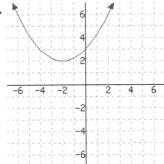

This function has domain ⬚ and range ⬚.

47.

This function has domain ⬚ and range ⬚.

48.

This function has domain ⬚ and range ⬚.

49.

This function has domain ⬚ and range ⬚.

50.

This function has domain ⬚ and range ⬚.

Finding Maximum and Minimum Values for Applications of Quadratic Functions

51. Consider two numbers where one number is 10 less than a second number. Find a pair of such numbers that has the least product possible. One approach is to let x represent the smaller number, and write a formula for a function of x that outputs the product of the two numbers. Then find its vertex and interpret it.

These two numbers are [⎕] and the least possible product is [⎕].

52. Consider two numbers where one number is 4 less than a second number. Find a pair of such numbers that has the least product possible. One approach is to let x represent the smaller number, and write a formula for a function of x that outputs the product of the two numbers. Then find its vertex and interpret it.

These two numbers are [⎕] and the least possible product is [⎕].

53. Consider two numbers where one number is 10 less than twice a second number. Find a pair of such numbers that has the least product possible. One approach is to let x represent the smaller number, and write a formula for a function of x that outputs the product of the two numbers. Then find its vertex and interpret it.

These two numbers are [⎕] and the least possible product is [⎕].

54. Consider two numbers where one number is 7 less than twice a second number. Find a pair of such numbers that has the least product possible. One approach is to let x represent the smaller number, and write a formula for a function of x that outputs the product of the two numbers. Then find its vertex and interpret it.

These two numbers are [⎕] and the least possible product is [⎕].

55. You will build a rectangular sheep enclosure next to a river. There is no need to build a fence along the river, so you only need to build on three sides. You have a total of 440 feet of fence to use. Find the dimensions of the pen such that you can enclose the maximum possible area. One approach is to let x represent the length of fencing that runs perpendicular to the river, and write a formula for a function of x that outputs the area of the enclosure. Then find its vertex and interpret it.

The length of the pen (parallel to the river) should be [⎕], the width (perpendicular to the river) should be [⎕], and the maximum possible area is [⎕].

56. You will build a rectangular sheep enclosure next to a river. There is no need to build a fence along the river, so you only need to build on three sides. You have a total of 460 feet of fence to use. Find the dimensions of the pen such that you can enclose the maximum possible area. One approach is to let x represent the length of fencing that runs perpendicular to the river, and write a formula for a function of x that outputs the area of the enclosure. Then find its vertex and interpret it.

The length of the pen (parallel to the river) should be [   ], the width (perpendicular to the river) should be [   ], and the maximum possible area is [   ].

57. You will build a rectangular sheep enclosure next to a river. There is no need to build a fence along the river, so you only need to build on three sides. You have a total of 470 feet of fence to use. Find the dimensions of the pen such that you can enclose the maximum possible area. One approach is to let x represent the length of fencing that runs perpendicular to the river, and write a formula for a function of x that outputs the area of the enclosure. Then find its vertex and interpret it.

The length of the pen (parallel to the river) should be [   ], the width (perpendicular to the river) should be [   ], and the maximum possible area is [   ].

58. You will build a rectangular sheep enclosure next to a river. There is no need to build a fence along the river, so you only need to build on three sides. You have a total of 480 feet of fence to use. Find the dimensions of the pen such that you can enclose the maximum possible area. One approach is to let x represent the length of fencing that runs perpendicular to the river, and write a formula for a function of x that outputs the area of the enclosure. Then find its vertex and interpret it.

The length of the pen (parallel to the river) should be [   ], the width (perpendicular to the river) should be [   ], and the maximum possible area is [   ].

59. You will build two identical rectangular enclosures next to a each other, sharing a side. You have a total of 408 feet of fence to use. Find the dimensions of each pen such that you can enclose the maximum possible area. One approach is to let x represent the length of fencing that the two pens share, and write a formula for a function of x that outputs the total area of the enclosures. Then find its vertex and interpret it.

The length of each (along the wall that they share) should be [   ], the width should be [   ], and the maximum possible area of each pen is [   ].

60. You will build two identical rectangular enclosures next to a each other, sharing a side. You have a total of 420 feet of fence to use. Find the dimensions of each pen such that you can enclose the maximum possible area. One approach is to let x represent the length of fencing that the two pens share, and write a formula for a function of x that outputs the total area of the enclosures. Then find its vertex and interpret it.

The length of each (along the wall that they share) should be [   ], the width should be [   ], and the maximum possible area of each pen is [   ].

61. You plan to build four identical rectangular animal enclosures in a row. Each adjacent pair of pens share a fence between them. You have a total of 312 feet of fence to use. Find the dimensions of each pen such that you can enclose the maximum possible area. One approach is to let x represent the length of fencing that adjacent pens share, and write a formula for a function of x that outputs the total area. Then find its vertex and interpret it.

The length of each pen (along the walls that they share) should be [_____], the width (perpendicular to the river) should be [_____], and the maximum possible area of each pen is [_____].

62. You plan to build four identical rectangular animal enclosures in a row. Each adjacent pair of pens share a fence between them. You have a total of 328 feet of fence to use. Find the dimensions of each pen such that you can enclose the maximum possible area. One approach is to let x represent the length of fencing that adjacent pens share, and write a formula for a function of x that outputs the total area. Then find its vertex and interpret it.

The length of each pen (along the walls that they share) should be [_____], the width (perpendicular to the river) should be [_____], and the maximum possible area of each pen is [_____].

63. Currently, an artist can sell 240 paintings every year at the price of $90.00 per painting. Each time he raises the price per painting by $15.00, he sells 5 fewer paintings every year.

   a. To obtain maximum income of [_____], the artist should set the price per painting at [_____].

   b. To earn $43,875.00 per year, the artist could sell his paintings at two different prices. The lower price is [_____] per painting, and the higher price is [_____] per painting.

64. Currently, an artist can sell 270 paintings every year at the price of $150.00 per painting. Each time he raises the price per painting by $5.00, he sells 5 fewer paintings every year.

   a. To obtain maximum income of [_____], the artist should set the price per painting at [_____].

   b. To earn $43,700.00 per year, the artist could sell his paintings at two different prices. The lower price is [_____] per painting, and the higher price is [_____] per painting.

9.3 Graphing Quadratic Functions

We have learned how to locate the key features of quadratic functions on a graph and find the vertex algebraically. In this section we'll explore how to find the intercepts algebraically and use their coordinates to graph a quadratic function. Then we will see how to interpret the key features in context and distinguish between quadratic and other functions.

Let's start by looking at a quadratic function that models the path of a baseball after it is hit by Ignacio, the batter. The height of the baseball, $H(t)$, measured in feet, after t seconds is given by $H(t) = -16t^2 + 75t + 4.7$. We know this quadratic function has the shape of a parabola and we want to know the initial height, the maximum height, and the amount of time it takes for the ball to hit the ground if it is not caught. These key features correspond to the vertical intercept, the vertex, and one of the horizontal intercepts.

The graph of this function is shown in Figure 9.3.2. We cannot easily read where the intercepts occur from the graph because they are not integers. We previously covered how to determine the vertex algebraically. In this section, we'll learn how to find the intercepts algebraically. Then we'll come back to this example and find the intercepts for the path of the baseball.

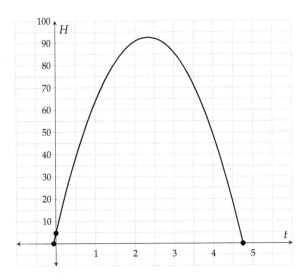

Figure 9.3.2: Graph of $H(t) = -16t^2 + 75t + 4.7$

9.3.1 Finding the Vertical and Horizontal Intercepts Algebraically

In List 9.2.13, we identified that the **vertical intercept** occurs where the graph of a function intersects the vertical axis. If we're using x and y as our variables, the x-value on the vertical axis is $x = 0$. We will substitute 0 for x to find the value of y. In function notation, we find $f(0)$.

The **horizontal intercepts** occur where the graph of a function intersects the horizontal axis. If we're using x and y as our variables, the y-value on the horizontal axis is $y = 0$, so we will substitute 0 for y and find the value(s) of x. In function notation, we solve the equation $f(x) = 0$.

Here is an example where we find the vertical and horizontal intercepts.

Example 9.3.3 Find the intercepts for the quadratic function $f(x) = x^2 - 4x - 12$ algebraically.

To determine the y-intercept, we find $f(0) = 0^2 - 4(0) - 12 = -12$. So the y-intercept occurs where $y = -12$. On a graph, this is the point $(0, -12)$.

To determine the x-intercept(s), we set $f(x) = 0$ and solve for x:

$$0 = x^2 - 4x - 12$$
$$0 = (x - 6)(x + 2)$$

$$x - 6 = 0 \qquad \text{or} \qquad x + 2 = 0$$
$$x = 6 \qquad \text{or} \qquad x = -2$$

The x-intercepts occur where $x = 6$ and where $x = -2$. On a graph, these are the points $(6, 0)$ and $(-2, 0)$.

Notice in Example 9.3.3 that the y-intercept was $(0, -12)$ and the value of $c = -12$. When we substitute 0 for x we will always get the value of c.

Fact 9.3.4. *The vertical intercept of a quadratic function occurs at the point $(0, c)$ because $f(0) = c$.*

Example 9.3.5 Algebraically determine any horizontal and vertical intercepts of the quadratic function $f(x) = -x^2 + 5x - 7$.

Explanation. To determine the vertical intercept, we find $f(0) - (0)^2 + 5(0) - 7 = -7$. Thus the y-intercept occurs at the point $(0, -7)$.

To determine the horizontal intercepts, we'll set $f(x) = 0$ and solve for x:

$$0 = -x^2 + 5x - 7$$

This equation cannot be solved using factoring so we'll use the quadratic formula:

$$x = \frac{-5 \pm \sqrt{5^2 - 4(-1)(-7)}}{2(-1)}$$
$$x = \frac{-5 \pm \sqrt{-3}}{-2}$$

The radicand is negative so there are no real solutions to the equation. This means there are no horizontal intercepts.

9.3.2 Graphing Quadratic Functions Using Their Key Features

To graph a quadratic function using its key features, we will algebraically determine the following: whether the function opens upward or downward, the vertical intercept, the horizontal intercepts and the vertex. Then we will graph the points and connect them with a smooth curve.

Example 9.3.6 Graph the function f where $f(x) = 2x^2 + 10x + 8$ by algebraically determining its key features.

To start, we'll note that this function will open upward, since the leading coefficient is positive.

To find the y-intercept, we evaluate $f(0) = 2(0)^2 + 10(0) + 8 = 8$. The y-intercept is $(0, 8)$.

Next, we'll find the horizontal intercepts by setting $f(x) = 0$ and solving for x:

$$2x^2 + 10x + 8 = 0$$
$$2(x^2 + 5x + 4) = 0$$
$$2(x + 4)(x + 1) = 0$$

$$x + 4 = 0 \qquad \text{or} \qquad x + 1 = 0$$
$$x = -4 \qquad \text{or} \qquad x = -1$$

The x-intercepts are $(-4, 0)$ and $(-1, 0)$.

Lastly, we'll determine the vertex. Noting that $a = 2$ and $b = 10$, we have:

$$h = -\frac{b}{2a}$$
$$h = -\frac{10}{2(2)}$$
$$h = -2.5$$

Using this x-value to find the y-coordinate, we have:

$$k = f(-2.5) = 2(-2.5)^2 + 10(-2.5) + 8$$
$$k = 12.5 - 25 + 8$$
$$k = -4.5$$

The vertex is the point $(-2.5, -4.5)$, and the axis of symmetry is the line $x = -2.5$.

We're now ready to graph this function. We'll start by drawing and scaling the axes so all of our key features will be displayed as shown in Figure 9.3.7. Next, we'll plot these key points as shown in Figure 9.3.8. Finally, we'll note that this parabola opens upward and connect these points with a smooth curve, as shown in Figure 9.3.9.

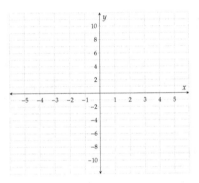

Figure 9.3.7: Setting up the grid.

Figure 9.3.8: Marking key features.

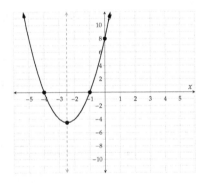

Figure 9.3.9: Completing the graph.

Example 9.3.10 Graph the function for which $y = -x^2 + 4x - 5$ by algebraically determining its key features.

To start, we'll note that this function will open downward, as the leading coefficient is negative.

To find the y-intercept, we'll substitute x with 0:

$$y = -(0)^2 + 4(0) - 5$$
$$y = -5$$

The y-intercept is $(0, -5)$.

Next, we'll find the horizontal intercepts by setting $y = 0$ and solving for x. We cannot use factoring to solve this equation so we'll use the quadratic formula:

$$-x^2 + 4x - 5 = 0$$

$$x = \frac{-4 \pm \sqrt{(4)^2 - 4(-1)(-5)}}{2(-1)}$$

$$x = \frac{-4 \pm \sqrt{16 - 20}}{-2}$$

$$x = \frac{-4 \pm \sqrt{-8}}{-2}$$

The radicand is negative, so there are no real solutions to the equation. This is a parabola that does not have any horizontal intercepts.

To determine the vertex, we'll use $a = -1$ and $b = 4$:

$$x = -\frac{4}{2(-1)}$$

$$x = 2$$

Using this x-value to find the y-coordinate, we have:

$$y = -(2)^2 + 4(2) - 5$$

$$y = -4 + 8 - 5$$

$$y = -1$$

The vertex is the point $(2, -1)$, and the axis of symmetry is the line $x = 2$.

Plotting this information in an appropriate grid, we have:

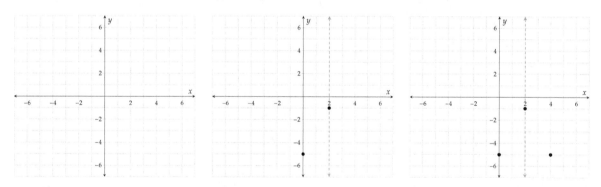

Figure 9.3.11: Setting up the grid.

Figure 9.3.12: Marking key features.

Figure 9.3.13: Using the axis of symmetry to determine one additional point.

Since we don't have any x-intercepts, we would like to have a few more points to graph. We will make a table with a few more values around the vertex, add these, and then draw a smooth curve. This is shown in Table 9.3.14 and Figure 9.3.15.

x	$y = -x^2 + 4x - 5$	Point
0	$-(0)^2 + 4(0) - 5 = -5$	$(0, -5)$
1	$-(1)^2 + 4(1) - 5 = -2$	$(1, -2)$
2	$-(2)^2 + 4(2) - 5 = -1$	$(2, -1)$
3	$-(3)^2 + 4(3) - 5 = -2$	$(3, -2)$
4	$-(4)^2 + 4(4) - 5 = -5$	$(4, -5)$

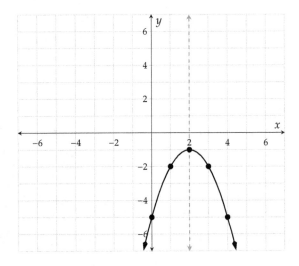

Table 9.3.14: Determine additional function values.

Figure 9.3.15: Completing the graph.

9.3.3 Applications of Quadratic Functions

Now we have learned how to find all the key features of a quadratic function algebraically. Here are some applications of quadratic functions so we can learn how to identify and interpret the vertex, intercepts and additional points in context. Let's look at a few examples.

Example 9.3.16 Returning to the path of the baseball in Figure 9.3.2, the function that represents the height of the baseball after Ignacio hit it, is $H(t) = -16t^2 + 75t + 4.7$. The height is is feet and the time, t, is in seconds. Find and interpret the following, in context.

    a. The vertical intercept.

    b. The horizontal intercept(s).

    c. The vertex.

    d. The height of the baseball 1 second after it was hit.

    e. The time(s) when the baseball is 80 feet above the ground.

Explanation.

    a. To determine the vertical intercept, we'll find $H(0) = -16(0)^2 + 75(0) + 4.7 = 4.7$. The vertical intercept occurs at $(0, 4.7)$. This is the height of the baseball at time $t = 0$, so the initial height of the baseball was 4.7 feet.

    b. To determine the horizontal intercepts, we'll solve $H(t) = 0$. Since factoring is not a possibility to solve this equation, we'll use the quadratic formula:

$$H(t) = 0$$
$$-16t^2 + 75t + 4.7 = 0$$
$$t = \frac{-75 \pm \sqrt{75^2 - 4(-16)(4.7)}}{2(-16)}$$

$$t = \frac{-75 \pm \sqrt{5925.8}}{-32}$$

Rounding these two values with a calculator, we obtain:

$$t \approx -0.06185, \ t \approx 4.749$$

The horizontal intercepts occur at approximately $(-0.06185, 0)$ and $(4.749, 0)$. If we assume that the ball was hit when $t = 0$, a negative time does not make sense. The second horizontal intercept tells us that the ball hit the ground after approximately 4.75 seconds.

c. The vertex occurs at $t = h = -\frac{b}{2a}$, and for this function $a = -16$ and $b = 75$. So we have:

$$h = -\frac{75}{2(-16)}$$
$$h = 2.34375$$

We can now find the output for this input:

$$H(2.34375) = -16(2.34375)^2 + 75(2.34375) + 4.7$$
$$\approx 92.59$$

Thus the vertex is $(2.344, 92.59)$.

The vertex tells us that the baseball reached a maximum height of approximately 92.6 feet about 2.3 seconds after Ignacio hit it.

d. To find the height of the baseball after 1 second, we can compute $H(1)$:

$$H(1) = -16(1)^2 + 75(1) + 4.7$$
$$= 63.7$$

The height of the baseball was 63.7 feet after 1 second.

e. If we want to know when the baseball was 80 feet in the air, then we set $H(t) = 80$ and we have:

$$H(t) = 80$$
$$-16t^2 + 75t + 4.7 = 80$$
$$-16t^2 + 75t - 75.3 = 0$$
$$t = \frac{-75 \pm \sqrt{75^2 - 4(-16)(-75.3)}}{2(-16)}$$
$$t = \frac{-75 \pm \sqrt{805.8}}{-32}$$

Rounding these two values with a calculator, we obtain:

$$t \approx 1.457, \ t \approx 3.231$$

The baseball was 80 feet above the ground at two times, at about 1.5 seconds on the way up and about 3.2 seconds on the way down.

Example 9.3.17 The profit that Keenan's manufacturing company makes for producing n refrigerators is given by $P = -0.01n^2 + 520n - 54000$, for $0 \le n \le 51{,}896$.

   a. Determine the profit the company will make when they produce 1,000 refrigerators.

   b. Determine the maximum profit and the number of refrigerators produced that yields this profit.

   c. How many refrigerators need to be produced in order for the company to "break even?" (In other words, for their profit to be $0.)

   d. How many refrigerators need to be produced in order for the company to make a profit of $1,000,000?

Explanation.

   a. This question is giving us an input value and asking for the output value. We will substitute 1000 for n and we have:

$$P = -0.01(1000)^2 + 520(1000) - 54000$$
$$P = 366000$$

If Keenan's company sells 1,000 refrigerators it will make a profit of $366,000.

   b. This question is asking for the maximum so we need to find the vertex. This parabola opens downward so the vertex will tell us the maximum profit and the corresponding number of refrigerators that need to be produced. Using $a = -0.01$ and $b = 520$, we have:

$$h = -\frac{b}{2a}$$
$$h = -\frac{520}{2(-0.01)}$$
$$h = 26000$$

Now we will find the value of P when $n = 26000$:

$$P = -0.01(26000)^2 + 520(26000) - 54000$$
$$P = 6706000$$

The maximum profit is $6,706,000, which occurs if 26,000 units are produced.

   c. This question is giving an output value of 0 and asking us to find the input(s) so we will be finding the horizontal intercept(s). We will set $P = 0$ and solve for n using the quadratic formula:

$$0 = -0.01n^2 + 520n - 54000$$
$$n = \frac{-520 \pm \sqrt{520^2 - 4(-0.01)(-54000)}}{2(-0.01)}$$
$$n = \frac{-520 \pm \sqrt{268240}}{-0.02}$$
$$n \approx 104, n \approx 51896$$

The company will break even if they produce about 104 refrigerators or 51,896 refrigerators. If the company produces more refrigerators than it can sell its profit will go down.

   d. This question is giving an output value and asking us to find the input. To find the number of refrigerators that need to be produced for the company to make a profit of $1,000,000, we will set

$P = 1000000$ and solve for n using the quadratic formula:

$$1000000 = -0.01n^2 + 520n - 54000$$
$$0 = -0.01n^2 + 520n - 1054000$$
$$n = \frac{-520 \pm \sqrt{520^2 - 4(-0.01)(-1054000)}}{2(-0.01)}$$
$$n \frac{-520 \pm \sqrt{228240}}{-0.02}$$
$$n \approx 2,113, n \approx 49,887$$

The company will make \$1,000,000 in profit if they produce about 2,113 refrigerators or 49,887 refrigerators.

Example 9.3.18 Maia has a remote-controlled airplane and she is going to do a stunt dive where the plane dives toward the ground and back up along a parabolic path. The height of the plane is given by the function H where $H(t) = 0.7t^2 - 23t + 200$, for $0 \leq t \leq 30$. The height is measured in feet and the time, t, is measured in seconds.

a. Determine the starting height of the plane as the dive begins.

b. Determine the height of the plane after 5 seconds.

c. Will the plane hit the ground, and if so, at what time?

d. If the plane does not hit the ground, what is the closest it gets to the ground, and at what time?

e. At what time(s) will the plane have a height of 50 feet?

Explanation.

a. This question is asking for the starting height which is the vertical intercept. We will find $H(0)$:

$$H(0) = 0.7(0)^2 - 23(0) + 200$$
$$H(0) = 200$$

When Maia begins the stunt, the plane has a height of 200 feet. Recall that we can also look at the value of $c = 200$ to determine the vertical intercept.

b. This question is giving an input of 5 seconds and asking for the output so we will find $H(5)$:

$$H(5) = 0.7(5)^2 - 23(5) + 200$$
$$H(5) = 102.5$$

After 5 seconds, the plane is 102.5 feet above the ground.

c. The ground has a height of 0 feet, so it is asking us to find the horizontal intercept(s) if there are any. We will set $H(t) = 0$ and solve for t using the quadratic formula:

$$H(t) = 0.7t^2 - 23t + 200$$
$$0 = 0.7t^2 - 23t + 200$$
$$t = \frac{23 \pm \sqrt{(-23)^2 - 4(0.7)(200)}}{2(0.7)}$$

$$t = \frac{23 \pm \sqrt{-31}}{1.4}$$

The radicand is negative so there are no real solutions to the equation $H(t) = 0$. That means the plane did not hit the ground.

d. This question is asking for the lowest point of the plane so we will find the vertex. Using $a = 0.7$ and $b = -23$, we have:

$$h = -\frac{b}{2a}$$

$$h = -\frac{(-23)}{2(0.7)}$$

$$h \approx 16.43$$

Now we will find the value of H when $t \approx 16.43$:

$$H(16.43) = 0.7(16.43)^2 - 23(16.43) + 200$$

$$H(16.43) \approx 11.07$$

The minimum height of the plane is about 11 feet, which occurs after about 16 seconds.

e. This question is giving us a height and asking for the time(s) so we will set $H(t) = 50$ and solve for t using the quadratic formula:

$$H(t) = 0.7t^2 - 23t + 200$$

$$50 = 0.7t^2 - 23t + 200$$

$$0 = 0.7t^2 - 23t + 150$$

$$t = \frac{23 \pm \sqrt{(-23)^2 - 4(0.7)(150)}}{2(0.7)}$$

$$t = \frac{23 \pm \sqrt{109}}{1.4}$$

$$t \approx 8.971, t \approx 23.89$$

Maia's plane will be 50 feet above the ground about 9 seconds and 24 seconds after the plane begins the stunt.

9.3.4 The Domain and Range of Quadratic Applications

Let's identify the domain and range in each of the applications of quadratic functions in this section.

Example 9.3.19 In the baseball example in Figure 9.3.2, Ignacio hit the ball at 0 seconds, and it lands on the ground at about 4.7 seconds. The domain is $[0, 4.7]$.

The baseball is at its lowest point when it hits the ground at 0 feet, and the vertex is its highest point at about 92.6 feet. The range is $[0, 92.6]$.

Example 9.3.20 Identify the domain and range in Keenan's refrigerator company application in Example 9.3.17. Write them in interval and set-builder notation.

Explanation. The domain is given in the model as $0 \leq n \leq 51{,}896$ refrigerators. Limits are often stated with a mathematical model because only part of the function fits the real-world situation. The domain is $[0, 51896]$ or $\{n \mid 0 \leq n \leq 51{,}896\}$.

When 0 units are produced, the profit is $-\$54{,}000$. The profit increases to a maximum value of $\$6{,}706{,}000$ at the vertex, and then goes back down to $\$0$ at 51,896 units produced. So the range is $[-54000, 6706000]$ or $\{P \mid -54{,}000 \leq P \leq 6{,}706{,}000\}$.

Example 9.3.21 Identify the domain and range of Maia's remote-controlled airplane application in Example 9.3.18. Write them in interval and set-builder notation.

Explanation. The domain is given in the model as $[0, 30]$ seconds, because this parabola opens upward and the plane cannot keep flying up forever. In set-builder notation the domain is $\{t \mid 0 \leq t \leq 30\}$.

When $t = 0$ seconds, the plane is 200 feet above the ground. It dives down to a height of about 11 feet and then flies up again. We need to know how high the plane is at 30 seconds to determine the range, so we find $H(30)$:

$$H(30) = 0.7(30)^2 - 23(30) + 200$$
$$H(30) = 140$$

The plane has returned to a height of 140 feet after 30 seconds. The starting point of 200 feet is still the highest point, so the range is $[11, 200]$ or $\{H(t) \mid 11 \leq H(t) \leq 200\}$.

9.3.5 Distinguishing Quadratic Functions from Other Functions and Relations

So far, we've seen that the graphs of quadratic functions are parabolas and have a specific, curved shape. We've also seen that they have the algebraic form of $y = ax^2 + bx + c$. Here, we will learn to tell the difference between quadratic functions and other relations and functions.

Example 9.3.22 Determine if each relation represented algebraically is a quadratic function.

a. $y + 5x^2 - 4 = 0$

b. $x^2 + y^2 = 9$

c. $y = -5x + 1$

d. $y = (x - 6)^2 + 3$

e. $y = \sqrt{x + 1} + 5$

Explanation.

a. As $y + 5x^2 - 4 = 0$ can be re-written as $y = -5x^2 + 4$, this equation represents a quadratic function.

b. The equation $x^2 + y^2 = 9$ cannot be re-written in the form $y = ax^2 + bx + c$ (due to the y^2 term), so this equation does not represent a quadratic function.

c. The equation $y = -5x + 1$ represents a linear function, not a quadratic function.

d. The equation $y = (x - 6)^2 + 3$ can be re-written as $y = x^2 - 12x + 39$, so this does represent a quadratic function.

e. The equation $y = \sqrt{x + 1} + 5$ does not represent a quadratic function as x is inside a radical, not squared.

Example 9.3.23 Determine if each function represented graphically *could* represent a quadratic function.

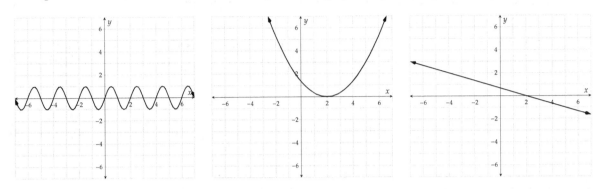

Figure 9.3.24: Function a.   **Figure 9.3.25:** Function b.   **Figure 9.3.26:** Function c.

Explanation.

a. Since this graph has multiple maximum points and minimum points, it is not a parabola and it is not possible that it represents a quadratic function.

b. This graph looks like a parabola, and it's possible that it represents a quadratic function.

c. This graph does not appear to be a parabola, but looks like a straight line. It's not likely that it represents a quadratic function.

Exercises

Review and Warmup   Solve the equation.

1. $x^2 + 14x + 48 = 0$ **2.** $x^2 + 11x + 28 = 0$ **3.** $x^2 - 20x + 100 = 0$ **4.** $x^2 - 22x + 121 = 0$

5. $x^2 - 100 = 0$ **6.** $x^2 - 64 = 0$ **7.** $41x^2 - 11 = 0$ **8.** $13x^2 - 17 = 0$

9. $7x^2 - 10x + 1 = 0$ **10.** $7x^2 + 10x + 1 = 0$ **11.** $4x^2 - 5x + 6 = 0$ **12.** $3x^2 - 3x + 5 = 0$

Finding the Intercepts of Quadratic Functions Algebraically

13. Find the y-intercept and any x-intercept(s) of the quadratic function $y = x^2 - 5x + 4$.

y-intercept:

x-intercept(s):

14. Find the y-intercept and any x-intercept(s) of the quadratic function $y = -x^2 + 2x + 15$.

y-intercept:

x-intercept(s):

15. Find the y-intercept and any x-intercept(s) of the quadratic function $y = x^2 - 16$.

y-intercept: [   ]

x-intercept(s): [   ]

16. Find the y-intercept and any x-intercept(s) of the quadratic function $y = -x^2 + 9$.

y-intercept: [   ]

x-intercept(s): [   ]

17. Find the y-intercept and any x-intercept(s) of the quadratic function $y = x^2 + 2x$.

y-intercept: [   ]

x-intercept(s): [   ]

18. Find the y-intercept and any x-intercept(s) of the quadratic function $y = -x^2 - x$.

y-intercept: [   ]

x-intercept(s): [   ]

19. Find the y-intercept and any x-intercept(s) of the quadratic function $y = x^2 + 8x + 16$.

y-intercept: [   ]

x-intercept(s): [   ]

20. Find the y-intercept and any x-intercept(s) of the quadratic function $y = x^2 + x + 7$.

y-intercept: [   ]

x-intercept(s): [   ]

21. Find the y-intercept and any x-intercept(s) of the quadratic function $y = x^2 + 4x + 8$.

y-intercept: [   ]

x-intercept(s): [   ]

22. Find the y-intercept and any x-intercept(s) of the quadratic function $y = x^2 + 3x + 9$.

y-intercept: [   ]

x-intercept(s): [   ]

23. Find the y-intercept and any x-intercept(s) of the quadratic function $y = x^2 + x + 10$.

y-intercept: [   ]

x-intercept(s): [   ]

24. Find the y-intercept and any x-intercept(s) of the quadratic function $y = x^2 + 5x - 9$.

y-intercept: [   ]

x-intercept(s): [   ]

25. Find the y-intercept and any x-intercept(s) of the quadratic function $y = x^2 - 2x - 6$.

y-intercept: [   ]

x-intercept(s): [   ]

26. Find the y-intercept and any x-intercept(s) of the quadratic function $y = x^2 - 10x - 4$.

y-intercept: [   ]

x-intercept(s): [   ]

27. Find the y-intercept and any x-intercept(s) of the quadratic function $y = x^2 + 4x - 2$.

y-intercept: _____

x-intercept(s): _____

28. Find the y-intercept and any x-intercept(s) of the parabola with equation $y = 4x^2 - 17x + 18$.

y-intercept: _____

x-intercept(s): _____

29. Find the y-intercept and any x-intercept(s) of the parabola with equation $y = 16x^2 - 8x + 1$.

y-intercept: _____

x-intercept(s): _____

30. Find the y-intercept and any x-intercept(s) of the parabola with equation $y = 25x^2 - 49$.

y-intercept: _____

x-intercept(s): _____

31. Find the y-intercept and any x-intercept(s) of the parabola with equation $y = -x + 4 - 5x^2$.

y-intercept: _____

x-intercept(s): _____

32. Find the y-intercept and any x-intercept(s) of the parabola with equation $y = -6x - 5x^2$.

y-intercept: _____

x-intercept(s): _____

Sketching Graphs of Quadratic Functions   Graph each curve by algebraically determining its key features.

33. $y = x^2 - 7x + 12$

34. $y = x^2 + 5x - 14$

35. $y = -x^2 - x + 20$

36. $y = -x^2 + 4x + 21$

37. $y = x^2 - 8x + 16$

38. $y = x^2 + 6x + 9$

39. $y = x^2 - 4$

40. $y = x^2 - 9$

41. $y = x^2 + 6x$

42. $y = x^2 - 8x$

43. $y = -x^2 + 5x$

44. $y = -x^2 + 16$

45. $y = x^2 + 4x + 7$

46. $y = x^2 - 2x + 6$

47. $y = x^2 + 2x - 5$

48. $y = x^2 - 6x + 2$

49. $y = -x^2 + 4x - 1$

50. $y = -x^2 - x + 3$

51. $y = 2x^2 - 4x - 30$

52. $y = 3x^2 + 21x + 36$

Applications of Quadratic Functions

53. An object was shot up into the air at an initial vertical speed of 320 feet per second. Its height as time passes can be modeled by the quadratic function f, where $f(t) = -16t^2 + 320t$. Here t represents the number of seconds since the object's release, and $f(t)$ represents the object's height in feet.

    a. After [   ], this object reached its maximum height of [   ].

    b. This object flew for [   ] before it landed on the ground.

    c. This object was [   ] in the air 12 s after its release.

    d. This object was 1584 ft high at two times: once [   ] after its release, and again later [   ] after its release.

54. An object was shot up into the air at an initial vertical speed of 384 feet per second. Its height as time passes can be modeled by the quadratic function f, where $f(t) = -16t^2 + 384t$. Here t represents the number of seconds since the object's release, and $f(t)$ represents the object's height in feet.

    a. After [   ], this object reached its maximum height of [   ].

    b. This object flew for [   ] before it landed on the ground.

    c. This object was [   ] in the air 7 s after its release.

    d. This object was 1520 ft high at two times: once [   ] after its release, and again later [   ] after its release.

55. From a clifftop over the ocean 200 m above sea level, an object was shot into the air with an initial vertical speed of 156.8 $\frac{m}{s}$. On its way down it fell into the ocean. Its height (above sea level) as time passes can be modeled by the quadratic function f, where $f(t) = -4.9t^2 + 156.8t + 200$. Here t represents the number of seconds since the object's release, and $f(t)$ represents the object's height (above sea level) in meters.

    a. After [   ], this object reached its maximum height of [   ].

    b. This object flew for [   ] before it landed in the ocean.

    c. This object was [   ] above sea level 26 s after its release.

    d. This object was 748.8 m above sea level twice: once [   ] after its release, and again later [   ] after its release.

56. From a clifftop over the ocean 160 m above sea level, an object was shot into the air with an initial vertical speed of 176.4 $\frac{m}{s}$. On its way down it fell into the ocean. Its height (above sea level) as time passes can be modeled by the quadratic function f, where $f(t) = -4.9t^2 + 176.4t + 160$. Here t represents the number of seconds since the object's release, and $f(t)$ represents the object's height (above sea level) in meters.

a. After [_____], this object reached its maximum height of [_____].

b. This object flew for [_____] before it landed in the ocean.

c. This object was [_____] above sea level 8 s after its release.

d. This object was 1742.7 m above sea level twice: once [_____] after its release, and again later [_____] after its release.

57. A remote control aircraft will perform a stunt by flying toward the ground and then up. Its height can be modeled by the function $h(t) = 1.2t^2 - 16.8t + 55.8$. The plane ($\square$ will $\quad \square$ will not) hit the ground during this stunt.

58. A remote control aircraft will perform a stunt by flying toward the ground and then up. Its height can be modeled by the function $h(t) = 0.1t^2 - 1.6t + 10.4$. The plane ($\square$ will $\quad \square$ will not) hit the ground during this stunt.

59. A submarine is traveling in the sea. Its depth can be modeled by $d(t) = -0.9t^2 + 16.2t - 72.9$, where t stands for time in seconds. The submarine ($\square$ will $\quad \square$ will not) hit the sea surface along this route.

60. A submarine is traveling in the sea. Its depth can be modeled by $d(t) = -1.6t^2 + 28.8t - 133.6$, where t stands for time in seconds. The submarine ($\square$ will $\quad \square$ will not) hit the sea surface along this route.

61. An object is launched upward at the height of 400 meters. It's height can be modeled by

$$h = -4.9t^2 + 90t + 400,$$

where h stands for the object's height in meters, and t stands for time passed in seconds since its launch. The object's height will be 420 meters twice before it hits the ground. Find how many seconds since the launch would the object's height be 420 meters. Round your answers to two decimal places if needed.

The object's height would be 420 meters the first time at [_____] seconds, and then the second time at [_____] seconds.

62. An object is launched upward at the height of 210 meters. It's height can be modeled by

$$h = -4.9t^2 + 70t + 210,$$

where h stands for the object's height in meters, and t stands for time passed in seconds since its launch. The object's height will be 250 meters twice before it hits the ground. Find how many seconds since the launch would the object's height be 250 meters. Round your answers to two decimal places if needed.

The object's height would be 250 meters the first time at [   ] seconds, and then the second time at [   ] seconds.

63. Currently, an artist can sell 220 paintings every year at the price of $60.00 per painting. Each time he raises the price per painting by $5.00, he sells 5 fewer paintings every year.

Assume he will raise the price per painting x times, then he will sell $220 - 5x$ paintings every year at the price of $60 + 5x$ dollars. His yearly income can be modeled by the equation:

$$i = (60 + 5x)(220 - 5x)$$

where i stands for his yearly income in dollars. If the artist wants to earn $18,375.00 per year from selling paintings, what new price should he set?

To earn $18,375.00 per year, the artist could sell his paintings at two different prices. The lower price is [   ] per painting, and the higher price is [   ] per painting.

64. Currently, an artist can sell 250 paintings every year at the price of $90.00 per painting. Each time he raises the price per painting by $15.00, he sells 5 fewer paintings every year.

Assume he will raise the price per painting x times, then he will sell $250 - 5x$ paintings every year at the price of $90 + 15x$ dollars. His yearly income can be modeled by the equation:

$$i = (90 + 15x)(250 - 5x)$$

where i stands for his yearly income in dollars. If the artist wants to earn $37,125.00 per year from selling paintings, what new price should he set?

To earn $37,125.00 per year, the artist could sell his paintings at two different prices. The lower price is [   ] per painting, and the higher price is [   ] per painting.

Challenge

65. Consider the function $f(x) = x^2 + nx + p$. Let n and p be real numbers. Give your answers as points.

   a. Suppose the function has two real x-intercepts. What are they?

   b. What is its y-intercept?

   c. What is its vertex?

9.4 Graphs of Quadratic Functions Chapter Review

9.4.1 Introduction to Functions

In Section 9.1 we covered the definitions of a relation, a function, and domain and range. We then discussed function notation and how to evaluate functions at a particular value as well as how to solve an equation with function notation. Last we introduced the vertical line test

Example 9.4.1 Introduction to Functions. One week in Portland, it was rainy. Shocking, I know. Shown is a diagram for the day of the week and how much rain fell on those days. Convert this diagram to a set of ordered pairs where the first coordinate is the day of the week and the second coordinate is the amount of rain that fell in Portland that day, in inches.

Figure 9.4.2: Diagram for the relation "Day of the week to rainfall in PDX"

Explanation. Since Monday has an arrow to 0.2, that must mean that it rained 0.2 inches of rain on Monday. That gives us the ordered pair (Monday, 0.2). Likewise, we also have (Tuesday, 0.4). To write down all of the ordered pairs, we will use a set:

$$\{(\text{Monday}, 0.2), (\text{Tuesday}, 0.4), (\text{Wednesday}, 0.0), (\text{Thursday}, 0.6), (\text{Friday}, 0.0)\}$$

Example 9.4.3 Functions as Predictors. Recall that a a function should predict the output perfectly if you know the input. In Example 9.4.1, we saw that the day of a particular week in Portland was related to the rainfall during that week. Was this relation a function?

Explanation. Yes, that relation was a function. If I tell you the day of the week, you can with certainty "predict" how much rain there was on that day by looking at the data. There was a single answer to the question "How much rain fell on Wednesday during that week in Portland?"

At this point, we should note that all historical weather data can be viewed as a function of some sort: On any particular day in the recent past, we can "predict" (by looking it up somewhere) how much rain fell in any city in the world.

However, the opposite question, "If it rained 0.0 inches, what day of the year was it?" is not going to represent a function because there will be more than one day of the year that it didn't rain at all. Even in Portland.

Example 9.4.4 Algebraic Functions.

   a. For the equation $y = 2x^2 + x$, will y be a function of x?

   b. For the equation $y^2 = x$, will y be a function of x?

Explanation.

a. This question is equivalent to the question "If you input any x value, will there be only one y value?" The answer is "yes." If you input any number for x you first square that number and multiply it by two, then add the number. For example, if we substitute -3, we get:

$$y = 2(x)^2 + (x)$$
$$y = 2(-3)^2 + (-3)$$
$$= 2(9) - 3$$
$$= 18 - 3$$
$$= 15$$

There is no way we could substitute a number for x and get more than one value for y.

b. This question is equivalent to the question "If you input any x value, will there be only one y value?" The answer is "no." If you input any number for x and solve for y, you might actually get two solutions. For example, if we substitute 1, we get:

$$y^2 = x$$
$$y^2 = 1$$

$$y = \sqrt{1} \qquad \text{or} \qquad y = -\sqrt{1}$$
$$y = 1 \qquad \text{or} \qquad y = -1$$

Since there is more than one y value for the x value 1, the relation $y^2 = x$ is *not* a function.

Example 9.4.5 Function Notation. For the function $V = g(x)$, what is the name of the function, what is the input variable, and what is the output variable?

Explanation. For the function $V = g(x)$, the name of the function is g, the input variable is x, and the output variable is V.

Example 9.4.6 Evaluating Functions.

Let $a(x) = x^2 - 2x$, let the function b be defined by the graph in Figure 9.4.7, and let the function c be defined by the table in Table 9.4.8.

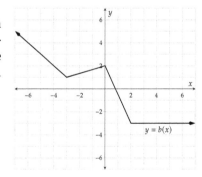

x	$c(x)$
-5	7
-3	-2
2	2
6	4
9	7

Figure 9.4.7: A graph of the function b

Table 9.4.8: The values of the function c

a. Evaluate $a(-3)$.   b. Evaluate $b(-3)$.   c. Evaluate $c(-3)$.

Explanation.

a. To evaluate $a(-3)$ we will substitute the value -3 in wherever we see x in the equation for $a(x)$.

$$a(x) = x^2 - 2x$$
$$a(-3) = (-3)^2 - 2(-3)$$
$$= 9 + 6$$
$$= 15$$

b. To evaluate $b(-3)$, we need to look at the graph of $y = b(x)$ and look for the one place on the graph where the x-value is -3. The y-value at this point is 1, so we would say that

$$b(-3) = 1$$

c. To evaluate $c(-3)$ we examine the table and find the place where the x-value is -3. We conclude that $c(-3) = -2$.

Example 9.4.9 Solving Equations That Have Function Notation.

Still let $a(x) = x^2 - 2x$, let the function b be defined by the graph in Figure 9.4.10, and let the function c be defined by the table in Table 9.4.11.

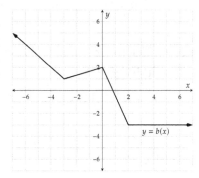

x	$c(x)$
-5	7
-3	-2
2	2
6	4
9	7

Figure 9.4.10: A graph of the function b

Table 9.4.11: The values of the function c

a. Solve the equation $a(x) = 0$.

b. Solve the equation $b(x) = 2$.

c. Solve the equation $c(x) = 7$.

Explanation.

a. To solve the equation $a(x) = 0$, we should set the formula for $a(x)$ equal to 0. In this case, factoring will then help.

$$x^2 - 2x = 0$$
$$x(x - 2) = 0$$

$$x = 0 \qquad \text{or} \qquad x - 2 = 0$$
$$x = 0 \qquad \text{or} \qquad x = 2$$

So, the solution set is $\{0, 2\}$.

b. To solve the equation $b(x) = 2$, we should examine the graph and find all of the locations where the y-value is 2. According to the graph, that happens twice, both when $x = -4$ and when $x = 0$. So the solution set is $\{-4, 0\}$.

c. To solve the equation $c(x) = 7$, we examine the table and find all the places where the function value is 7. According to the table, that happens twice: once when $x = -5$ and again when $x = 9$. So the solution set is $\{-5, 9\}$.

Example 9.4.12 Domain and Range.

Still let the function b be defined by the graph in Figure 9.4.13, and let the function c be defined by the table in Table 9.4.14.

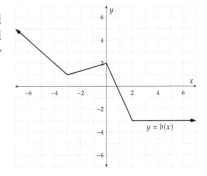

x	$c(x)$
-5	7
-3	-2
2	2
6	4
9	7

Figure 9.4.13: A graph of the function b

Table 9.4.14: The values of the function c

a. Write the domain of b in interval notation.

b. Write the domain of c in interval notation.

c. Write the range of b in interval notation.

d. Write the range of c in interval notation.

Explanation. Recall the definitions of domain and range.

a. To find the the domain of b, look at the graph of b and see what x-values are used for the graph. In other words, what x-values could you input and expect to get a y-value? It looks like the graph extends forever to the right as well as the left. Therefore, the domain is $(-\infty, \infty)$.

b. To find the the domain of c, we list the x-values in the table in a set. That would be $\{-5, -3, 2, 6, 9\}$.

c. To find the the range of b, look at the graph of b and see what y-values are on the graph. The lowest y-value is -3 and all higher y-values are possible outputs. Therefore, the range is $[-3, \infty)$.

d. To find the the range of c, we list the y-values in the table in a set. That would be $\{7, -2, 2, 4\}$.

Example 9.4.15 Determining if a Relation is a Function. Is y a function of x in the following graphs?

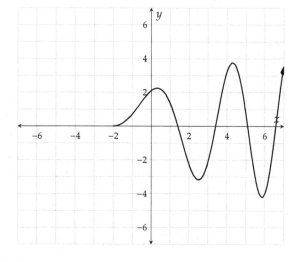

Figure 9.4.16: Is y a function of x?   Figure 9.4.17: Is y a function of x?

Explanation. The graph in Figure 9.4.16 fails the vertical line test and so y is not a function of x. The graph in Figure 9.4.17 passes the vertical line test and so y is a function of x.

9.4.2 Properties of Quadratic Functions

In Section 9.2 we covered the definition of a quadratic function, and how to determine the direction, vertex, axis of symmetry, and vertical and horizontal intercepts. We then learned how to algebraically find the vertex, how to graph a parabola using a table, and how to find the domain and range of quadratic functions.

Example 9.4.18 Finding the Vertex and Axis of Symmetry. Algebraically find the vertex of the parabola described by the quadratic function $f(x) = 3x^2 + 8x - 7$.

Explanation. To find the vertex of a parabola algebraically, we use the formula $h = -\frac{b}{2a}$ to find the axis of symmetry first. For our equation, $a = 3$ and $b = 8$, so:

$$h = -\frac{8}{2(3)} = -\frac{4}{3}$$

Next, we evaluate $f\left(-\frac{4}{3}\right)$ to find the y-coordinate of the vertex.

$$f\left(-\frac{4}{3}\right) = 3\left(-\frac{4}{3}\right)^2 + 8\left(-\frac{4}{3}\right) - 7$$

$$= 3\left(\frac{16}{9}\right) - \frac{32}{3} - 7$$

$$= \frac{16}{3} - \frac{32}{3} - \frac{21}{3}$$

$$= \frac{37}{3}$$

So, the parabola has its axis of symmetry at $x = -\frac{4}{3}$ and has its vertex at the point $\left(-\frac{4}{3}, -\frac{37}{3}\right)$.

Example 9.4.19 Properties of Quadratic Functions.

Identify the key features of the quadratic function $y = -2x^2 - 4x + 6$ shown in Figure 9.4.20.

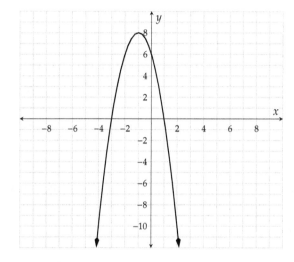

Figure 9.4.20: Graph of $y = -2x^2 - 4x + 6$

Explanation.

First, we see that this parabola opens downward because the leading coefficient is negative.

Then we locate the vertex which is the point $(-1, 8)$. The axis of symmetry is the vertical line $x = -1$.

The vertical intercept or y-intercept is the point $(0, 6)$.

The horizontal intercepts are the points $(-3, 0)$ and $(1, 0)$.

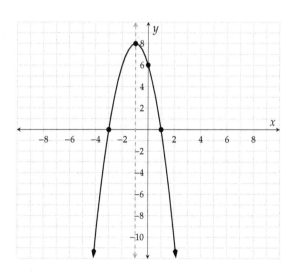

Figure 9.4.21: Graph of $y = -2x^2 - 4x + 6$

Example 9.4.22 Graphing Quadratic Functions by Making a Table. Determine the vertex and axis of symmetry for the quadratic function $g(x) = 2x^2 + 8x + 2$. Then make a table of values and sketch the graph of the function.

Explanation. To determine the vertex of $g(x) = 2x^2 + 8x + 2$, we want to find the x-value of the vertex first. We will use $h = -\frac{b}{2a}$ for $a = 2$ and $b = 8$:

$$h = -\frac{b}{2a}$$
$$h = -\frac{8}{2(2)}$$
$$h = -2$$

To find the vertex, we evaluate $g(-2)$.

$$g(-2) = 2(-2)^2 + 8(-2) + 2$$
$$= 2(4) - 16 + 2$$
$$= 8 - 16 + 2$$
$$= -6$$

Now we know that our axis of symmetry is the line $x = -2$ and the vertex is the point $(-2, -6)$. We will set up our table with two values on each side of $x = -2$. We choose $x = -4, -3, -2, -1$, and 0. Then, we'll determine the y-coordinates by replacing x with each value and we have the complete table as shown in Table 9.4.23. Notice that each pair of y-values on either side of the vertex match. This helps us to check that our vertex and y-values are correct.

x	$g(x) = 2x^2 + 8x + 2$	Point
-4	$g(-4) = 2(-4)^2 + 8(-4) + 2 = 2$	$(-4, 2)$
-3	$g(-3) = 2(-3)^2 + 8(-3) + 2 = -4$	$(-3, -4)$
-2	$g(-2) = 2(-2)^2 + 8(-2) + 2 = -6$	$(-2, -6)$
-1	$g(-1) = 2(-1)^2 + 8(-1) + 2 = -4$	$(-1, -4)$
0	$g(0) = 2(0)^2 + 8(0) + 2 = 2$	$(0, 2)$

Table 9.4.23: Function values and points for $g(x) = 2x^2 + 8x + 2$

Now that we have our table, we will plot the points and draw in the axis of symmetry as shown in Figure 9.4.24. We complete the graph by drawing a smooth curve through the points and drawing an arrow on each end as shown in Figure 9.4.25

Figure 9.4.24: Plot of the points and axis of symmetry

Figure 9.4.25: Graph of $g(x) = 2x^2 + 8x + 2$

Example 9.4.26 The Domain and Range of Quadratic Functions. Use the graph of $g(x) = 2x^2 + 8x + 2$ in from Example 9.4.22 shown in Figure 9.4.25 to identify the domain and range of g in set-builder and interval notation.

Explanation. For the domain, we look horizontally and see the graph is a continuous curve and one arrow points to the left and the other arrow points to the right. The domain is $\{x \mid x \text{ is a real number}\}$ which is equivalent to $(-\infty, \infty)$.

For the range we look up and down on the graph, which opens upward. Both arrows point upward and the lowest point on the graph is the vertex at the point $(-2, -6)$. The range is $\{y \mid y \geq -6\}$ which is equivalent to $[-6, \infty)$.

Example 9.4.27 Applications of Quadratic Functions Involving the Vertex. The value of FedEx stock, in dollars, between December 28, 2017 and February 9, 2018 can be very closely approximated by a quadratic function, $F(x) = -0.07x^2 + 2.7x + 248$. Find the vertex of this parabola and interpret it in the context of the situation.

Explanation. To find the vertex of $F(x) = -0.07x^2 + 2.7x + 248$, we first find the axis of symmetry using the formula $h = -\frac{b}{2a}$. For our equation, $a = -0.07$ and $b = 2.7$, so:

$$h = -\frac{2.7}{2(-0.07)}$$
$$h = -\frac{2.7}{-0.14}$$
$$h \approx 19.3$$

Next, we substitute the value of the axis of symmetry, 19.3, into the formula for the parabola.

$$F(x) = -0.07x^2 + 2.7x + 248$$
$$F(19.3) = -0.07(19.3)^2 + 2.7(19.3) + 248$$
$$\approx 274$$

So, the parabola has its axis of symmetry at $x = 19.3$ and has its vertex at about the point $(19.3, 274)$. In addition, since $a = -0.07$, the parabola will be opening downward. Given this information, the vertex of this parabola is its maximum.

In the reality of the situation, the 19.3 indicates the number of days after December 28, 2017 which would be sometime on January 16. The 274 indicates the stock price in dollars. In conclusion, FedEx's stock peaked (between December 28, 2017 and February 9, 2018) on January 16 at a value of $274.

9.4.3 Graphing Quadratic Functions

In Section 9.3 we covered how to find the vertical and horizontal intercepts of a quadratic function algebraically, how to make a graph of a parabola using key features, applications of quadratic functions and what their real world domains and ranges mean, and how to tell a quadratic function from other types of functions.

Example 9.4.28 Finding the Vertical and Horizontal Intercepts Algebraically.

a. Algebraically determine the vertical and horizontal intercepts of the quadratic function $h(x) = 6x^2 - 13x + 6$.

b. Algebraically determine the vertical and horizontal intercepts of the quadratic function $k(x) = 2x^2 - 2x - 5$.

Explanation.

a. To find the vertical intercept, we evaluate the function when $x = 0$.

$$h(0) = 6(0)^2 - 13(0) + 6$$
$$= 6$$

So, the vertical intercept of h is $(0, 6)$.

Next, to find the horizontal intercepts, we set the function equal to zero. To solve that equation we will use Algorithm 8.6.2. For this particular example, we will practice factoring using the AC method. Here, $ac = 36$ and factor pairs that add up to -13 are -9 and -4, as you will see.

$$h(x) = 6x^2 - 13x + 6$$
$$0 = 6x^2 - 13x + 6$$
$$0 = 6x^2 - 9x - 4x + 6$$
$$0 = 3x(2x - 3) - 2(2x - 3)$$
$$0 = (3x - 2)(2x - 3)$$

$$0 = 3x - 2 \qquad \text{or} \qquad 0 = 2x - 3$$
$$x = \frac{2}{3} \qquad \text{or} \qquad x = \frac{3}{2}$$

So, the horizontal intercepts are $\left(\frac{2}{3}, 0\right)$ and $\left(\frac{3}{2}, 0\right)$.

b. To find the vertical intercept, we have to evaluate the function when $x = 0$.

$$k(0) = 2(0)^2 - 2(0) - 5$$
$$= -5$$

So, the vertical intercept of k is $(0, -5)$.

Next, to find the horizontal intercepts, we set the function equal to zero. To solve that equation we will use Algorithm 8.6.2. For this particular example, we will use the quadratic formula because the square root method will not work (because there is a linear term) and factoring fails.

$$k(x) = 2x^2 - 2x - 5$$
$$0 = 2x^2 - 2x - 5$$

We identify that $a = 2$, $b = -2$, and $c = -5$

$$x = \frac{-b \pm \sqrt{b^2 - 4ac}}{2a}$$
$$x = \frac{-(-2) \pm \sqrt{(-2)^2 - 4(2)(-5)}}{2(2)}$$
$$x = \frac{2 \pm \sqrt{4 + 40}}{4}$$

$$x = \frac{2 \pm \sqrt{44}}{4}$$

$$x = \frac{2 \pm \sqrt{4 \cdot 11}}{4}$$

$$x = \frac{2 \pm 2\sqrt{11}}{4}$$

$$x = \frac{2}{4} \pm \frac{2\sqrt{11}}{4}$$

$$x = \frac{1}{2} \pm \frac{\sqrt{11}}{2}$$

$$x = \frac{1 \pm \sqrt{11}}{2}$$

So, the horizontal intercepts are $\left(\frac{1+\sqrt{11}}{2}, 0\right)$ and $\left(\frac{1-\sqrt{11}}{2}, 0\right)$. If you wanted to graph these points, you would need to approximate them as $(2.16, 0)$ and $(-1.16, 0)$.

Example 9.4.29 Graphing Quadratic Functions Using their Key Features. Graph the function $j(x) = -2x^2 + 6x + 8$ by algebraically determining its key features.

Explanation. To start, we'll note that this function will open downward, as the leading coefficient is negative.

To find the y-intercept, we'll evaluate $j(0)$:

$$j(0) = -2(0)^2 + 6(0) + 8$$
$$= 8$$

The y-intercept is $(0, 8)$.

Next, we'll find the horizontal intercepts by setting $j(x) = 0$ and solving for x:

$$-2x^2 + 6x + 8 = 0$$
$$-2(x^2 - 3x - 4) = 0$$
$$2(x - 4)(x + 1) = 0$$

$$x - 4 = 0 \qquad \text{or} \qquad x + 1 = 0$$
$$x = 4 \qquad \text{or} \qquad x = -1$$

The x-intercepts are $(4, 0)$ and $(-1, 0)$.

Lastly, we'll determine the vertex. Noting that $a = -2$ and $b = 6$, we have:

$$x = -\frac{b}{2a}$$
$$x = -\frac{6}{2(-2)}$$
$$= 1.5$$

Using the x-value 1.5 to find the y-coordinate, we have:

$$j(1.5) = -2(1.5)^2 + 6(1.5) + 8$$
$$= 12.5$$

The vertex is the point $(1.5, 12.5)$, and the axis of symmetry is the line $x = 1.5$.

We're now ready to graph this function. We'll start by drawing and scaling the axes so all of our key features will be displayed as shown in Figure 9.4.30. Next, we'll plot these key points as shown in Figure 9.4.31. Finally, we'll note that this parabola opens downward and connect these points with a smooth curve, as shown in Figure 9.4.32.

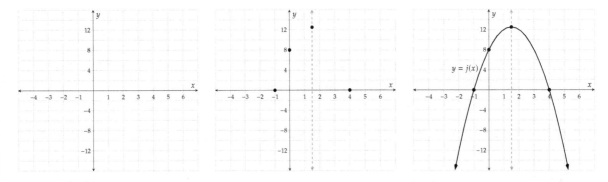

Figure 9.4.30: Setting up the grid.

Figure 9.4.31: Marking key features.

Figure 9.4.32: Completing the graph.

Example 9.4.33 Applications of Quadratic Functions. The Five-hundred-meter Aperture Spherical Telescope[a] (FAST) located in SW China is the worlds largest radio telescope. The name is actually incorrect because it is not spherical: it is parabolic! Parabolic telescopic dishes are very common because parabolas have the unique feature that they focus light coming in at a single point where a collecting instrument is placed.

If a scientist somehow managed to climb to the rim of the dish and roll a ball from the top down toward the center, her ball would travel along the parabola $h(x) = 0.0018x^2 - 0.9x$, where x is the horizontal distance from the rim of the dish where the ball was released, in meters, and $h(x)$ is the height of the ball above above the rim, in meters (so negative height would mean *below* the rim). Note that this is an approximation based on real data.

    a. Find and interpret the vertical intercept of the graph.

    b. Find and interpret the horizontal intercepts of the graph.

    c. Find and interpret the vertex of the graph.

Explanation.

    a. The vertical intercept is found when $x = 0$.

$$h(x) = 0.0018x^2 - 0.9x$$
$$h(0) = 0.0018(0)^2 - 0.9(0)$$
$$= 0$$

So the vertical intercept is $(0, 0)$ which means that the ball was released 0 meters from the rim (on the rim) at a height of 0 meters above the rim (again, on the rim).

b. The horizontal intercepts are found when $h(x) = 0$. This will be a quadratic equation that we can solve using factoring.

$$h(x) = 0$$
$$0.0018x^2 - 0.9x = 0$$
$$x(0.0018x - 0.9) = 0$$

$x = 0$	or	$0.0018x - 0.9 = 0$
$x = 0$	or	$0.0018x = 0.9$
$x = 0$	or	$x = 500$

The horizontal intercepts are $(0, 0)$ and $(500, 0)$. The interpretation of $(0, 0)$ is the same as before. The interpretation of $(500, 0)$ is that the other side of the rim of the dish is 500 meters away. If the ball were to somehow roll all the way down, and then back up the other side, it would pop up at the opposite rim exactly 500 meters horizontally away.

c. The vertex is found when the x-value is at the axis of symmetry. To find that, we use the formula $x = -\frac{b}{2a}$ where $a = 0.0018$ and $b = -0.9$

$$x = -\frac{b}{2a}$$
$$x = -\frac{-0.9}{2(0.0018)}$$
$$= 250$$

Then, to find the y-value, we substitute that x-value, 250 into the original function.

$$h(x) = 0.0018x^2 - 0.9x$$
$$h(250) = 0.0018(250)^2 - 0.9(250)$$
$$= -112.5$$

The vertex is the point $(250, -112.5)$. Since the vale of a is positive, we know that this parabola opens upward. This means that the vertex is the lowest point on the graph. So the lowest point of the telescope dish is 250 meters from the rim (horizontally) and 112.5 meters below the height of the rim. That is one big dish!

---

[a]en.wikipedia.org/wiki/Five_hundred_meter_Aperture_Spherical_Telescope

Example 9.4.34 The Domain and Range of Quadratic Applications. In the FAST telescope example, find the real world domain and range of the function h.

Explanation. Real world domain and range problems are sometimes subjective. The domain in this case would represent all possible x-values that make sense in the reality of the situation. Recall that x is the horizontal distance from the rim of the dish where the ball was released, in meters. Since we found out that the dish is 500 meters across we could say that the domain is $[0, 500]$.

The range will represent all possible y values, that make sense in reality. Recall that $h(x)$ is the height above above the rim, in meters. Since the ball won't roll any higher than the rim of the dish, 0 will be the largest y-value. We found out that the lowest point on the graph is -112.5 meters below the rim. So the range is [-112.5,0].

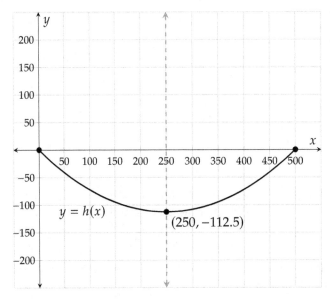

Figure 9.4.35: A diagram of the parabolic function representing the telescope's dish

Example 9.4.36 Distinguishing Quadratic Functions from Other Functions and Relations. Decide if the equations represent quadratic functions or something else.

a. $y - 1 = 3(x - 2) - 5$ b. $y = 3(x - 2)^2 - 5$ c. $y = \sqrt{x - 2} - 5$ d. $y^2 = x^2 - 5$

Explanation. Recall that Definition 9.2.2 says that a quadratic function has the form $f(x) = ax^2 + bx + c$ where a, b, and c are real numbers, and $a \neq 0$.

a. The equation $y - 1 = 3(x - 2) - 5$ *is not* quadratic. It is in fact a linear equation.

b. The equation $y = 3(x - 2)^2 - 5$ *is* quadratic. We could simplify the right hand side and would have something of the form $y = ax^2 + bx + c$.

c. The equation $y = \sqrt{x - 2} - 5$ *is not* quadratic. The x is inside a radical, not squared, so it cannot be converted into the form $y = ax^2 + bx + c$.

d. The equation $y^2 = x^2 - 5$ *is not* quadratic. It cannot be re-written in the form $y = ax^2 + bx + c$ (due to the y^2 term).

Exercises

Introduction to Functions

1. Which of the following graphs show y as a function of x?

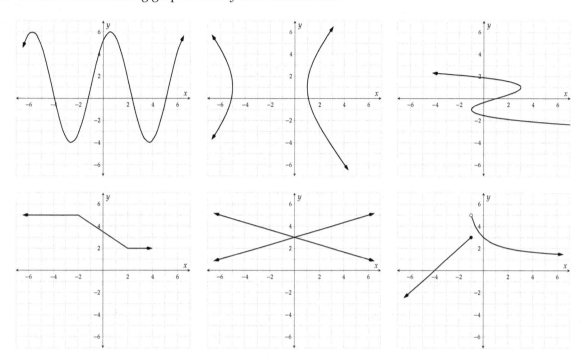

2. Which of the following graphs show y as a function of x?

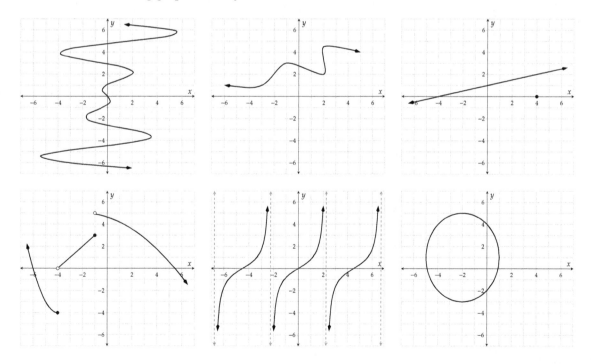

3. Evaluate the function at the given values.

$F(x) = -5x + 4$

   a. $F(1) =$ ☐

   b. $F(-3) =$ ☐

   c. $F(0) =$ ☐

4. Evaluate the function at the given values.

$G(x) = -3x + 8$

   a. $G(5) =$ ☐

   b. $G(-3) =$ ☐

   c. $G(0) =$ ☐

5. Evaluate the function at the given values.

$H(x) = \dfrac{7x}{-6x + 3}$

   a. $H(4) =$ ☐.

   b. $H(-6) =$ ☐.

6. Evaluate the function at the given values.

$K(x) = \dfrac{2x}{3x - 3}$

   a. $K(4) =$ ☐.

   b. $K(-1) =$ ☐.

7. Evaluate the function at the given values.

$K(x) = -2x^2 - 5x - 6$

   a. $K(2) =$ ☐

   b. $K(-4) =$ ☐

8. Evaluate the function at the given values.

$f(x) = -3x^2 + 3x - 1$

   a. $f(4) =$ ☐

   b. $f(-5) =$ ☐

9. Solve for x, where $g(x) = x^2 - 2x - 82$.

If $g(x) = -2$, then $x =$ ☐.

10. Solve for x, where $h(x) = x^2 - x - 17$.

If $h(x) = -5$, then $x =$ ☐.

11. A function is graphed.

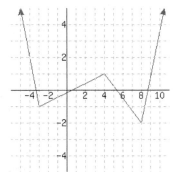

This function has domain ☐ and range ☐.

12. A function is graphed.

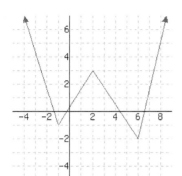

This function has domain ☐ and range ☐.

13. Function f is graphed.

a. $f(2) =$ ☐

b. Solve $f(x) = 1$.

14. Function f is graphed.

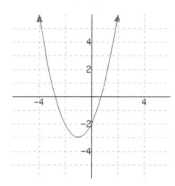

a. $f(0) =$ ☐

b. Solve $f(x) = 1$.

15. Heather started saving in a piggy bank on her birthday. The function $f(x) = 5x + 3$ models the amount of money, in dollars, in Heather's piggy bank. The independent variable represents the number of days passed since her birthday.

Interpret the meaning of $f(2) = 13$.

⊙ A. Two days after Heather started her piggy bank, there were $13 in it.

⊙ B. The piggy bank started with $2 in it, and Heather saves $13 each day.

⊙ C. Thirteen days after Heather started her piggy bank, there were $2 in it.

⊙ D. The piggy bank started with $13 in it, and Heather saves $2 each day.

16. Haley started saving in a piggy bank on her birthday. The function $f(x) = 4x + 3$ models the amount of money, in dollars, in Haley's piggy bank. The independent variable represents the number of days passed since her birthday.

Interpret the meaning of $f(4) = 19$.

⊙ A. Nineteen days after Haley started her piggy bank, there were $4 in it.

⊙ B. The piggy bank started with $4 in it, and Haley saves $19 each day.

⊙ C. Four days after Haley started her piggy bank, there were $19 in it.

⊙ D. The piggy bank started with $19 in it, and Haley saves $4 each day.

The following figure has the graph $y = d(t)$, which models a particle's distance from the starting line in feet, where t stands for time in seconds since timing started.

17.

18.

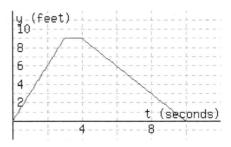

a. $d(2) =$ ☐

b. Interpret the meaning of $d(2)$:

    ⊙ A. In the first 8 seconds, the particle moved a total of 2 feet.

    ⊙ B. In the first 2 seconds, the particle moved a total of 8 feet.

    ⊙ C. The particle was 8 feet away from the starting line 2 seconds since timing started.

    ⊙ D. The particle was 2 feet away from the starting line 8 seconds since timing started.

c. Solve $d(t) = 4$ for t. $t =$ ☐

d. Interpret the meaning of part c's solution(s):

    ⊙ A. The article was 4 feet from the starting line 1 seconds since timing started, or 8 seconds since timing started.

    ⊙ B. The article was 4 feet from the starting line 8 seconds since timing started.

    ⊙ C. The article was 4 feet from the starting line 1 seconds since timing started.

    ⊙ D. The article was 4 feet from the starting line 1 seconds since timing started, and again 8 seconds since timing started.

a. $d(5) =$ ☐

b. Interpret the meaning of $d(5)$:

    ⊙ A. The particle was 5 feet away from the starting line 7.5 seconds since timing started.

    ⊙ B. The particle was 7.5 feet away from the starting line 5 seconds since timing started.

    ⊙ C. In the first 7.5 seconds, the particle moved a total of 5 feet.

    ⊙ D. In the first 5 seconds, the particle moved a total of 7.5 feet.

c. Solve $d(t) = 3$ for t. $t =$ ☐

d. Interpret the meaning of part c's solution(s):

    ⊙ A. The article was 3 feet from the starting line 8 seconds since timing started.

    ⊙ B. The article was 3 feet from the starting line 1 seconds since timing started, and again 8 seconds since timing started.

    ⊙ C. The article was 3 feet from the starting line 1 seconds since timing started, or 8 seconds since timing started.

    ⊙ D. The article was 3 feet from the starting line 1 seconds since timing started.

The function C models the the number of customers in a store t hours since the store opened.

19.

t	0	1	2	3	4	5	6	7
$C(t)$	0	41	75	97	97	73	48	0

20.

t	0	1	2	3	4	5	6	7
$C(t)$	0	43	78	97	95	81	43	0

19.

a. $C(1) = \boxed{}$

b. Interpret the meaning of $C(1)$:

⊙ A. There were 41 customers in the store 1 hour after the store opened.

⊙ B. In 1 hour since the store opened, the store had an average of 41 customers per hour.

⊙ C. In 1 hour since the store opened, there were a total of 41 customers.

⊙ D. There was 1 customer in the store 41 hours after the store opened.

c. Solve $C(t) = 97$ for t. $t = \boxed{}$

d. Interpret the meaning of Part c's solution(s):

⊙ A. There were 97 customers in the store 3 hours after the store opened.

⊙ B. There were 97 customers in the store 4 hours after the store opened.

⊙ C. There were 97 customers in the store either 4 hours after the store opened, or 3 hours after the store opened.

⊙ D. There were 97 customers in the store 4 hours after the store opened, and again 3 hours after the store opened.

20.

a. $C(3) = \boxed{}$

b. Interpret the meaning of $C(3)$:

⊙ A. There were 97 customers in the store 3 hours after the store opened.

⊙ B. There were 3 customers in the store 97 hours after the store opened.

⊙ C. In 3 hours since the store opened, there were a total of 97 customers.

⊙ D. In 3 hours since the store opened, the store had an average of 97 customers per hour.

c. Solve $C(t) = 43$ for t. $t = \boxed{}$

d. Interpret the meaning of Part c's solution(s):

⊙ A. There were 43 customers in the store 1 hours after the store opened, and again 6 hours after the store opened.

⊙ B. There were 43 customers in the store either 1 hours after the store opened, or 6 hours after the store opened.

⊙ C. There were 43 customers in the store 1 hours after the store opened.

⊙ D. There were 43 customers in the store 6 hours after the store opened.

Properties of Quadratic Functions

21. Find the axis of symmetry and vertex of the quadratic function.

$y = 4x^2 + 32x - 3$

Axis of symmetry: $\boxed{}$

Vertex: $\boxed{}$

22. Find the axis of symmetry and vertex of the quadratic function.

$y = x^2 - 4x + 5$

Axis of symmetry: $\boxed{}$

Vertex: $\boxed{}$

23. Find the axis of symmetry and vertex of the quadratic function.

$y = 3x^2 - 15x + 2$

Axis of symmetry: $\boxed{}$

Vertex: $\boxed{}$

24. Find the axis of symmetry and vertex of the quadratic function.

$y = 4x^2 + 12x - 4$

Axis of symmetry: ☐

Vertex: ☐

25. For $y = 4x^2 - 8x + 5$, determine the vertex, create a table of ordered pairs, and then graph the function.

26. For $y = 2x^2 + 4x + 7$, determine the vertex, create a table of ordered pairs, and then graph the function.

27. For $y = -x^2 + 4x + 2$, determine the vertex, create a table of ordered pairs, and then graph the function.

28. For $y = -x^2 + 2x - 5$, determine the vertex, create a table of ordered pairs, and then graph the function.

29. You will build a rectangular sheep enclosure next to a river. There is no need to build a fence along the river, so you only need to build on three sides. You have a total of 500 feet of fence to use. Find the dimensions of the pen such that you can enclose the maximum possible area. One approach is to let x represent the length of fencing that runs perpendicular to the river, and write a formula for a function of x that outputs the area of the enclosure. Then find its vertex and interpret it.

The length of the pen (parallel to the river) should be ☐, the width (perpendicular to the river) should be ☐, and the maximum possible area is ☐.

30. You will build a rectangular sheep enclosure next to a river. There is no need to build a fence along the river, so you only need to build on three sides. You have a total of 400 feet of fence to use. Find the dimensions of the pen such that you can enclose the maximum possible area. One approach is to let x represent the length of fencing that runs perpendicular to the river, and write a formula for a function of x that outputs the area of the enclosure. Then find its vertex and interpret it.

The length of the pen (parallel to the river) should be ☐, the width (perpendicular to the river) should be ☐, and the maximum possible area is ☐.

Graphing Quadratic Functions

31. Find the y-intercept and any x-intercept(s) of the quadratic function $y = x^2 + 4x + 3$.

y-intercept: ☐

x-intercept(s): ☐

32. Find the y-intercept and any x-intercept(s) of the quadratic function $y = -x^2 - 3x - 2$.

y-intercept: ☐

x-intercept(s): ☐

33. Find the y-intercept and any x-intercept(s) of the quadratic function $y = x^2 + x + 5$.

y-intercept: ☐

x-intercept(s): ☐

34. Find the y-intercept and any x-intercept(s) of the quadratic function $y = x^2 + 8x + 4$.

y-intercept: ☐

x-intercept(s): ☐

35. Find the y-intercept and any x-intercept(s) of the parabola with equation $y = 16x^2 - 8x + 1$.

y-intercept: ☐

x-intercept(s): ☐

36. Find the y-intercept and any x-intercept(s) of the parabola with equation $y = 25x^2 - 1$.

y-intercept: ☐

x-intercept(s): ☐

Graph each curve by algebraically determining its key features.

37. $y = x^2 - 7x + 12$

38. $y = x^2 + 5x - 14$

39. $y = -x^2 - x + 20$

40. $y = -x^2 + 4x + 21$

41. $y = x^2 + 6x$

42. $y = x^2 - 8x$

43. $y = x^2 + 4x + 7$

44. $y = x^2 - 2x + 6$

45. $y = 2x^2 - 4x - 30$

46. $y = 3x^2 + 21x + 36$

47. From a clifftop over the ocean 170 m above sea level, an object was shot into the air with an initial vertical speed of 264.6 $\frac{m}{s}$. On its way down it fell into the ocean. Its height (above sea level) as time passes can be modeled by the quadratic function f, where $f(t) = -4.9t^2 + 264.6t + 170$. Here t represents the number of seconds since the object's release, and $f(t)$ represents the object's height (above sea level) in meters.

a. After ☐ , this object reached its maximum height of ☐ .

b. This object flew for ☐ before it landed in the ocean.

c. This object was ☐ above sea level 47 s after its release.

d. This object was 3663.7 m above sea level twice: once ☐ after its release, and again later ☐ after its release.

48. From a clifftop over the ocean 130 m above sea level, an object was shot into the air with an initial vertical speed of 294 $\frac{m}{s}$. On its way down it fell into the ocean. Its height (above sea level) as time passes can be modeled by the quadratic function f, where $f(t) = -4.9t^2 + 294t + 130$. Here t represents the number of seconds since the object's release, and $f(t)$ represents the object's height (above sea level) in meters.

a. After ☐ , this object reached its maximum height of ☐ .

b. This object flew for ☐ before it landed in the ocean.

c. This object was ☐ above sea level 18 s after its release.

d. This object was 967.9 m above sea level twice: once ☐ after its release, and again later ☐ after its release.

Index

FOIL, 485

absolute value, 27
    as a grouping symbol, 35
    formal definition, 27
    of a real number, 63
Addition
    of fractions with different denominators, 17
    of fractions with the same denominator, 17
addition
    additive identity, 158
    associative property of, 159
    commutative property of, 159
    inverses, 158
    of polynomials, 473
    property of equality, 111
    property of inequality, 123
addition method for solving systems of linear
    equations, *see* elimination method for
    solving systems of linear equations
area, 80
associative property
    of addition, 159
    of multiplication, 159
axis
    horizontal, 255
    vertical, 255
axis of symmetry, 676

base, 153
binomial, 473

Cartesian coodrinate system, 253
Cartesian Coordinate System, 255
circumference, 83
coefficient, 471
common denominator, 17
commutative property
    of addition, 159
    of multiplication, 159
complex number, 625
conjugates, 497
constant, 279
constant term, 473
coordinate plane, 255

coordinates, 254, 255
cross-multiplication, 227

degree, 472
denominator, 12
difference of squares, 497
discriminant, 616
distributive property, 159
dividing a polynomial by a monomial, 503
division
    property of equality, 111
    property of inequality, 123

elimination method for solving systems of linear
    equations, 423
empty set, 399
equation, 99
equivalent equation, 111
evaluate, 70
exponent, 62, 153
    base, 153
    introductory rules, 155
    negative exponent rule, 453
    power, 153
    power to a power rule, 154
    product rule, 153
    product to a power rule, 155
    quotient of powers rule, 449
    quotient to a power rule, 449
    zero power rule, 451
expression, 70

factor, 96
factoring
    by grouping, 525
    by the AC method, 538
    decision tree, 556
    difference of squares, 546
    in stages, 533
    perfect square trinomials, 547
    sum and difference of cubes, 549
    trinomials with higher powers, 532
    with factor pairs, 529
false statement, 239
form

point-slope, 315
slope-intercept, 296
standard, 331
formula, 73
fraction
common denominator, 17
denominator, 12
improper, 19
mixed number, 18
numerator, 12
reduction, 13
fraction bar
as a grouping symbol, 35
function, 650
domain, 658
informal definition, 649
notation, 653
range, 658

G.C.F., 517
generic rectangles, 483
graph
of an equation, 260
greater-than symbol, 53
greater-than-or-equal-to symbol, 54
greatest common factor, 517

horizontal intercept, 694
horizontal line, 346

identity, 239
additive, 158
multiplicative, 158
imaginary numbers, 624
imaginary unit, 624
improper fraction, 19
inequality, 99
integers, 45
intercept, 295
horizontal, 694
vertical, 694
x, 694
y, 694
interval notation, 56
inverses
additive, 158
multiplicative, 158
irrational numbers, 45

less-than symbol, 55
like term, 94
linear, 275

linear equation, 99
linear relationship, 279

maximum, 676
minimum, 676
mixed number, 18
monomial, 473
Multiplication
of fractions, 15
multiplication
associative property of, 159
commutative property of, 159
inverses, 158
multiplicative identity, 158
property of equality, 111
property of inequality, 123
Multiplication Properties of Square Roots, 590

natural numbers, 6, 45
negative, 1
number line, 1
numerator, 12

order of magnitude, 463
ordered pair, 254, 255
origin, 255

parabola, 675
axis of symmetry, 676
horizontal intercept, 676
opening up or down, 676
vertex, 676
vertical intercept, 676
parallel lines, 351
perfect square trinomial, 494
perfect squares, 28
perimeter, 80
perpendicular lines, 352
point-slope form, 315
polynomial, 471
addition and subtraction, 473
binomial, 473
coefficients of, 471
constant term of, 473
degree of a polynomial, 472
degree of a term, 472
divided by a monomial, 503
leading coefficient, 473
leading term, 472
monomial, 473
prime, 521
terms of, 471

trinomial, 473
positive, 1
power, 153
prime polynomial, 521
product of sum and difference of two terms, 497
properties
    associative property of addition, 159
    associative property of multiplication, 159
    commutative property of addition, 159
    commutative property of multiplication, 159
    distributive property, 159
    of equivalent equations, 111
    square root property, 602
    zero product property, 578
proportion, 226
Pythagorean theorem, 605

quadrant, 255
quadratic, 532
    equation, 577
    equation in standard form, 577
    function, 675
    graph, 675
    vertex, 679

radical, 28, 589
    as a grouping symbol, 35
radicand, 589
rate of change, 273
ratio, 226
rational numbers, 45
rationalizing the denominator, 595
real numbers, 45
reciprocal, 158
reduced fraction, 13
relation, 649
    domain, 649
    range, 649
relationship
    linear, 279
rise over run, 280
root, 589

scientific notation, 463
set, 44
set notation, 44
set-builder notation, 55
signed numbers, 1
slope, 279
    formula, 288
    given intercepts, 335
    review, 377

triangle, 280
slope-intercept form, 296
solution, 100
    to a system of linear equations, 393
solving
    linear and quadratic equations, 617
    linear equations
        multi-step, 186
        one-step, 112
    linear inequalities
        multistep, 201
        one-step, 123
    quadratic equations
        by factoring, 579
        factoring, 579
        quadratic formula, 613
        using the square root property, 602
    system of linear equations
        elimination method, 423
        graphing, 393
        substitution method, 405
special cases of multiplication formulas, 498
square root, 589
    of a real number, 63
square root property, 602
squaring a binomial, 494, 495
standard form, 331
standard form of a quadratic equation, 577
substitution method for solving a system of
        linear equations, 405
subtraction
    property of equality, 111
    property of inequality, 123
system of linear equations, 393

term, 94, 471
trinomial, 473
types of numbers, 45

variable, 69
vertex, 676
vertical line, 347
vertical line test, 662
vertical-intercept, 694
volume, 85
    cylinder, 87
    prism, 86

whole numbers, 45

x-axis, 255
x-coordinate, 255

x-intercept, 694

y-axis, 255
y-coordinate, 255

y-intercept, 295, 694

zero product property, 578

CPSIA information can be obtained
at www.ICGtesting.com
Printed in the USA
LVHW062334140319
610751LV00004B/53/P

9 781724 271570